International Vocational Education Bilingual Textbook Series

国际化职业教育双语系列教材

Technology of Steel Bar Production
棒材生产技术

Wang Lei

王 磊 主编

Beijing

Metallurgical Industry Press

2020

内 容 提 要

本书主要介绍了棒材生产技术，内容分为两个项目，项目1棒材生产，包括任务1.1棒材生产认知、任务1.2原料准备与加热、任务1.3棒材的轧制、任务1.4棒材的精整；项目2螺纹钢棒材生产，包括任务2.1坯料准备、任务2.2坯料的加热、任务2.3轧材轧制、任务2.4控制冷却工艺及设备、任务2.5螺纹钢的精整、任务2.6产品缺陷及质量控制。

本书既可作为职业院校冶金相关专业的国际化教学用书，也可作为冶金企业员工的培训教材和有关专业人员的参考书。

图书在版编目(CIP)数据

棒材生产技术=Technology of Steel Bar Production：汉、英/王磊主编. —北京：冶金工业出版社，2020.9
国际化职业教育双语系列教材
ISBN 978-7-5024-8529-0

Ⅰ.①棒… Ⅱ.①王… Ⅲ.①棒材—生产技术—双语教学—教材—汉、英 Ⅳ.①TB3

中国版本图书馆 CIP 数据核字(2020)第 149795 号

出 版 人　陈玉千
地　　址　北京市东城区嵩祝院北巷39号　邮编　100009　电话　(010)64027926
网　　址　www.cnmip.com.cn　电子信箱　yjcbs@cnmip.com.cn
责任编辑　俞跃春　王　颖　美术编辑　郑小利　版式设计　孙跃红
责任校对　李　娜　责任印制　李玉山
ISBN 978-7-5024-8529-0
冶金工业出版社出版发行；各地新华书店经销；三河市双峰印刷装订有限公司印刷
2020年9月第1版，2020年9月第1次印刷
787mm×1092mm　1/16；17.25印张；412千字；255页
49.00元

冶金工业出版社　投稿电话　(010)64027932　投稿信箱　tougao@cnmip.com.cn
冶金工业出版社营销中心　电话　(010)64044283　传真　(010)64027893
冶金工业出版社天猫旗舰店　yjgycbs.tmall.com

(本书如有印装质量问题，本社营销中心负责退换)

Editorial Board of International Vocational Education Bilingual Textbook Series

Director Kong Weijun (Party Secretary and Dean of Tianjin Polytechnic College)

Deputy Director Zhang Zhigang (Chairman of Tiantang Group, Sino-Uganda Mbale Industrial Park)

Committee Members Li Guiyun, Li Wenchao, Zhao Zhichao, Liu Jie, Zhang Xiufang, Tan Qibing, Liang Guoyong, Zhang Tao, Li Meihong, Lin Lei, Ge Huijie, Wang Zhixue, Wang Xiaoxia, Li Rui, Yu Wansong, Wang Lei, Gong Na, Li Xiujuan, Zhang Zhichao, Yue Gang, Xuan Jie, Liang Luan, Chen Hong, Jia Yanlu, Chen Baoling

国际化职业教育双语系列教材编委会

主　任　孔维军（天津工业职业学院党委书记、院长）

副主任　张志刚（中乌姆巴莱工业园天唐集团董事长）

委　员　李桂云　李文潮　赵志超　刘　洁　张秀芳

　　　　　谭起兵　梁国勇　张　涛　李梅红　林　磊

　　　　　葛慧杰　王治学　王晓霞　李　蕊　于万松

　　　　　王　磊　宫　娜　李秀娟　张志超　岳　刚

　　　　　玄　洁　梁　娈　陈　红　贾燕璐　陈宝玲

Foreword

With the proposal of the 'Belt and Road Initiative', the Ministry of Education of China issued *Promoting Education Action for Building the Belt and Road Initiative* in 2016, proposing cooperation in education, including 'cooperation in human resources training'. At the Forum on China-Africa Cooperation (FOCAC) in 2018, President Xi proposed to focus on the implementation of the 'Eight Actions', which put forward the plan to establish 10 Luban Workshops to provide skills training to African youth. Draw lessons from foreign advanced experience of vocational education mode, China's vocational education has continuously explored and formed the new mode of vocational education with Chinese characteristics. Tianjin, as a demonstration zone for reform and innovation of modern vocational education in China, has started the construction of 'Luban Workshop' along the 'Belt and Road Initiative', to export high-quality vocational education achievements.

The compilation of these series of textbooks is in response to the times and it's also the beginning of Tianjin Polytechnic College to explore the internationalization of higher vocational education. It's a new model of vocational education internationalization by Tianjin, response to the 'Belt and Road Initiative' and the 'Going Out' of Chinese enterprises. Tianjin Polytechnic College and Uganda Technical College-Elgon reached a cooperation intention to establish the Luban Workshop to carry out vocational education cooperation on mechatronics technology and ferrous metallurgy technology major in 2019. The establishment of Luban Workshop is conducive to strengthen the cooperation between China and Uganda in vocational education, promote the export of high-quality higher vocational education resources, and serve Chinese enterprises in Uganda and Ugandan local enterprises. Exploring and standardizing the overseas operation of Chinese colleges, the expansion of international influences of China's higher vocational education is also one of the purposes.

The construction of 'Luban Workshop' in Uganda is mainly based on the EPIP (Engineering, Practice, Innovation, Project) project, and is committed to cultivating high-quality talents with innovative spirit, creative ability and entrepreneurial spirit. To meet the learning needs of local teachers and students accurately, the compilation of these international vocational skills bilingual textbooks is based on the talent demand of Uganda and the specialty and characteristics of Tianjin Polytechnic.

These textbooks are supporting teaching material, referring to Chinese national professional standards and developing international professional teaching standards. The internationalization of the curriculums takes into account the technical skills and cognitive characteristics of local students, to promote students' communication and learning ability. At the same time, these textbooks focus on the enhancement of vocational ability, rely on professional standards, and integrate the teaching concept of equal emphasis on skills and quality. These textbooks also adopted project-based, modular, task-driven teaching model and followed the requirements of enterprise posts for employees.

In the process of writing the series of textbooks, Wang Xiaoxia, Li Rui, Wang Zhixue, Ge Huijie, Yu Wansong, Wang Lei, Li Xiujuan, Gong Na, Zhang Zhichao, Jia Yanlu, Chen Baoling and other chief teachers, professional teams, English teaching and research office have made great efforts, receiving strong support from leaders of Tianjin Polytechnic College. During the compilation, the series of textbooks referred to a large number of research findings of scholars in the field, and we would like to thank them for their contributions.

Finally, we sincerely hope that the series of textbooks can contribute to the internationalization of China's higher vocational education, especially to the development of higher vocational education in Africa.

Principal of Tianjin Polytechnic College Kong Weijun
May, 2020

序

随着"一带一路"倡议的提出，2016年中华人民共和国教育部发布了《推进共建"一带一路"教育行动》，提出了包括"开展人才培养培训合作"在内的教育合作。2018年习近平主席在中非合作论坛上提出，要重点实施"八大行动"，明确要求在非洲设立10个鲁班工坊，向非洲青年提供技能培训。中国职业教育在吸收和借鉴发达国家先进职教发展模式的基础上，不断探索和形成了中国特色职业教育办学模式。天津市作为中国现代职业教育改革创新示范区，开启了"鲁班工坊"建设工作，在"一带一路"沿线国家搭建"鲁班工坊"平台，致力于把优秀职业教育成果输出国门与世界分享。

本系列教材的编写，契合时代大背景，是天津工业职业学院探索高职教育国际化的开端。"鲁班工坊"是由天津率先探索和构建的一种职业教育国际化发展新模式，是响应国家"一带一路"倡议和中国企业"走出去"，创建职业教育国际合作交流的新窗口。2019年天津工业职业学院与乌干达埃尔贡技术学院达成合作意向，共同建立"鲁班工坊"，就机电一体化技术专业、黑色冶金技术专业开展职业教育合作。此举旨在加强中乌职业教育交流与合作，推动中国优质高等职业教育资源"走出去"，服务在乌中资企业和乌干达当地企业，探索和规范我国职业院校"鲁班工坊"建设和境外办学，扩大中国高等职业教育的国际影响力。

中乌"鲁班工坊"的建设主要以工程实践创新项目（EPIP：Engineering, Practice, Innovation, Project）为载体，致力于培养具有创新精神、创造能力和创业精神的"三创"复合型高素质技能人才。国际化职业教育双语系列教材的编写，立足于乌干达人才需求和天津工业职业学院专业特色，是为了更好满足当地师生学习需求。

本系列教材采用中英双语相结合的方式，主要参照中国专业标准，开发国际化专业教学标准，课程内容国际化是在专业课程设置上，结合本地学生的技术能力水平与认知特点，合理设置双语教学环节，加强学生的学习与交流能力。同时，教材以提升职业能力为核心，以职业标准为依托，体现技能与质量并重的教学理念，主要采用项目化、模块化、任务驱动的教学模式，并结合企业岗位对员工的要求来撰写。

本系列教材在撰写过程中，王晓霞、李蕊、王治学、葛慧杰、于万松、王磊、李秀娟、宫娜、张志超、贾燕璐、陈宝玲等主编老师、专业团队、英语教研室付出了辛勤劳动，并得到了学院各级领导的大力支持，同时本系列教材借鉴和参考了业界有关学者的研究成果，在此一并致谢！

最后，衷心希望本系列教材能为我国高等职业教育国际化，尤其是高等职业教育走进非洲、支援非洲高等职业教育发展尽绵薄之力。

天津工业职业学院书记、院长　孔维军

2020 年 5 月

Preface

Tianjin Polytechnic College and Uganda Technical College-Elgon reached a cooperation intention to establish the Luban Workshop to carry out vocational education cooperation on mechatronics technology and ferrous metallurgy technology major in 2019. In order to strengthen the cooperation between China and Uganda in vocational education, the two colleges plan to compile a series of international vocational skills bilingual textbooks.

This book is one of the international vocational skills bilingual textbooks.

'Technology of Steel Bar Production' is a supporting teaching material for the construction project of Luban Workshop in Uganda, Tianjin Institute of Technology. According to the characteristics of the course design, such as wide range of subjects and strong practicality, this book introduces the bar production technology in a simple way, avoids the tedious theoretical derivation, enhances the practical application, and comprehensively introduces the bar production process and technology. This book introduces the new achievements and progress in the field of bar production at home and abroad as far as possible, and fully embodies the purpose of higher vocational education to cultivate applied talents. With reference to the national vocational standards for metal rolling workers, this book follows the requirements of enterprise posts for employees, relies on Vocational standards, focuses on improving professional ability, focuses on working process operation, and adopts task driven teaching mode.

This book mainly takes the production process and equipment of round steel and screw steel as the carrier, mainly introduces the theoretical knowledge and practical skills of relevant rolling mill and auxiliary equipment, detection system, console and product quality control. This book is the result of school enterprise cooperation. The editor, together with the technical experts in the production line, tracks the technical development trend on the basis of the industry experts and graduates' job research, and compiles this book according to the cultivation of Applied Talents in bar production enterprises. This book was jointly completed by Wang Lei and Gong Na of Tianjin Polytechnic College. The Chinese part of Project 2 was written by Gong Na, and the rest was written by Wang Lei. Many of the author's documents are cited in the book, and I would like to express my thanks.

Due to the limited level of editors, it is inevitable that there are some defects or omissions in the book. Please criticize and correct them.

The editor
May, 2020

前　言

2019年天津工业职业学院与乌干达埃尔贡技术学院达成合作意向，共同建立"鲁班工坊"，就机电一体化技术专业、黑色冶金技术专业开展职业教育合作，双方计划编撰国际化职业教育双语系列教材。

本书是国际化职业教育双语系列教材之一。

本书是天津工业职业学院乌干达鲁班工坊建设项目配套教材。本书根据棒材课程设计的学科面广、实践性强等特点，深入浅出地介绍了棒材生产技术，避免了烦琐的理论推导，增强了实际应用，对棒材的生产工艺与技术进行了全面介绍。该书尽可能多的介绍了国内外棒材生产领域的新成果、新进展，并且参照金属轧制工国家职业标准，遵循企业岗位对就业人员的要求，以职业标准为依托，以提升职业能力为核心，以工作过程操作为重点，采用任务驱动的教学模式来进行编写，充分体现了高等职业教育培养应用型人才的宗旨。

本书以圆钢和螺纹钢的生产工艺及设备为载体，主要介绍了相关轧机和辅助设备、检测系统、主控台以及产品质量控制等方面的理论知识和实践技能。本书是校企合作的成果，是编者在与行业专家、生产一线的技术人员一起，以及毕业生工作岗位调研的基础上，跟踪技术发展趋势，根据棒材生产企业应用型人才的培养完成编写本书。

本书由天津工业职业学院王磊担任主编，宫娜参编，其中项目2的中文部分由宫娜编写，其余部分由王磊编写。书中引用了许多作者的文献资料，在此一并表示感谢。

由于编者水平所限，书中不妥之处，恳请读者批评指正。

编　者

2020年5月

Contents

Project 1 Production of Steel Bar ·· 1

  Task **1.1** Cognition of Steel Bar Production ··· 1

    1.1.1 Product Outline and Product Standard ·· 1

    1.1.2 The Process of Steel Bar Production ··· 5

  Task **1.2** Raw Material Preparation and Heating ·· 10

    1.2.1 Preparation of Billet ·· 10

    1.2.2 Heating of Raw Materials ·· 15

  Task **1.3** Rolling of Steel Bar ··· 28

    1.3.1 Theoretical Basis of Rolling ··· 29

    1.3.2 Rolling Process and Equipment ··· 38

    1.3.3 Rolling Pass System ·· 50

    1.3.4 Rollers and Guides ·· 62

  Task **1.4** Finishing of Steel Bars ··· 86

    1.4.1 Process of Finishing Area ·· 87

    1.4.2 Multiple Shear of Flying Shear ·· 87

    1.4.3 Fixed Length Shear ··· 91

    1.4.4 Process and Equipment in Cooling Bed Area ·································· 92

Project 2 Production of HRB ··· 96

  Task **2.1** Preparation of the Raw Material ··· 96

    2.1.1 Selection of the Billet ··· 96

    2.1.2 Continuous Casting Billet ··· 97

    2.1.3 Hot Delivery and Hot Charging of Billet ·· 98

Task **2.2** Heating of the Billet ································· 106
  2.2.1 Heating Equipment ································· 107
  2.2.2 Fuel Selection ································· 115
  2.2.3 Heating Process of Billet ································· 118
  2.2.4 Thermal System ································· 131

Task **2.3** Rolling ································· 137
  2.3.1 Rolling Process System ································· 137
  2.3.2 Pass System ································· 143
  2.3.3 Guide Device ································· 156
  2.3.4 Common Rolling Defects and Prevention ································· 170

Task **2.4** Control Cooling Process and Equipment ································· 176
  2.4.1 Theoretical Basis of Controlled Cooling Technology after Rolling ················· 176
  2.4.2 Controlled Cooling Process ································· 183
  2.4.3 Process Equipment of Controlled Cooling Process ································· 189

Task **2.5** Finishing of Threaded Steel ································· 197
  2.5.1 Overview ································· 197
  2.5.2 Shear ································· 197
  2.5.3 Cooling ································· 202
  2.5.4 On Line Sizing and Shearing of Bar ································· 213
  2.5.5 Steel Packaging (Bundling) ································· 214
  2.5.6 Typical Finishing Process of Continuous Rolling Bar ································· 217

Task **2.6** Product Defects and Quality Control ································· 234
  2.6.1 Quality Requirements for Use of Threaded Steel ································· 234
  2.6.2 Production Characteristics and Quality Control of Screw Steel ················· 240
  2.6.3 Inspection and Inspection of Product Quality ································· 245

References ································· 256

目 录

项目1 棒材生产 ... 1

任务1.1 棒材生产认知 ... 1
1.1.1 产品大纲及产品标准 ... 1
1.1.2 棒材生产工艺流程 ... 5

任务1.2 原料准备与加热 ... 10
1.2.1 坯料准备 ... 10
1.2.2 原料加热 ... 15

任务1.3 棒材的轧制 ... 28
1.3.1 轧制理论基础 ... 29
1.3.2 轧制工艺及设备 ... 38
1.3.3 轧制孔型系统 ... 50
1.3.4 轧辊与导卫 ... 62

任务1.4 棒材的精整 ... 86
1.4.1 精整区工艺流程 ... 87
1.4.2 飞剪倍尺剪切 ... 87
1.4.3 定尺剪 ... 91
1.4.4 冷床区工艺及设备 ... 92

项目2 螺纹钢棒材生产 ... 96

任务2.1 坯料准备 ... 96
2.1.1 坯料的选择 ... 96
2.1.2 连铸坯 ... 97
2.1.3 坯料的热送热装 ... 98

任务 2.2　坯料的加热	106
2.2.1　加热设备	107
2.2.2　燃料选择	115
2.2.3　坯料的加热工艺	118
2.2.4　热工制度	131
任务 2.3　轧材轧制	137
2.3.1　轧制工艺制度	137
2.3.2　孔型系统	143
2.3.3　导卫装置	156
2.3.4　常见的轧制缺陷及其预防	170
任务 2.4　控制冷却工艺及设备	176
2.4.1　钢材轧后控制冷却技术的理论基础	176
2.4.2　控制冷却工艺	183
2.4.3　控冷工艺的工艺设备	189
任务 2.5　螺纹钢的精整	197
2.5.1　概述	197
2.5.2　剪切	197
2.5.3　冷却	202
2.5.4　棒材在线定尺剪切	213
2.5.5　钢材的包装（打捆）	214
2.5.6　连轧棒材典型精整工艺	217
任务 2.6　产品缺陷及质量控制	234
2.6.1　螺纹钢使用的质量要求	234
2.6.2　螺纹钢的生产特点及质量控制	240
2.6.3　产品质量的检查、检验	245
参考文献	256

Project 1　Production of Steel Bar
项目1　棒材生产

Task 1.1　Cognition of Steel Bar Production
任务1.1　棒材生产认知

Mission objectives
任务目标

(1) Understand the round bar steel products, and understand its characteristics, use and quality requirements;

(1) 认识棒材圆钢产品,并了解其特点、用途和质量要求;

(2) Master the production process of bar round steel.

(2) 掌握棒材圆钢的生产工艺。

1.1.1　Product Outline and Product Standard
1.1.1　产品大纲及产品标准

Bar is one of the hot-rolled strip steel. Generally, bar is defined as the hot-rolled bar shaped steel cut according to the fixed length. According to its section shape, it is generally divided into round steel, square steel flat steel, hexagonal riveting, octagonal steel, etc., there are also periodic section steel such as screw steel. In special cases, there are also rolled bars, not straight bars. For example, Korean POSCO iron and steel company produces $\phi14 \sim 42$mm bars in high-speed wire rod workshop, which are rolled by Garrett type coiler.

棒材是热轧的条状钢材中的一种,一般把棒材定义为热轧成棒状的按定尺长度切断而成的钢材。根据它的断面形状一般分为圆钢、方钢扁钢、六角铆、八角钢等,也有周期断面型钢,如螺纹钢。在特殊情况下,也有卷状棒材,不是直条形式,如韩国浦项钢铁公司在高速线材车间生产 $\phi14 \sim 42$mm 的棒材,由加勒特式卷取机卷取,就是卷状的。

Bars can be divided into large, medium and small bars according to their size. The corresponding production workshops are also large, medium and small workshops. In the international standard ISO1035, the size (diameter) of hot-rolled bar is $8 \sim 220$mm, the size (side length) of square steel is 8~120mm, and the size of flat steel is 20mm×5mm~150mm×50mm.

棒材视其尺寸大小分为大型、中型和小型棒材。其对应的生产车间也分别为大型、中型、小型车间。在国际标准 ISO1035 中，热轧棒材的圆钢尺寸（直径）范围为 8~220mm，方钢尺寸（边长）范围为 8~120mm，扁钢尺寸范围为 20mm×5mm~150mm×50mm。

Bars are widely used in industrial fields such as machinery, automobiles, ships, and construction. Some are directly used for concrete, such as light round steel bars and threaded steel bars, and some can be used to make various shafts, gears, bolts, females, anchor chains, springs, etc. through secondary processing. The bars produced by a certain factory mainly include reinforced concrete bars, ordinary carbon-bonded round steel, high-quality carbon cable structural steel, spring steel, and so on.

棒材广泛用于机械、汽车、船舶、建筑等工业领域。有的直接用于混凝土，如光圆钢筋和螺纹钢筋，有的可通过二次加工来制作各种轴、齿轮、螺栓、螺母、锚链、弹簧等。某厂生产的棒材主要有钢筋混凝土用钢筋、普通碳素结合圆钢、优质碳素结构钢、弹簧钢等。

Because of its wide range of uses, bars account for a certain percentage of the total steel in various countries, and the situation in different countries is different. This is true even in Western countries. For example, in 1989, Britain accounted for 13.0%, Japan accounted for 23.5%, Italy accounted for 37.9%, and China has been 35% to 40% in recent years.

棒材因其用途广泛，在各国的钢材总量中都占有一定比例，各国情况有所不同。即使是西方国家，也是如此，如 1989 年，英国占 13.0%，日本占 23.5%，意大利占 37.9%，我国近些年来一直为 35%~40%。

Full continuous small bar continuous rolling mill is the current development direction of small bar production. The rolling speed of a fully continuous small bar mill is usually 15~22m/s, and the highest speed is 30m/s (Gallet coil type). Rolling mills that only produce round steel and rebar, the production capacity can reach 750,000 tons/year. The yield rate is generally above 95%. Modern continuous rolling small bar production lines generally have the following characteristics:

全连续式小型棒材连轧机则是目前小型棒材生产的发展方向。全连续式小型棒材轧机的轧制速度通常为 15~22m/s，最高可达 30m/s（加勒特卷取型）。只生产圆钢和螺纹钢筋的轧机，生产能力可达 75 万吨/年。成材率一般在 95% 以上。现代化的连轧小型棒材生产线一般都具有如下特点：

(1) Taking the long continuous casting slab as the raw material, the step furnace is used for heating, and the hot delivery and charging are realized as far as possible. It can increase the yield and reduce the energy consumption.

(1) 以长尺连铸坯为原料，采用步进炉加热，并尽可能实现热送热装炉。使得成材率提高，能耗降低。

(2) Most of the production lines are single line rolling without torsion, that is, the rolling mill is arranged alternately in horizontal and vertical direction, so as to reduce production accidents.

(2) 大多数生产线为单线无扭轧制，即轧机平立交替布置，从而使生产事故减少。

(3) The short stress line and high rigidity mill are used in the medium and finishing mills, which improves the dimensional accuracy of the finished products.

(3) 中、精轧机采用短应力线、高刚度轧机，成品尺寸精度提高。

(4) Slitting rolling. Slitting rolling can reduce rolling passes, reduce energy consumption, and increase the hourly output of small specifications. Generally, the output of small-sized machine hour is only half of that of large-sized machine hour, but it accounts for more than half of the market demand. The capacity of the rolling mill is matched with that of the heating furnace through the slitting rolling technology.

(4) 切分轧制。切分轧制可以减少轧制道次，降低能耗，提高小规格的机时产量。通常小规格的机时产量只有大规格的一半，但却占市场需要的一半以上。通过切分轧制技术，使轧机能力与加热炉能力匹配。

(5) Control rolling and cooling technology are adopted. Grain refinement can be achieved by low temperature rolling or the strength can be improved by post rolling heat treatment.

(5) 采用控轧控冷技术。通过低温轧制实现晶粒细化或通过轧后热处理提高强度。

Especially the on-line heat treatment technology after rolling. After rolling, the screw steel is cooled by water-cooling device, quenched by high temperature and tempered by central waste heat. The tempered martensite structure is formed on the surface of the steel and the pearlite structure is refined on the core, which improves the comprehensive mechanical properties of the steel and reduces the amount of secondary iron oxide skin.

尤其是轧后在线热处理技术。螺纹钢轧后经水冷装置穿水冷却，利用轧件高温快速淬火，中心余热回火。钢材表面形成回火马氏体组织，芯部形成细化的珠光体组织，从而提高了钢材的综合机械性能，同时减少二次氧化铁皮量。

(6) It adopts DC motor drive and process computer control. Due to the large speed range, fast response and stable speed of DC motor, the computer can control all links in production, and realize micro tension rolling in rough and medium rolling. Through looper, no tension rolling is realized. It creates conditions for ensuring the dimensional accuracy of products.

(6) 采用直流电机传动及过程计算机控制。由于直流电机调速范围大、响应快、速度稳定，计算机可以控制生产中的各个环节，在粗、中轧中实现了微张力轧制。中精轧中通过活套，实现了无张力轧制。为保证产品尺寸精度创造了条件。

(7) Long length cooling is adopted, and the finishing process is mechanized and automatic. In the modern small bar production line, the length of the cooling bed is generally about 1201m. Before the rolling piece is put on the cooling bed, it is cut into hot multiple length by flying shear. The sizing ratio and yield are increased. The bar can be bundled by counter according to the number of bars, which creates a good condition for negative tolerance rolling.

(7) 采用长尺冷却，精整工序机械化、自动化。现代化小型棒材生产线上，冷床长度一般在1201m左右，轧件上冷床之前用飞剪切成热倍尺。提高了定尺率和成材率。棒材可通过计数器按根数打捆，为负公差轧制创造了好的条件。

1.1.1.1 Common Production Specifications and Steel Grades for Steel Bar Production
1.1.1.1 棒材生产常见生产规格和钢种

The common production specifications and steel types of small bar production plants are shown in Table 1-1.

小棒材生产厂常见的生产规格和钢种主要见表 1-1。

Table 1-1 Product outline for bar production
表 1-1 棒材生产产品大纲

Serial number 序号	Product 产品	Specification/mm 规格/mm	Steel 钢种
1	Round steel (tube blank) 圆钢（管坯）	$\phi 12 \sim 60$	Low carbon steel, spring steel, carbon structural steel, alloy structural steel, cold heading steel 低碳钢、弹簧钢、碳素结构钢、合金结构钢、冷镦钢
2	HRB 螺纹钢	$\phi 10 \sim 50$	Low alloy steel 低合金钢

1.1.1.2 Product Executive Standard of Bar Mill
1.1.1.2 棒材生产厂产品执行标准

The product executive standards of the bar mill are shown in Table 1-2.

棒材生产厂产品执行标准见表 1-2。

Table 1-2 Executive standard of bar products
表 1-2 棒材产品执行标准

Steel 钢种	Executive standard 执行标准
HRB 螺纹钢	GB1499—1998
Round steel 圆钢	GB 1222, GB/T 3077, GB/T 699, YB/T 5222, GB 702—86

1.1.1.3 Bar Product Precision
1.1.1.3 棒材产品精度

The precision of bar products is:
棒材产品精度为:

(1) Roundness of round steel: no more than $\phi 40$mm, roundness $\leq 50\%$ of nominal diameter tolerance;
Greater than $\phi 40 \sim 85$mm, out of roundness $\leq 70\%$ of the nominal diameter tolerance.
(1) 圆钢不圆度：不大于 $\phi 40$mm，不圆度≤公称直径公差的 50%；
大于 $\phi 40 \sim 85$mm，不圆度≤公称直径公差的 70%。

(2) Bending degree: the bending degree per meter shall not exceed 4mm, and the total bending degree shall not exceed 0.4% of the delivery length.

（2）弯曲度：每米弯曲度不超过4mm，总弯曲度不超过交货长度的0.4%。

(3) The packaging standard shall be in accordance with GB/T 2101—1989 and relevant regulations.

（3）包装标准执行 GB/T 2101—1989 及相关规定。

1.1.2 The Process of Steel Bar Production
1.1.2 棒材生产工艺流程

According to the production organization form, the whole plant is divided into four areas: heating furnace area, rolling area, finishing area and peripheral auxiliary area. The production process is shown in Figure 1-1.

棒材厂按生产组织形式全厂划分为加热炉区、轧制区、精整区、外围辅助区四个区域。生产工艺流程如图 1-1 所示。

1.1.2.1 Heating Furnace Area
1.1.2.1 加热炉区

The main equipment includes feeding platform, loading roller table, walking beam furnace, discharging roller table and stripping device. The hourly output of the full beam walking beam furnace is 100t, and the walking mechanism can realize continuous walking and retreating, which solves some defects that other kinds of furnaces can not meet the requirements of the heating process in the past. A roller table is also arranged at the tapping side of the heating furnace and the opposite side of the rolling, which is very convenient to deal with production accidents. This kind of heating furnace is widely used in the new advanced bar rolling mill in recent years.

主要设备有上料台架、装钢辊道、步进式加热炉、出钢辊道及剔除装置。全梁式步进加热炉小时产量为100t，步进机构可实现连续步进和后退，解决了过去其他各式加热炉不能满足加热工艺要求的一些缺陷。加热炉出钢侧在与轧制反方向侧也设置了辊道，处理生产事故十分方便。这种加热炉在我国近几年新建的较先进的棒材轧钢厂中被广泛应用。

1.1.2.2 Rolling Zone
1.1.2.2 轧制区

There are 18 rolling mills in the rolling area, which are arranged horizontally and vertically in a staggered way to realize torsion free rolling.

轧制区共 18 架轧机，平立交错布置，实现无扭轧制。

(1) One to six rough rolling mills are cantilever type high rigidity mills, including three $\phi 685mm$ (No. 1 and No. 3 are horizontal mills, No. 2 are vertical mills), three $\phi 585mm$ (No. 4 and No. 6 are vertical mills, and No. 5 are horizontal mills).

（1）粗轧机 1~6 架，为悬臂式高刚度轧机，其中 3 架 $\phi 685mm$（1 号、3 号为水平轧机，2 号为立式轧机），3 架 $\phi 585mm$（4 号、6 号为立式轧机，5 号为水平轧机）。

(2) There are 6~12 medium rolling mills, all of which are $\phi 470mm$ short stress line high rigidity rolling mills (No. 7, No. 9 and No. 11 are horizontal rolling mills, No. 8 and No. 10 are vertical rolling mills, and No. 12 are horizontal vertical conversion rolling mills). This kind of rolling mill is commonly used in small-scale rolling mills such as bar mills, with different

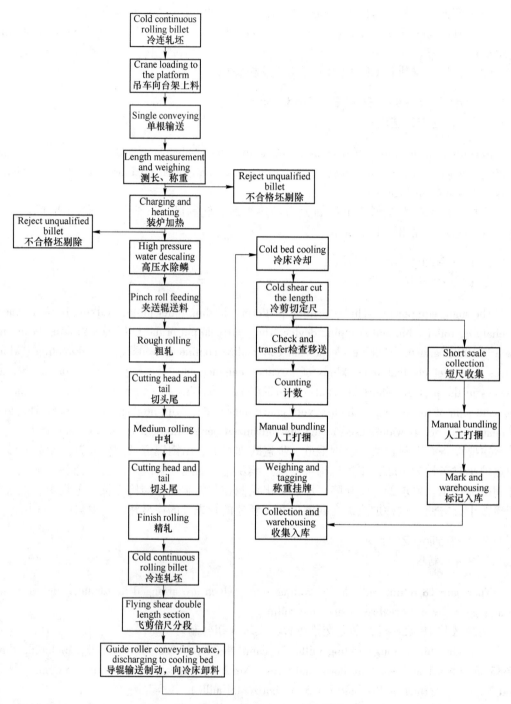

Figure1-1　Production process of steel bar mill
图 1-1　棒材厂生产工艺流程

forms. The rolling mills manufactured by manufacturers all over the world have their own characteristics. Danieli's rolling mill design and manufacture have become a series, which is widely used in Europe and America.

（2）中轧机6~12架，全部为φ470mm短应力线高刚度轧机（7号、9号、11号为水平轧机，8号、10号为立式轧机，12号为平立转换轧机）。这种轧机常见于棒材等小型轧机上，形式各有不同，世界各国制造商制造的此种轧机均有各自的特点，达涅利的轧机设计制造已成系列化，在欧美被广泛采用。

（3）There are 12~18 finishing mills, 3 of which are φ470mm and 3 of which are φ370mm short stress line high rigidity mills, just like the medium mill. Among them, three vertical rolling mills can be vertically and horizontally converted for slitting rolling. The rolling process is controlled automatically by computer program and can also be operated manually.

（3）精轧机12~18架，其中3架φ470mm，3架φ370mm短应力线高刚度轧机，与中轧机一样。其中3架立式轧机可立平转换，用以切分轧制。轧制过程全部由计算机程序自动控制，也可单体手动操作。

（4）Hot shear: there is one hot shear at the back of each rolling mill, and the bypass of the third hot shear is provided with four pieces of shear. The details are as follows:

（4）热剪：每组轧机后边均有一台热剪，第3台热剪的旁路设置了一台碎断剪，共4台。具体内容如下：

Shear 1——start stop crank shear, shear section 5200mm^2

1号剪——启停式曲柄剪，剪切断面5200mm^2

Shear 2——start stop flying shear, shear section 2600mm^2

2号剪——启停式飞剪，剪切断面2600mm^2

Shear 3——start stop flying shear, with two pairs of blades, shear section 2000mm^2

3号剪——启停式飞剪，有两对刀片，剪切断面2000mm^2

Shear 4——start stop flying shear, shear section 800mm^2

4号剪——启停式飞剪，剪切断面800mm^2

No. 1 and No. 2 shears are mainly used for cutting head, tail and breaking. No. 3 shears are mainly used for cutting multiple length. No. 4 shears are mainly used for cutting bar in case of accident and optimizing cutting.

1号、2号剪主要用于切头、切尾和碎断，3号剪主要用于剪切倍尺，4号剪主要用于事故时切碎棒材，优化剪切。

1.1.2.3　Finishing Area
1.1.2.3　精整区

Bar finishing refers to the process of cold bed, cold shear, inspection, bundling, packing, weighing, collection and warehousing on the rolling line. The equipment mainly includes:

棒材精整是指在轧制线上从冷床、冷剪、检查、成捆、打包、称重、收集入库的工序。其设备主要包括：

（1）Cold bed. The step rack cooling bed is 96m long and 12.5m wide. The step mechanism adopts eccentric wheel type and operates stably and accurately. The main drive adopts the motor to drive the worm gear reducer. Because the bed body is too long, two sets of drives are used to drive the eccentric axle, which requires high synchronous accuracy.

（1）冷床。步进式齿条冷床长96m，有效宽12.5m，步进机构采用偏心轮式，运行平稳动作准确，主传动采用电机带动蜗杆减速机。由于床体过长，采用两套传动分别驱动偏心轮轴，要求同步精度较高。

(2) Cold shear. The shear capacity is 350t, which is swing type flying shear. Because of continuous shear, the operation rate is increased, and the huge fixed length movable baffle equipment is banned. The operation is normal during the thermal load test run.

(2) 冷剪。剪切能力 350t，为摆动式飞剪，因为是连续剪切，提高了作业率，并取缔了庞大的定尺活动挡板设备。热负荷试车时运转正常。

(3) Baler. The baling diameter is 150~500mm, with small hydraulic station.

(3) 打捆机。打捆直径 150~500mm，自身带有小液压站。

The baler used to bundle bar has different forms. Its main feature is that 8 small column plugs are used to evenly arrange along the periphery of the feeding slot, and block the binding line. With the pulling action, the baler is opened in turn, and the binding is tight, and the structure is simple and compact. The tie with the same degree of tightness is less than the loop tension of other forms of balers.

用于捆棒材的打捆机形式各有不同，其主要特点是采用 8 个小柱塞沿喂线槽周边均匀布置，并挡住捆线，随着抽线动作依次打开，捆得紧，结构简单紧凑。同样松紧程度的捆结，较其他形式的打捆机回线拉力要小一些。

Other equipment: there are lifting apron plate, roller table, transport trolley, chain bed, collection basket and other equipment in the finishing area.

其他设备：精整区域还有升降裙板、辊道、运输小车、链床、收集筐等一些设备。

1.1.2.4 Hydraulic Lubrication Equipment
1.1.2.4 液压润滑设备

Hydraulic lubrication equipment area mainly includes:

液压润滑设备区域主要有：

(1) Hydraulic equipment. It mainly includes heating furnace hydraulic station, rough rolling mill hydraulic station, common hydraulic station of medium and finishing mill and finishing area hydraulic station. The details are as follows:

(1) 液压设备。主要包括加热炉液压站，粗轧机液压站，中、精轧机共用的液压站和精整区液压站。具体内容如下：

1) Heating furnace hydraulic station, oil tank volume $3.5m^3$, flow 1100L/min, pressure 14MPa.

1) 加热炉液压站，油箱容积 $3.5m^3$，流量 1100L/min，压力 14MPa。

2) Hydraulic station of rough rolling mill, oil tank volume $0.5m^3$, flow 70L/min, pressure 12MPa.

2) 粗轧机液压站，油箱容积 $0.5m^3$，流量 70L/min，压力 12MPa。

3) Medium and finishing rolling hydraulic station, oil tank volume $1m^3$, flow 200L/min, pressure 12MPa.

3）中、精轧液压站，油箱容积 1m³，流量 200L/min，压力 12MPa。

4）Finishing hydraulic station, oil tank volume 5m³, flow 1000L/min, pressure 12MPa.

4）精整液压站，油箱容积 5m³，流量 1000L/min，压力 12MPa。

（2）Lubrication equipment. Lubrication equipment mainly includes rough rolling mill lubrication station, medium rolling mill lubrication station, finishing mill lubrication station, cold shear lubrication station, dry oil lubrication system and oil steam lubrication system. It includes：

（2）润滑设备。润滑设备主要有粗轧机润滑站、中轧机润滑站、精轧机润滑站、冷剪润滑站、干油润滑系统和油—汽润滑系统。具体包括：

1）At the lubrication station of rough rolling mill, the volume of oil tank is 25m³, the flow of lubricating oil for oil supply film bearing is 180L/m³, and the pressure is 0.4MPa：the flow of lubricating oil for other parts is 540L/min, and the pressure is 0.2MPa.

1）粗轧机润滑站，油箱容积 25m³，供油膜轴承的润滑油流量 180L/m³，压力 0.4MPa；供其他部位润滑的油流量 540L/min，压力 0.2MPa。

2）Lubricating station of medium rolling mill, oil tank volume 25m³, oil flow 900L/min, pressure 0.2MPa.

2）中轧机润滑站，油箱容积 25m³，油流量 900L/min，压力 0.2MPa。

3）Lubrication station of finishing mill, oil tank capacity 25m³, oil flow 900L/min, pressure 0.2MPa.

3）精轧机润滑站，油箱容积 25m³，油流量 900L/min，压力 0.2MPa。

4）Centralized dry oil system, oil tank volume 0.04m³, flow 0.13L/min, pressure 50MPa.

4）集中干油系统，油箱容积 0.04m³，流量 0.13L/min，压力 50MPa。

5）Oil gas lubrication system, oil tank volume 0.5m³, total flow 3L/min, pressure 16MPa.

5）油气润滑系统，油箱容积 0.5m³，总流量 3L/min，压力 16MPa。

1.1.2.5　Accessory Equipment
1.1.2.5　附属设备

Auxiliary equipment includes air compressor, crane, etc.
附属设备包括空压机、起重机等。

（1）Air compressor. Three are 60Nm³/h air compressors.

（1）空压机。三台 60Nm³/h 的空压机。

（2）Crane. There are 4 sets of 16t + 16t electromagnetic cranes for slab collapse, 2 sets of 20t+20t for main rolling, 2 sets of bridge cranes and 2 sets of 10t bridge cranes, 4 sets of 10t+10t electromagnetic cranes for finishing collapse, one set of 10t bridge cranes for machine repair room and one set of 10t bridge cranes for production preparation.

（2）起重机。钢坯跨作用 16t+16t 的电磁吊 4 台，主轧跨 20t+20t，桥吊和 10t 桥吊各两台，精整跨使用 10t+10t 电磁吊 4 台，机修间 10t 一台，生产准备 10t 桥吊一台。

Task 1.2　Raw Material Preparation and Heating
任务1.2　原料准备与加热

Mission objectives
任务目标

(1) Be able to select the type and technical parameters of raw materials according to the actual production, be able to identify the defects of raw materials;

(1) 能够根据实际生产选择原料种类及技术参数,能够识别原料缺陷;

(2) Be able to master the billet heating process, master the formation mechanism of common heating defects in the heating process, master the basic process of billet heating operation.

(2) 能够掌握坯料加热工艺,掌握加热过程中常见加热缺陷的形成机理,掌握坯料加热操作基本流程。

1.2.1　Preparation of Billet
1.2.1　坯料准备

1.2.1.1　Selection of Billet
1.2.1.1　坯料的选择

Round steel production requires high quality of billet, and its selection and acceptance criteria include:

圆钢生产对钢坯的质量要求较高,其选择和验收的标准包括:

(1) Requirements for steel grades and chemical composition.

(1) 对钢种与化学成分的要求。

The grade and chemical composition of the billet shall conform to the relevant standards and regulations, and there are corresponding requirements for the residual element content of different steel grades.

钢坯的牌号和化学成分应符合有关标准规定,对不同的钢种在钢的残余元素含量上有相应的要求。

(2) Requirements for billet size and quality.

(2) 对钢坯尺寸和质量的要求。

1) The requirements for the shape, dimension and quality of the billet include: the section shape and allowable deviation of the billet, the length of the fixed length, the shortest length and proportion of the short ruler, bending, twisting, etc. These requirements are determined by considering comprehensive factors such as giving full play to the production capacity of rolling mill,

ensuring the smooth heating and rolling, and considering the possibility and rationality of slab supply. Allowable deviation of square billet size and side length:

1) 对钢坯的外形、尺寸、质量的要求有：钢坯断面形状及允许偏差、定尺长度、短尺的最短长度及比例、弯曲、扭转等。这些要求是考虑了充分发挥轧机生产能力，保证加热与轧制顺利并考虑供坯的可能性和合理性等综合因素确定的。方坯尺寸和边长允许偏差：

Side length of square billet: 130±4mm (YB/T 001—91)

Fillet radius: $R>8$mm (YB/T 001—91)

Diagonal length difference: ≤6mm (YB/T 001—91)

Standard for continuous casting square billet and rectangular billet: YB 2011—83

方坯边长：130±4mm (YB/T 001—91)

圆角半径：$R>8$mm (YB/T 001—91)

对角线长度差：≤6mm (YB/T 001—91)

连铸方坯和矩形坯标准：YB 2011—83

2) The standard length of billets is 10000mm; the minimum length of short billets is 8000mm; the weight of each batch (furnace) of short billets is not more than 10% of the weight of all billets; the allowable deviation of length is 10000 + 70, 8000 - 0mm.

2) 钢坯标准长度为10000mm；短尺钢坯最短长度为8000mm；每批（炉）短尺钢坯的重量不大于全部钢坯重量的10%；长度允许偏差为10000+70，8000-0mm。

3) The bending degree of billet shall not be greater than 100mm when it is 10000mm; however, it is not allowed to be at both ends of billet, with the maximum of 50mm at both ends. For steel billets requiring fluorescent magnetic particle inspection, such as spring steel, bearing steel and cold heading steel, the bending per meter shall not be more than 6mm.

3) 钢坯的弯曲度在10000mm不大于100mm；但不允许在钢坯两端，两端最大50mm。需荧光磁粉探伤的钢坯，如弹簧钢、轴承钢、冷镦钢等，每米弯曲不大于6mm。

4) The maximum torsion of billet is 6° within 10000mm.

4) 钢坯扭转在10000mm内最大为6°。

5) The local spread caused by shear deformation at the end of billet shall not be greater than 10% of the side length. The burr of the cutting head shall be removed. The local bending of the end due to shear deformation shall not be more than 20mm. The shear end face shall be perpendicular to the axis of the billet length direction; the end face deflection shall not be greater than 6% of the side length.

5) 钢坯端部因剪切变形而造成的局部宽展不大于边长的10%。切头飞边应清除。端部因剪切变形造成的局部弯曲不得大于20mm。剪切端面应与钢坯长度方向轴线垂直；端面歪斜量不得大于边长的6%。

(3) Requirements for surface quality and internal quality.

(3) 对表面质量与内部质量的要求。

1) Billet.

1) 钢坯。

The requirements for the surface quality of the billet are: the end face of the billet shall be free of shrinkage cavity, tail hole and stratification; the surface of the billet shall be free of defects such as cracks, folds, ears, scabs, pull cracks and inclusions; the surface defects of the billet must be removed along the longitudinal processing, and the removal place shall be smooth and free of edges and corners, the removal width shall not be less than 5 times of the removal depth, and the surface cleaning depth shall not be greater than 8% of the nominal thickness.

对钢坯表面质量的要求是：钢坯端面不得有缩孔、尾孔和分层；钢坯表面应无裂纹、折叠、耳子、结疤、拉裂和夹杂等缺陷；钢坯表面缺陷必须沿纵向加工清除，清除处应圆滑、无棱角，清除宽度不得小于清除深度的5倍，表面清理深度不大于公称厚度的8%。

The requirements for the internal quality of the billet are as follows: the macrostructure of the billet shall be free of visible shrinkage, stratification, bubbles, cracks, white spots, etc.; for the high-quality carbon structural steel and spring steel, bearing steel, cold heading steel and other alloy steel, according to the requirements of the demander, the high-power inspection can be conducted to check the decarburization layer, check the non-metallic inclusions in the steel, and check whether the grain size meets the specified requirements.

对钢坯内部质量的要求是：钢坯低倍组织不得有肉眼可见的缩孔、分层、气泡、裂纹、白点等；对优质碳素结构钢和弹簧钢、轴承钢、冷镦钢等合金钢种，根据需方要求，可以做高倍检验，检查脱碳层，检查钢中非金属夹杂，检查晶粒度是否达到规定的要求。

2) Continuous casting slab.

2) 连铸坯。

The most common surface defects of continuous casting slab are pinhole and oxidation scar. The pinhole shall be worn away, and it shall be scrapped if it is serious. 100mm×100mm high carbon steel square billet is allowed to have 2mm deep pinholes. The oxidation scar can be removed by local polishing.

连铸坯最常见的表面缺陷是针孔及氧化结疤。针孔要求磨去，严重时报废。100mm×100mm的高碳钢方坯允许有2mm深的针孔。氧化结疤局部磨光即可清除。

The internal quality of continuous casting slab is often judged by the presence and severity of segregation, center porosity and cracks.

连铸坯的内部质量常以偏析、中心疏松和裂纹的有无和轻重为判断依据。

1.2.1.2　Inspection of Billet

1.2.1.2　坯料检查

(1) Billet inspection content.

(1) 坯料检查内容。

Billet inspection mainly includes:

坯料检查主要包括：

Blank specification. The continuous casting billet is 150mm×150mm, the length is 8000~

12000mm, and the short ruler shorter than 12000mm is not more than 10%.

坯料规格。连铸方坯 150mm×150mm，长度 8000~12000mm，短于 12000mm 的短尺不超过 10%。

The continuous casting billet standard shall be in accordance with continuous casting billet and rectangular billet (YB 2011—83), high quality carbon structural steel and alloy steel continuous casting billet and rectangular billet (YB/T 154—1999) and relevant regulations issued by the technical center.

连铸坯标准执行《连续铸钢方坯和矩形坯》（YB 2011—83）、《优质碳素结构钢和合金钢连铸方坯和矩形坯》（YB/T 154—1999）及技术中心下发的有关规定。

Treatment of unqualified billet. After the unqualified billet is removed at the removal device, the external inspector shall be informed to confirm and return it to the raw material warehouse. The weight of single billet shall be calculated according to the average weight of the heat number. When picking up the material in the raw material warehouse, if it is found that there are many unqualified billets in a certain heat number, the heat number can be replaced for picking up.

不合格钢坯的处理。不合格钢坯在剔除装置处剔除后，通知外检人员确认，退回原料库，其单根坯料重量按所属炉号平均重量计。到原料库领料时，如发现某一炉号的钢坯中不合格坯料数量较多，可更换炉号领料。

（2）Billet inspection method.

（2）钢坯检查方法。

In order to ensure the quality of round steel and smooth production, it is very important to check and clean the quality of billet. The quality inspection methods are as follows:

为确保圆钢材的质量及生产的顺利进行，对钢坯进行质量检查及清理是十分重要的。质量检查的方法有下述几种：

1) Visual inspection refers to manual inspection of surface defects of billet. This method is inefficient and unreliable, and only obvious defects can be detected. However, the equipment of this inspection method is few, only the bench and running roller table are needed.

1）目视检查即人工检查钢坯的表面缺陷。这种方法效率低，也不太可靠，只能检查出较明显的缺陷。但这种检查方法的设备少，只需要台架和运行辊道即可。

2) Electromagnetic induction flaw detection is used to check the surface defects of billet. There are many methods of electromagnetic induction testing, such as dry fluorescent magnetic particle testing, eddy current testing and magnetic recording testing.

2）电磁感应探伤检查钢坯的表面缺陷。电磁感应探伤的方法较多，例如干荧光磁粉探伤法、涡流探伤法、录磁探伤法。

3) Ultrasonic flaw detection is used to check the internal defects of square billet. This method is basically based on the principle of pulse echo (pulse reflection principle) by using the detection probe with transmitter and receiver.

3）超声波探伤检查方坯的内部缺陷。这种方法基本上是用带有发射和接收器的检测探头通过脉冲回波原理（脉冲反射原理）进行工作的。

1.2.1.3 Billet Cleaning
1.2.1.3 钢坯清理

The methods to clean the surface defects of billet include: manual flame gun cleaning, flame cleaner cleaning, manual air shovel cleaning, manual grinding wheel grinding and grinding machine cleaning. Artificial flame cleaning and air shovel cleaning have been widely used in the early stage of industrialization. They are the main cleaning methods to clean the surface defects of ingots and billets in China's iron and steel enterprises from 1950s to 1970s. At present, the main method of cleaning billet is grinding machine cleaning, a small number of local defects are assisted by manual grinding wheel cleaning.

清理钢坯表面缺陷的方法有：人工火焰枪清理、火焰清理机清理、人工风铲清理、人工砂轮修磨清理、修磨机清理。人工火焰清理和人工风铲清理在工业化初期曾被广泛用过，是20世纪50年代至70年代我国钢铁企业清理钢锭和钢坯表面缺陷的主要清理方法。现在清理钢坯的主要方法是修磨机清理，少量的局部缺陷辅助以人工砂轮修磨清理。

The old type of billet grinding machine adopts the structure of swinging grinding wheel. Now it has been replaced by the fixed grinding machine with fixed grinding wheel and trolley moving and controlled billet lifting and turning.

老式的钢坯修磨机采用的是摆动砂轮的结构，现在已为砂轮固定而小车移动并可控制钢坯升降和翻转的固定修磨机所代替。

Inspection and grinding line of billet.
钢坯的检查修磨作业线。

The continuous billet inspection and grinding line integrates shot blasting, flaw detection and grinding, and the middle is connected by track, with high automation, high efficiency and few operators.

连续式的钢坯检查修磨作业线将抛丸、探伤、修磨连成一体，中间以轨道联接，自动化程度高、效率高、操作人员少。

Generally, the steel billets are hoisted to the shot blasting feeding platform in groups by a crane, and the steel billets are moved into the V-shaped groove of the detection roller table by one action of the chain type steel moving machine, the push out device and the pull in device of the feeding platform. The track will send the billet into the shot blasting machine in a diamond state, and the shot blasting machine will spray steel shot at four sides of the billet at the same time to remove the oxide scale on the surface.

一般用吊车将钢坯成组地吊运至抛丸上料台架上，上料台架的链式移钢机、推出装置、拨入装置一次动作将钢坯移入检测辊道V形槽内。轨道将钢坯以菱形状态送入抛丸机内，抛丸机同时向钢坯四个面喷射钢丸，以清除表面的氧化铁皮。

After shot blasting, the steel billet is sent to the magnetizing device by the track, and the fluorescent magnetic powder liquid is sprayed on the surface of the steel billet. The magnetic force line at the surface defect is leaked to produce a magnetic force to absorb the magnetic powder to form a pattern, and then it is irradiated by the ultraviolet lamp in the darkroom to emit fluorescence, which is inspected manually and marked at the defect.

钢坯经抛丸处理后由轨道送进磁化装置,荧光磁粉液喷淋到钢坯表面,表面缺陷处的磁力线外泄产生磁力将磁粉吸住形成花纹,再经暗室中的紫外灯照射发出荧光,人工目视检查并在缺陷处做出标记。

The billets marked with defects shall be transported by rail to the billet repair mill to repair and grind the defective parts. For some steel with high quality requirements, such as stainless steel and valve steel, it is necessary to conduct 100% peeling and grinding, so it is not necessary to conduct surface flaw detection, but directly sent to the grinding machine for grinding.

标出缺陷的钢坯由轨道运至钢坯修磨机对缺陷部位进行修磨。对某些质量要求高的钢种如不锈钢、阀门钢等要进行100%的扒皮修磨,则不必再进行表面探伤,直接送至修磨机修磨。

1.2.2 Heating of Raw Materials
1.2.2 原料加热

1.2.2.1 Heating System
1.2.2.1 加热制度

(1) Heating steel.

(1) 加热钢种。

The heating steel types are shown in Table 1-3.

加热钢种见表1-3。

Table 1-3 The heating steel types
表1-3 加热钢种

Steel 钢种	Representative steel number 代表钢号
Carbon structural steel 碳素结构钢	Q195~Q235, Q255
High-quality carbon structural steel 优质碳素结构钢	45, 20
Low-alloy steel 低合金钢	HRB335, HRB400
Spring steel 弹簧钢	65Mn, 60Si2Mn, 55SiMnVB
Alloy structural steel 合金结构钢	40Cr, 40CrV, 20CrMo, 20CrMoA, 20Cr

(2) Billet heating temperature.

(2) 钢坯加热温度。

The heating temperature settings of different steel grades are different, including:

不同钢种加热温度设定不同,具体有:

1) Heating temperature setting of carbon structural steel, high quality carbon structural steel and low alloy steel. The tapping temperature is 1050~1150℃; the rolling temperature is 950~1050℃, as shown in Table 1-4.

1) 碳素结构钢、优质碳素结构钢、低合金钢加热温度设定。出钢温度1050~1150℃;开轧温度950~1050℃,见表1-4。

Table 1-4 Heating temperature of carbon structural steel, high quality
carbon structural steel and low alloy steel

表 1-4 碳素结构钢、优质碳素结构钢、低合金钢加热温度

Serial number 序号	Product specification 产品规格	Output/t·h^{-1} 产量/t·h^{-1}	Temperature setting value of heating section/℃ 加热段温度设定值/℃	Temperature setting value of soaking section/℃ 均热段温度设定值/℃
1	HRBϕ10×3 螺纹钢 ϕ10×3	92.5	1050	1160
2	HRBϕ12×3 螺纹钢 ϕ12×3	129.6	1110	1170
3	HRBϕ14×2 螺纹钢 ϕ14×2	140.2	1120	1170
4	ϕ14	73.6	1030	1150
5	ϕ15	83.8	1050	1150
6	ϕ16	94.6	1070	1160
7	ϕ17	105.9	1090	1160
8	ϕ18	117.8	1090	1170
9	ϕ19	130.0	1110	1170
10	ϕ20	142.7	1120	1170
11	ϕ21	150	1120	1180
12	Other specifications 其他规格	150	1120	1180

2) Heating temperature setting of alloy structural steel. The tapping temperature is 1050~1120℃; the rolling temperature is 950~1020℃, as shown in Table 1-5.

2) 合金结构钢加热温度设定。出钢温度 1050~1120℃；开轧温度 950~1020℃，具体见表 1-5。

Table 1-5 Heating temperature of alloy structural steel

表 1-5 合金结构钢加热温度

Serial number 序号	Product specification 产品规格	Output/t·h^{-1} 产量/t·h^{-1}	Temperature setting value of heating section/℃ 加热段温度设定值/℃	Temperature setting value of soaking section/℃ 均热段温度设定值/℃
1	ϕ14	73.6	1000	1130
2	ϕ15	83.8	1020	1130
3	ϕ16	94.6	1040	1140
4	ϕ17	105.9	1060	1140
5	ϕ18	117.8	1080	1150
6	ϕ19	130.0	1080	1150
7	ϕ20	142.7	1090	1150

Continued Table 1-5

Serial number 序号	Product specification 产品规格	Output/t·h⁻¹ 产量/t·h⁻¹	Temperature setting value of heating section/℃ 加热段温度设定值/℃	Temperature setting value of soaking section/℃ 均热段温度设定值/℃
8	φ21	150	1090	1160
9	Other specifications 其他规格	150	1090	1160

3) Spring steel heating temperature setting. The tapping temperature is 1050~1120℃; the rolling temperature is 950~1020℃, as shown in Table 1-6.

3) 弹簧钢加热温度设定。出钢温度1050~1120℃；开轧温度950~1020℃，具体见表1-6。

Table 1-6　Heating temperature of spring steel
表 1-6　弹簧钢加热温度

Serial number 序号	Output/t·h⁻¹ 产量/t·h⁻¹	Temperature setting value of heating section/℃ 加热段温度设定值/℃	Temperature setting value of soaking section/℃ 均热段温度设定值/℃
1	75	1000	1140
2	85	1010	1140
3	95	1030	1140
4	105	1050	1140
5	120	1070	1150
6	130	1090	1150
7	140	1100	1150
8	150	1110	1150

（3）Heating speed and time of billet.

（3）钢坯加热速度和加热时间。

Steel 钢种	Heating speed 加热速度
Low-carbon steel 低碳钢	6~9min/cm
Low-alloy steel 低合金钢	9~12min/cm
High-carbon steel 高碳钢	12~18min/cm
High alloy steel 高合金钢	18~24min/cm

The heating time of billet is the time of billet in furnace, that is, the time of billet from heating in furnace to heating out of furnace, that is, the sum of preheating time, heating time and soaking time. Although the heating time of billet can be calculated, but in the actual production process, the in furnace time of billet is affected by many process factors. Therefore, in the heating design and operation, the most basic heating time should be guaranteed, and the hot working parameters should be adjusted at any time according to the actual situation, so that the billet can be fully heated in the shortest possible time.

钢坯的加热时间,就是钢坯的在炉时间,即钢坯从进炉加热直到出炉的时间,也就是预热时间、加热时间、均热时间的总和。虽然钢坯的加热时间是可以计算的,但在实际生产过程中,钢坯的在炉时间受到诸多工艺因素的影响,因此,在加热设计和操作时,要保证最基本的加热时间,而且应根据实际情况,随时调整热工参数,使钢坯在尽量短的时间内得到充分的加热。

(4) Cooling system to be rolled.

(4) 待轧降温制度。

If the normal production is temporarily unavailable for some reason, the specific cooling range shall be determined according to the time to be rolled, as shown in Table 1-7.

因故而暂时不能正常生产时,应根据待轧时间而确定具体的降温幅度,规定见表1-7。

Table 1-7 Cooling system
表1-7 降温制度

Waiting time 待轧时间	Soaking section/℃ 均热段/℃	Heating section/℃ 加热段/℃
<10min	可不降温	可不降温
10~30min	980~1180	900~1150
0.5~1h	950~1150	850~1100
1~1.5h	900~1100	800~1050
1.5~2h	800~1100	750~1000
2~4h	800~1100	700~950

After rolling, it is necessary to raise the temperature again for rolling. The stoker shall follow the following system.

待轧后,需要重新升温而进行轧制,司炉工应遵循以下制度。

1) It takes 2 hours for furnace temperature to rise from 800℃ to 1200℃, and 15 minutes for steel tapping from 1200℃ to 1250℃.

1) 炉温从800℃升到1200℃需2h,从1200℃升到1250℃出钢需15min。

2) It takes 6 hours for the furnace temperature to rise from 500℃ to 1250℃.

2) 炉温从500℃升到1250℃出钢需6h。

3) It takes 12 hours for the furnace temperature to rise from 250℃ to 1250℃.

3) 炉温从250℃升到1250℃出钢需12h。

4) It takes 20~24 hours to raise the temperature from cold furnace to 1250℃.

4) 从冷炉提升到1250℃出钢需20~24h。

1.2.2.2 Heating Equipment
1.2.2.2 加热设备

(1) Heating furnace.
(1) 加热炉。

The continuous heating furnace is mainly used in the round steel rolling mill, and the side exit mode is often used. There are two ways to put billet into the furnace: side entry and end entry. The side entry door is small, which is easy to ensure the tightness of the furnace, but not as easy to arrange billets as end entry, so both ways are adopted.

圆钢材轧机所使用的主要是连续式加热炉，多采用侧出方式。钢坯入炉有侧入、端入两种方式，侧入式炉门小，易保证炉子的严密性，但不如端入式容易排列坯料，所以两种方式均有采用。

The continuous reheating furnace can be divided into push type and step type according to the operation mode of billet in the furnace. The walking beam furnace is divided into walking beam furnace, walking bottom furnace and walking beam bottom combined furnace. In recent years, most of the rolling mills have been built with walking beam furnace, because the walking beam furnace is more suitable for round steel production process and product quality requirements.

连续式加热炉按钢坯在炉内运行方式分为推钢式和步进式。步进式中又分为步进梁式炉、步进底式炉和步进梁底组合式炉。最近几年所建方钢材轧机大都选用步进式连续加热炉，因为步进式加热方式更适合圆钢材生产工艺和产品质量要求。

The main technical performance parameters of the furnace are:
炉子主要技术性能参数为：

Rated capacity of furnace: 90t/h
炉子额定能力：90t/h

Effective bottom strength: 500kg/(m^2·h)
有效炉底强度：500kg/(m^2·h)

Heating steel: ordinary carbon steel and low alloy steel such as Q235 and 20MnSi
加热钢种：Q235、20MnSi 等普碳钢和低合金钢

Billet size: 150mm×150mm×10m
坯料尺寸：150mm×150mm×10m

Billet heating temperature: 1050~1080℃, maximum 1150℃
钢坯加热温度：1050~1080℃，最高 1150℃

Air:
空气：

Preheating temperature: 450℃
预热温度：450℃

Working pressure: 4~6kPa
工作压力：4~6kPa

Maximum air consumption (α=1.1): 29600m^3/h

最大空气消耗量（α=1.1）：29600m³/h

Fuel: Mixed gas of blast furnace and coke oven

燃料：高焦炉混合煤气

Design calorific value: 7531kJ/m³

设计发热值：7531kJ/m³

Workshop gas contact pressure: 8000Pa

车间煤气接点压力：8000Pa

Maximum fuel consumption: 16000m³/h

最大燃料消耗量：16000m³/h

Hearth area: 190.8m²

炉底面积：190.8m²

Exhaust gas temperature: 300~350℃

排烟温度：300~350℃

Unit fuel consumption: under the condition of charging temperature of 20℃, discharging temperature of 1050~1150℃, output of 90t/h, air preheating temperature of 400~450℃, and 100% thermal insulation of walking beam, the unit heat consumption of furnace is 1.34GJ/t (45.7kg standard coal/T).

单位燃耗：在装料温度 20℃，出料温度 1050~1150℃，产量 90t/h，空气预热温度 400~450℃，步进梁 100% 绝热情况下，炉子的单位热耗为 1.34GJ/t（45.7kg 标煤/t）。

Compared with pusher furnace, the walking furnace has the following remarkable characteristics:

步进式炉与推钢式炉相比较，有以下显著特点：

1) In the pusher furnace, the operation of the billet depends on the thrust of the pusher to slide on the slide rail. Therefore, scratches often occur on the lower surface of the billet, which has a negative impact on the surface quality of the steel. However, in the step furnace, the operation of the billet is completed by the step beam and the bottom supporting→advancing→putting down, so there is no scratch.

1) 在推钢式炉内，钢坯的运行是靠推钢机的推力在滑轨上滑行的，因此，钢坯下表面往往产生划痕，对钢材表面质量带来不利影响。但在步进炉内，钢坯的运行是靠步进梁、底托起→前进→放下来完成的，所以不产生划痕。

2) In the push-type furnace, the temperature of the part of the billet in contact with the water-cooled slide rail is relatively low, and the "black mark" is serious, which greatly affects the dimensional deviation of the rolled product. Although the walking beam furnace has a water beam, the billet does not continuously contact the water beam, but intermittently and alternately contacts the water beam. The "black mark" phenomenon is lighter and the temperature difference is smaller, which greatly improves the effect on product size deviation.

2) 在推钢式炉内，钢坯接触水冷滑轨部分温度较低，"黑印"严重，对轧件的尺寸偏差影响很大，但步进底式炉无水冷滑轨不产生"黑印"，步进梁式炉虽有水梁，钢坯并不连续接触水梁，而是间断、交替地接触水梁，"黑印"现象较轻，温差较小，对产品尺寸偏差的影响大有改善。

3) In the pusher furnace, the billets are close together, which is easy to produce the phenomenon of "sticking steel" at high temperature, and can only be heated on one side or both sides, the heating speed is slow, and the temperature is not uniform enough, but in the walking furnace, there is a large gap between each group of billets, which not only avoids the phenomenon of "sticking steel", but also realizes the heating of three sides and four sides, with fast heating speed and uniform temperature.

3) 在推钢式炉内，钢坯是紧紧靠在一起的，高温下易产生"粘钢"现象，并且只能单面或双面受热，加热速度慢，温度不够均匀，但在步进炉内，每组钢坯间都留有较大的间隙，不仅避免了"粘钢"现象，而且实现三面、四面加热，加热速度快，温度均匀。

4) The effective length of pusher is limited by the section size of billet and the capacity of pusher, but the length of walking beam furnace is not limited by this limit, so the residual heat of flue gas can be fully used to preheat billet.

4) 推钢式炉在推钢时易发生拱钢事故，炉子有效长度受钢坯断面尺寸和推钢机能力的限制，但步进式炉其长度不受此限制，可充分利用烟气的余热预热钢坯。

5) The pusher type furnace can't be empty, so the flexibility of adjusting and repairing the empty furnace for different steel grades and different heating processes is very poor, but the step-by-step furnace is convenient, the step-by-step operation is flexible, and the adaptability of heating various kinds of feeds is strong.

5) 推钢式炉不能空炉，因此，对不同钢种不同加热工艺的调整、检修空炉等灵活性很差，但步进炉空炉方便，步进操作灵活，加热各种钢种的适应性强。

6) The step furnace has the advantages of fast heating speed, uniform temperature and flexible operation, so the burning loss of billet is reduced. The oxide scale of pusher furnace accounts for 1%~1.5% of the total weight of billet, and that of step furnace only accounts for 0.5%~0.8%, which reduces the labor intensity of slag cleaning.

6) 步进炉加热速度快，温度均匀，操作灵活，因此，减少了钢坯的烧损，推钢式炉的氧化铁皮占钢坯总重的1%~1.5%，步进式炉的仅占0.5%~0.8%，减轻了清渣劳动强度。

7) The total cooling surface area of the fixed beam, walking beam and supporting pipe of the walking beam furnace is about twice as large as that of the bottom pipe of the pusher type furnace, so its heat consumption is 15%~20% larger than that of the pusher type furnace, and its water consumption is about 60% larger than that of the pusher type furnace. In recent years, a series of measures have been taken abroad to improve the structure of stepping mechanism, heat insulation structure and sealing device, so as to reduce the water consumption of walking beam furnace, and its unit heat consumption is close to that of pusher furnace.

7) 步进梁式炉的固定梁、步进梁和支撑管总的冷却表面积，约比推钢式炉底管的冷却表面积大一倍，故其热耗比推钢式炉的大15%~20%，水耗比推钢式炉的大60%左右。近几年，国外采取一系列措施改进步进机构、隔热结构和密封装置的结构，使步进梁式炉的水耗下降，其单位热耗已接近推钢式炉的单位热耗。

8) The total maintenance cost of pusher furnace is 50% higher than that of walking beam furnace due to the vibration of pusher, low service life of insulation layer of furnace bottom pipe, and heavy maintenance work of full bottom soaking pit bed. However, the maintenance cost of driving mechanism of walking beam is higher than that of steel furnace.

8) 推钢式炉由于推钢发生振动，炉底管隔热层的使用寿命低，加之实底均热炉床的维护工作量大，其总的维护费用比步进梁式炉的大 50%，但其中步进梁传动机构的维护费用比推钢式炉大。

9) In terms of investment, in terms of furnace body, the step-by-step furnace is 10% ~ 15% higher than the pusher furnace. However, as for the whole system of furnace body, foundation, pusher furnace and building, due to the small investment in pusher and equipment foundation of walking beam furnace, the proportion of investment difference is relatively reduced.

9) 在投资方面，就炉体而言，步进式炉的比推钢式炉的高 10% ~ 15%。但就炉体、基础、推钢炉和建筑物整个系统而言，因步进式炉的推钢机及设备基础投资少，投资的差额比例相对缩小。

The walking beam furnace is divided into walking beam furnace, walking beam furnace and walking beam bottom combined furnace according to the different walking mechanism of the furnace. The furnace bottom of walking beam furnace is composed of two parts: fixed bottom and preliminary bottom, which can be arranged for upper heating; walking beam furnace has fixed beam and walking beam, which can be arranged for upper and lower heating; walking beam bottom combined furnace not only has fixed bottom and walking bottom, but also has fixed beam and walking beam, some pre heating sections are designed as walking beam, heating and soaking sections are designed as walking bottom, some preheating sections are designed as walking beam. The heating and soaking sections are designed as walking beam type, which can be arranged with upper heating and lower heating.

步进式加热炉根据炉子的步进机构不同分为步进底式炉、步进梁式炉和步进梁底组合式炉。步进底式炉的炉底由固定底和步进底两部分组成，可布置上加热；步进梁式炉有固定梁和步进梁，可布置上、下加热；步进梁底组合式炉不仅有固定底和步进底，而且有固定梁和步进梁，有的预热段设计成步进梁式，加热和均热段设计成步进底式，有的预热段设计成步进底式，加热、均热段设计成步进梁式，可布置上加热和下加热。

The stepping mechanism of walking heating furnace is composed of driving system, stepping frame and control system. The stepping system can be generally divided into two types: electric type and hydraulic type. The driving of the moving part is generally realized by the hydraulic system. Some of the lifting parts adopt the electric cam type, and some adopt the hydraulic curved rod type or the hydraulic inclined rail type. The electric cam type and the hydraulic curved bar type mechanism adopt the single-layer step frame, the hydraulic inclined rail type adopts the double-layer step frame, and the step frame slides on the roller through the track. The constant power pump with flow and pressure compensation and the full hydraulic material rail stepper with large capacity proportional valve operate stably and reliably. When the lifting hydraulic cylinder drives the lifting frame to move up and down on the inclined rail, the stepping bottom and the stepping beam move forward and backward. The action and movement speed of hydraulic cylinder are controlled by the opening, closing and opening degree of hydraulic valve, while the action of valve is controlled by PLC program. The signals of each step action point are sent out by the contact switch and programmable travel switch installed on the corresponding parts.

步进式加热炉的步进机构由驱动系统、步进框架和控制系统组成。步进系统一般可分为电动式和液压式两种，行进部分的驱动一般靠液压系统实现，升降部分有的采用电动凸轮式，也有的采用液压曲杆式或液压斜轨式。电动凸轮式和液压曲杆式机构采用单层步进框架，液压斜轨式采用双层步进框架，步进框架通过轨道在滚轮上滑动。采用流量、压力补偿的恒功率泵和大容量比例阀的全液压料轨式步进机运行稳定可靠。升降液压缸带动升降框架在斜轨上做升降运动时，步进底和步进梁便随着作前进和后退动作。液压缸的动作和其运动速度是由液压阀门的开、闭和开启度的大小来控制的，而阀门的动作则由PLC程序控制。步进各动作点的信号是由安装在相应部位上的接触开关和可编程行程开关发出。

The step bottom and the step beam move up, forward, down and backward periodically with the movement of the step frame. The action sequence is different according to the position of the starting point. Generally, the starting point is set at the position of the step beam falling down and backward. Then the step beam moves from the rising to the original starting position after the action, and completes one cycle after moving forward, down and backward step by step. The trace of step action curve is rectangular. In the process of ascending, the walking bottom and walking beam lift the billet from the fixed bottom and fixed beam, advance one step, put the billet back to the fixed bottom and fixed beam during descending, and then return to the original position. In this way, when the step bottom and the step beam complete the step action of each cycle, the billet will be transported forward by one step, and in the process of continuous action, the billet will be transported forward step by step. In order to eliminate the impact of stepping action, its running speed is strictly controlled by proportional valve in each step action. There is a slow acceleration process at the beginning of each action, and a deceleration process at the end. There is a short deceleration process when the billet is to be lifted and lowered during the rise and fall.

步进底和步进梁随着步进框架的运动作上升、前进、下降、后退的周期性动作，其动作顺序根据起始点位置的不同而不同，一般将起始点位置设在步进梁下降和后退的位置上，则步进梁的动作起动后由上升开始，经前进、下降，后退到原来的起始位置，完成一个周期的步进动作。步进动作曲轨迹呈矩形。步进底和步进梁在上升过程中，从固定底和固定梁上抬起钢坯，前进一个步距，在下降的过程中将钢坯放回到固定底和固定梁上，然后回到原位。这样，步进底和步进梁每完成个周期的步进动作，将钢坯向前输送一个步距，在连续动作的过程中将钢坯一步一步向前输送。为了消除步进动作时的冲击，在每个步进动作中其运行速度是用比例阀严格控制的，每个动作开始有缓慢的加速过程，结束时又有减速过程，在上升和下降时，将要抬起钢坯和放下钢坯时，有短时间的减速过程。

(2) Heating furnace area equipment.

(2) 加热炉区设备。

1) Mechanical equipment.

1) 机械设备。

The mechanical equipment in the furnace area includes charging platform, steel separator, knockout device, roller table in front of the furnace, two lifting baffles, weighing device, charging furnace door, charging cantilever roller table in the furnace, aligned pusher, furnace bottom machinery, discharging furnace door, discharging cantilever roller table in the furnace, fan, etc.

炉区机械设备包括上料台架、分钢机、剔除装置、炉前辊道、两个升降挡板、称重装置、装料炉门、炉内装料悬臂辊道、对齐推钢机、炉底机械、出料炉门、炉内出料悬臂辊道、风机等。

2) Hydraulic system equipment.

2) 液压系统设备。

Hydraulic system equipment includes heating furnace hydraulic station, weighing device hydraulic cylinder, pusher hydraulic steel, furnace bottom mechanical translation and lifting hydraulic cylinder, relevant hydraulic pipeline, etc.

液压系统设备包括加热炉液压站、称重装置液压缸、推钢机液压钢、炉底机械平移与升降液压缸、相关液压管路等。

1.2.2.3 Oxidation and Burning Loss of Billet
1.2.2.3 钢坯的氧化烧损

In the heating process of billet, the iron element on the surface of billet reacts with the oxidizing gas in the furnace gas to generate iron oxide, which causes the loss of metal. This phenomenon is called the oxidation burning loss of billet. The oxidation burning loss of billet is expressed by burning loss rate, which means the percentage of the weight of burned metal in the total weight of billet. Oxidation burning loss will reduce the yield of copper billet. If the oxide scale on the surface of billet cannot fall off during rolling, it will cause surface indentation and directly affect the quality of steel after rolling. When the oxidation and burning in the furnace are serious, the iron oxide scale accumulates in the furnace and makes the furnace bottom higher, which brings difficulties to the maintenance of furnace bottom and the operation of billet in the furnace. In the step furnace, the step operation of the billet is blocked after the fixed bottom is increased, and the discharging is difficult after the bottom slagging in the discharging section, which directly affects the production. When the continuous casting slab is directly used as wire rod billet, the surface of the continuous casting slab has a lot of oxide scales and is very rough. Therefore, the oxidation and burning loss of the continuous casting slab is large, which brings great harm to the furnace. Obviously, reducing oxidation and burning loss is of great significance to wire rod production.

在钢坯的加热过程中，钢坯表面的铁元素与炉气中的氧化性气体发生氧化反应，生成铁的氧化物，造成金属的损失，这种现象称为钢坯的氧化烧损。钢坯的氧化烧损用烧损率来表示，其意义就是烧损掉的金属重量在钢坯总重量中所占的百分比。氧化烧损必然降低铜坯的成材率。如果钢坯表面的氧化铁皮在轧制过程中不能脱落，轧后，氧化铁皮脱落则会造成钢坯表面压痕，直接影响钢材质量。当炉内的氧化烧损严重时，氧化铁皮在炉内堆积使炉底增高，给炉底的维护和钢坯在炉内的运行带来困难。在步进式炉中，固定底增高后钢坯的步进运行受阻，出料段炉底结渣后侧出料困难，直接影响生产。当连铸坯直接作为线材用坯时，由于连铸坯表面氧化铁皮多，又很粗糙，因此，连铸坯的氧化烧损大，给炉子带来的危害也大。显然，减少氧化烧损，对线材生产具有重要的意义。

There are many factors influencing the oxidation loss, among which the main ones are the atmosphere in the furnace, the heating temperature and the heating time. The oxidation capacity of the furnace gas depends on the ratio of the concentration of the oxidizing atmosphere to the concentration of the reducing atmosphere. The more the content of the oxidizing gas is, the stronger the oxidation capacity of the furnace gas is SO_2, H_2O, O_2 and CO_2 in the furnace have strong oxidation capacity, of which SO_2 has the strongest oxidation capacity. When there is SO_2 in the furnace gas, over 1100℃, low melting point compound [FeO·FeS] is produced on the surface of billet, and the new surface of billet is exposed continuously, resulting in intense oxidation. When the content of SO_2 reaches 0.1%~0.2%, the burning loss increases 1~2 times. The oxidation capacity of H_2O in furnace gas is also strong, which is twice of 2 and 2. The influence of excess air on the oxidation loss is not obvious. When the excess air coefficient increases from 0.9, the oxidation loss increases. When the excess air coefficient exceeds 1.2, the oxidation loss does not increase significantly, because the oxidation speed mainly depends on the diffusion speed of oxygen.

影响氧化烧损的因素较多，其中炉内气氛和加热温度及加热时间的影响是主要的。炉气的氧化能力取决于氧化性气氛的浓度与还原性气氛的浓度之比，氧化性气体含量越多，炉气的氧化能力越强。炉中的 SO_2、H_2O、O_2 和 CO_2 具有较强的氧化能力，其中 SO_2 的氧化能力最强。当炉气中有 SO_2 时，在 1100℃ 以上，钢坯表面产生 [FeO·FeS] 低熔点化合物，钢坯不断暴露新表面，造成激烈的氧化。当 SO_2 的含量（质量分数）达到 0.1%~0.2%时，烧损增加 1~2 倍。炉气中 H_2O 的氧化能力也较强，是 CO_2 和 O_2 的 2 倍。过剩空气对氧化烧损的影响不明显，空气过剩系数由 0.9 逐步增加时，氧化烧损有所增加，超过 1.2 时烧损并不显著增加，因为，此时氧化速度主要取决于氧气的扩散速度。

The oxide scale formed in reducing atmosphere has a compact texture, which is not easy to fall off on the surface of billet and easy to cause surface defects of bar. When the content of some alloy components such as Cr, Ni, Mo in steel increases, the process of iron oxide scale formation is intensified. The oxidation rate of billet is accelerated with the increase of temperature. Below 700℃, the oxidation rate is not significant. At 900℃, the oxidation rate begins to increase. At 1100℃, the oxidation rate rises sharply. If the oxidation loss of billet at 900℃ is 1, then the oxidation loss at 1000℃ is 2, 1100℃ is 4, 1200℃ is 8~9, and 1300℃, which can reach more than ten or more. The scale formed below 1270℃ is easy to fall off, and the scale formed above 1300℃ is not easy to fall off. The heating time of billet (i.e. in furnace time) is also very important to the oxidation burning loss. The longer the heating time is, the more oxide scale is generated, and the thicker the scale is, and it is not easy to fall off. Especially in the high temperature section, the longer the residence time, the more serious the oxidation and burning loss. At 1100℃, the oxidation loss rate increases by more than 0.5% for every 1 hour.

在还原性气氛中生成的氧化铁皮其质地致密，粘着在钢坯表面不易脱落，容易造成棒材表面缺陷。当钢内有些合金成分如 Cr、Ni、Mo 的含量增高时，生成氧化铁皮的过程加剧。钢坯的氧化速度随着温度的升高而加快，在 700℃ 以下，氧化不显著，900℃ 开始加剧，1100℃ 时氧化量急剧上升。若以钢坯在 900℃ 时的氧化烧损量为 1，则 1000℃ 时的为 2，1100℃ 时的为 4，1200℃ 时的为 8~9，1300℃ 时的可达十几或更多。在 1270℃ 以下生成的氧化铁皮易脱落，在 1300℃ 以上生成的氧化铁皮不易脱落。钢坯的加热时间（即在炉时间）对氧化烧损的影响也很重要，加热时间越长，生成的氧化铁皮越多，铁皮层也越厚，又不易脱落。尤其在高温段停留的时间越长氧化烧损越严重。在 1100℃ 温度下每多停留 1h，氧化烧损率增加 0.5% 以上。

In a word, the formation of oxide scale not only affects the yield and steel quality, but also brings great difficulties to the maintenance of furnace bottom. The heating furnaces for bar production are very wide, and the maintenance of furnace bottom is very difficult. Therefore, it is of great significance to reduce the oxidation loss to keep the furnace condition normal and the production smooth. In order to reduce the oxidation and burning loss of billet, it is necessary to reduce the heating temperature of billet as much as possible, implement rapid heating, shorten the furnace time of billet and control the oxidation atmosphere. The oxidation loss rate of walking beam furnace is lower than that of pusher furnace. The burning loss rate of the step-by-step wire heating furnace abroad is 0.4%.

总之，氧化铁皮的生成不仅影响成材率和钢材质量，而且给炉底的维护带来很大困难。棒材生产用的加热炉，都很宽，炉底的维护难度很大。因此，降低氧化烧损对保持炉况正常，生产顺利，具有重要意义。为了减少钢坯的氧化烧损，应在可能的情况下尽力降低钢坯的加热温度，实行快速加热，缩短钢坯的在炉时间，控制氧化气氛。步进式加热炉的氧化烧损率低于推钢式加热炉。国外步进式线材加热炉的氧化烧损率达到 0.4%。

1.2.2.4　Billet Heating Defects
1.2.2.4　钢坯加热缺陷

(1) Decarburization of billet surface.
(1) 钢坯表面脱碳。

During the heating process, the carbon element on the surface of billet is oxidized, which reduces the carbon content on the surface of billet. This phenomenon is called decarburization of billet surface. Steel with carbon content more than 0.35%~0.4% has decarburization tendency. Due to decarburization, the carbon content of steel surface and inside is inconsistent, which reduces the strength of steel and affects the service performance of steel. The main reactions of decarbonization process are as follows:

钢坯在加热过程中，其表面的碳元素被氧化，使钢坯表面含碳量减少，这种现象被称为钢坯的表面脱碳。含碳量（质量分数）大于 0.35%~0.4% 的钢都具有脱碳倾向。由于脱碳，钢材表面与内部的含碳量不一致，降低了钢材的强度，影响了钢材的使用性能。脱碳过程的主要反应如下：

$$Fe_3C + H_2O \longrightarrow 3Fe + CO + H_2$$

$$Fe_3C+CO_2 \longrightarrow 3Fe+2CO$$
$$2Fe_3C+O_2 \longrightarrow 6Fe+2CO$$
$$Fe_3C+2H_2 \longrightarrow 3Fe+CH_4$$

Decarburization and oxidation occur at the same time, and decarburization is easy to take place in oxidizing atmosphere. Decarburization is slow when the billet temperature is below 700℃. When the temperature reaches 800~850℃, decarburization begins to be serious. With the intense oxidation, decarburization slows down. The longer the heating time, the more serious the decarburization. High carbon steel and high silicon steel are easy to decarburize, and the elements such as Al, CO and W in the steel accelerate decarburization, and the elements such as Cr and Mn inhibit decarburization. The steel grades prone to decarburization should be heated slowly in the low temperature section and rapidly in the high temperature section, so as to shorten the retention time in the high temperature section as much as possible.

脱碳过程与氧化过程是同时发生的，在氧化性气氛下容易发生脱碳反应。钢坯温度低于700℃时脱碳缓慢。温度达到800~850℃时脱碳开始严重起来，随着氧化的激烈，脱碳有所减缓。加热时间越长，脱碳越严重。高碳钢、高硅钢容易脱碳，钢中Al、Co、W等元素加快脱碳，Cr、Mn等元素抑制脱碳。容易发生脱碳的钢种宜在低温段缓慢加热，而在高温段快速加热，尽量缩短在高温段的滞留时间。

（2）Overheating and overburning of billets.

（2）钢坯的过热和过烧。

The overheating and overburning of steel in the heating process mean that the crystal structure of steel has changed. When the billet is heated at high temperature for a long time, the grain of the steel grows continuously, when the grain grows to a certain extent. The bond strength between grains is weakened and the plasticity of steel is deteriorated. This phenomenon is the overheating of steel.

钢在加热过程中的过热和过烧都意味着钢的结晶组织发生了变化。钢坯在高温下长时间加热时，钢的晶粒不断长大，当晶粒长大到一定程度时。晶粒间结合力减弱，钢的塑性变坏。这种现象就是钢的过热。

The overheated billet cracks in the rolling process, making the product scrapped. If the billet temperature continues to rise and reaches the liquidus of the iron carbon equilibrium diagram, the grain boundary of the steel begins to melt. In the solidification process of steel, the solidification point of non-metallic inclusions is low, which is left between the metal grains and finally solidified. When the temperature increases, the inclusions with low melting point melt first. Once the grain boundary begins to melt, the crystal structure of the steel will be destroyed and the plasticity and strength of the metal will be lost. This phenomenon is called overburning of steel. The billet is easy to break and break after over burning, which will cause pushing accident when it is fed into the rolling mill.

过热的钢坯在轧制过程中产生裂纹，使产品报废。如果钢坯温度继续上升，达到铁碳平衡图的液相线时，钢的晶粒边界便开始熔化。因为，钢在凝固过程中，非金属夹杂的凝固点较低，被留在金属晶粒之间最后凝固，当温度升高时，熔点低的夹杂先熔化。一旦晶粒边界开始熔化,则钢的结晶组织遭到破坏,失去金属应具有的塑性和强度。这种现象称为钢的过烧。钢坯过烧后易折断和碎裂,喂入轧机轧制时便造成堆钢事故。

Overheating and overburning are serious heating quality accidents. Overburning occurs on the basis of overheating, which is more serious. If the overheated billet is not rolled, it can be cooled to below 700℃ and then reheated for use. However, the overheated billet cannot be restored to its original structure state and can only be scrapped. The melting point of high carbon steel is lower than that of low carbon steel. If the maximum heating temperature is not properly controlled, the phenomenon of sodium billet overheating and overburning often occurs. When the rolling production line suddenly breaks down and stops rolling, the furnace temperature is not controlled in time, which is easy to cause overheating or overburning. In the process of billet heating, as long as the furnace temperature and billet heating temperature are strictly controlled, and the furnace condition is adjusted in time when the rolling line breaks down, all kinds of heating defects can be completely avoided.

钢坯过热和过烧都是严重的加热质量事故,过烧是在过热的基础上发生的,情况更为严重。过热的钢坯若未经轧制,可将其冷却至700℃以下。然后重新加热使用,而过烧的钢坯无法恢复原来的组织状态,只能报废。高碳钢较低碳钢熔点低,若最高加热温度控制不当,往往发生钠坯过热和过烧现象,当轧制作业线突然出现故障停轧时,对炉温控制不及时,很容易造成过热或过烧。在钢坯的加热过程中,只要严格控制炉子温度和钢坯的加热温度,并在轧制作业线出现故障时及时调整炉况,则各种加热缺陷是完全可以避免的。

Task 1.3　Rolling of Steel Bar
任务1.3　棒材的轧制

Mission objectives
任务目标

(1) Master the rolling theoretical basis of bar round steel and pass design of round steel;

(1) 掌握棒材圆钢的轧制理论基础,掌握圆钢的孔型设计;

(2) Master the structure and application of the guide and guard device in the rolling of bar and round steel, master the basic knowledge of roll;

(2) 掌握棒材圆钢轧制中的导卫装置的构造和用途,掌握轧辊的基本知识;

(3) Be able to arrange, install, dismantle and adjust the guide rail on the rolling line;

(3) 能够布置轧线上的导卫,安装、拆卸导卫并调整;

(4) Be able to dismantle and replace the roll, operate the rolling mill for rolling production.

(4) 能够拆卸和更换轧辊,操作轧机进行轧制生产。

1.3.1 Theoretical Basis of Rolling
1.3.1 轧制理论基础

1.3.1.1 Rolling and Conditions for Its Realization
1.3.1.1 轧制及其实现的条件

Rolling, also known as rolling, refers to the pressure processing process in which metal is compressed between rotating rolls to produce plastic deformation.

轧制又称压延,是指金属通过旋转的轧辊间受到压缩而产生塑性变形的压力加工过程。

(1) Purpose of rolling.

(1) 轧制的目的。

The two tasks of rolling process are precise forming and improvement of structure and performance, so rolling is a central link to ensure the quality of products.

轧钢工序的两个任务是精确成型和改善组织、性能,因此轧制是保证产品实物质量的一个中心环节。

In the aspect of precise forming, it is required that the product shape is correct, the size is accurate and the surface is complete and smooth. Pass design and mill adjustment are the decisive factors for precise forming. Deformation temperature, speed regulation (through the influence of deformation resistance) and wear of roll tools also have a very important influence on precise forming. In order to improve the accuracy of product size, it is necessary to strengthen the process control, which not only requires the pass design to be reasonable, but also to maintain the stability of rolling deformation conditions as much as possible, mainly the stability of temperature, speed, front and rear tension and other conditions.

在精确成型方面,要求产品形状正确,尺寸精确,表面完整光洁。对精确成型有决定性影响的因素是孔型设计和轧机调整,变形温度、速度规程(通过对变形抗力的影响)和轧辊工具的磨损等也对精确成型有很重要的影响。为了提高产品尺寸的精确度,必须加强工艺控制,不仅要求孔型设计合理,而且也要尽可能保持轧制变形条件稳定,主要是温度、速度及前后张力等条件的稳定。

In improving the properties of steel, the decisive factors are the thermal dynamic factors of deformation, mainly the deformation temperature, deformation speed and deformation degree.

在改善钢材性能方面,有决定性影响的因素是变形的热动力因素,主要是变形温度、变形速度和变形程度。

Generally speaking, the larger the deformation degree is, the stronger the three-dimensional pressure state is, the more favorable the microstructure and properties of hot rolled steel are. This is because:

变形程度与应力状态对产品组织性能的影响,一般来说,变形程度越大,三向压力状态越强,对于热轧钢材的组织性能越为有利。这是因为:

1) It is advantageous to break the dendrite segregation and carbide of the alloy composition

in the metal and to change its as cast structure. Therefore, it is necessary to use rolling or forging to process with large total deformation degree, so as to fully break the casting structure, make the steel structure compact and even carbide distribution.

1) 变形程度大，应力状态强，有利于破碎金属内部合金成分的枝晶偏析及碳化物，且有利于改变其铸态组织。因此，需采用轧制或锻造，以较大的总变形程度进行加工，才能充分破碎铸造组织，使钢材组织致密，碳化物分布均匀。

2) In order to improve the mechanical properties, it is necessary to improve the casting structure of the metal and make the steel structure compact, that is, to ensure a certain degree of total deformation, that is, to ensure a certain compression ratio.

2) 为改善力学性能，必须改善金属的铸造组织，使钢材组织致密，即要保证一定的总变形程度，也就是保证一定的压缩比。

3) When the total deformation degree is certain, the distribution of the deformation amount of each pass also has a certain impact on the product quality. This is to consider the recrystallization characteristics of the steel. If the fine and uniform grain size is required, it is necessary to avoid falling into the critical reduction range of coarse grains.

3) 在总变形程度一定时，各道变形量的分配对产品质量也有一定的影响。这是考虑钢种再结晶的特性，如果是要求细致均匀的晶粒度，就必须避免落入使晶粒粗大的临界压下量范围内。

The rolling temperature schedule shall be determined according to the relevant data of plasticity, deformation resistance and steel characteristics, so as to ensure that the products are correctly formed without cracks, the structure and performance are qualified and the force and energy consumption is small. The determination of rolling temperature mainly includes the determination of start rolling temperature and finish rolling temperature. The determination of the start rolling temperature must be based on the guarantee of the final rolling temperature; the final rolling temperature varies with the steel type, which mainly depends on the structure and properties specified in the product technical requirements.

轧制温度规程要根据有关塑性、变形抗力和钢种特性等数据来确定，以保证产品正确成型而不出现裂纹，组织、性能合格及力能消耗少。轧制温度的确定主要包括开轧温度和终轧温度的确定。开轧温度的确定必须以保证终轧温度为依据；终轧温度因钢种不同而不同，它主要取决于产品技术要求中规定的组织性能。

The deformation speed or rolling speed mainly affects the output of rolling mill, so improving the rolling speed is one of the main ways to improve the productivity of modern rolling mill. However, the improvement of rolling speed is limited by a series of equipment and process factors such as motor capacity, rolling mill equipment and temperature, mechanical automation level, bite condition and billet specification. The influence of rolling speed or deformation speed through hardening and recrystallization also has a certain impact on the microstructure and properties of steel. In addition, the change of rolling speed through the influence of friction coefficient, also often affects the quality index of steel size accuracy.

变形速度或轧制速度主要影响到轧机产量，因此提高轧制速度是现代轧机提高生产率的主要途径之一。但轧制速度的提高受到电机能力、轧机设备及温度、机械自动化水平以及咬入条件和坯料规格等一系列设备和工艺因素的限制,轧制速度或变形速度通过硬化和再结晶的影响也对钢材组织性能产生一定的影响。此外,轧制速度的变化通过摩擦系数的影响,还经常影响到钢材尺寸精确度等质量指标。

(2) Conditions for realizing rolling process.

(2) 实现轧制过程的条件。

1) Bite condition.

1) 咬入条件。

Depending on the friction between the rotating roll and the rolled piece, the phenomenon that the roll drags the roll is called biting. In order to realize the plastic deformation between the roll and the work piece, the roller must have the same horizontal force on the work piece as the rolling direction.

依靠旋转的轧辊与轧件之间的摩擦力,轧辊将轧件拖入轧辊之间的现象称为咬入。为使轧件进入轧辊之间实现塑性变形,轧辊对轧件必须有与轧制方向相同的水平作用力。

The biting process of the rolled piece is shown in Figure 1-2.

轧件的咬入过程如图1-2所示。

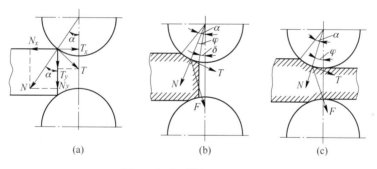

Figure 1-2 Biting process
图 1-2 咬入过程

The rolling piece first contacts two points on the circumference of the roll [Shown in Figure 1-2 (a)], which is affected by the force V exerted by the roll. At the same time, due to the friction between the two, the friction between the roll and the rolling piece tries to drag the rolling piece between the rolls, which has friction T on its effect. Obviously, in order for the roll to bite into the rolled piece, it must meet the following requirements:

轧件首先与轧辊在圆周上的两点接触[见图1-2 (a)],受到轧辊对其的作用力 V,同时由于两者之间存在摩擦,轧辊对轧件的摩擦力试图将轧件拖入轧辊之间,对其作用有摩擦力 T。显然,欲使轧辊咬入轧件,必须满足:

$$\sum F_x = T_x - N_x > 0$$

Because:

因为:

$$N_x = N\sin\alpha; \quad T_x = Nf\cos\alpha$$

Where α——Biting angle, (°);
　　　f——Friction coefficient between roll and rolled piece.
式中 α——咬入角, (°);
　　　f——轧辊与轧件间的摩擦系数。

Substitute for:
代入得:

$$Nf\cos\alpha - N\sin\alpha > 0$$

Simplify to get:
简化可得:

$$\tan\alpha < f$$

And:
而:

$$f = \tan\beta$$

Where β——Friction angle, (°).
式中 β——摩擦角, (°)。

Available: $\alpha < \beta$
可得: $\alpha < \beta$

It can be concluded that the natural biting angle of the rolled piece by the roll should meet the condition that the biting angle α is less than the friction angle β.

由此可得出结论: 轧件被轧辊自然咬入应满足咬入角 α 小于摩擦角 β 的条件。

2) Stable rolling conditions.

2) 稳定轧制条件。

When the rolled piece is bitten by the roll, it begins to fill the roll gap gradually, shown in Figure 1-2 (b)、(c). In the process of filling the roll gap, the included angle δ between the front end of the rolled piece and the axis of the roll decreases continuously. When the rolled piece is fully filled with the roll gap, $\delta = 0$, it begins to enter the stable rolling stage.

当轧件被轧辊咬入后开始逐渐充填辊缝, 见图 1-2 (b)、(c), 在轧件充填辊缝的过程中, 轧件前端与轧辊轴心连线间的夹角 δ 不断减小, 当轧件完全充满辊缝时, $\delta = 0$, 开始进入稳定轧制阶段。

3) Ways to change biting conditions.

3) 改变咬入条件的途径。

According to the biting condition $\alpha < \beta$, it can be concluded that all factors that can improve the angle and all factors that can reduce the angle are conducive to biting.

根据咬入条件 $\alpha < \beta$ 可得出: 凡是能提高 β 角的一切因素和降低角 α 的一切因素都有利于咬入。

Because:
由于:

$$\alpha = \arccos(1 - \Delta h/D)$$

Where Δh——Reduction, mm;
　　　D——Roll diameter, mm.

式中 Δh——压下量,mm;
D——轧辊直径,mm。

Then, in order to reduce the angle, the diameter D of the roll can be increased and the reduction of the reduction value Δh can be reduced.

那么,为降低 α 角,就可以增大轧辊直径 D 和减小压下量 Δh。

The commonly used methods to reduce the angle α in actual production are as follows:

实际生产中常用的降低 α 角的方法有:

①At this time, the corresponding biting angle is small.

①以小头先送入轧辊或以带楔形的钢坯进行轧制,此时对应的咬入角较小。

②Force to bite, that is, to force the rolled piece into the roll with external force, such as using pinch roll. The external force flattens the front end of the rolled piece, which is equivalent to reducing the contact angle, thus improving the biting condition.

②强迫咬入,即用外力将轧件强制送入轧辊中,如利用夹送辊。外力作用使轧件前端压扁,相当于减小接触角,从而改善咬入条件。

It is more complicated to improve the friction coefficient or friction angle, because under rolling conditions, the friction coefficient depends on many factors, such as the surface state and chemical composition of tools and deformed metals, unit pressure of contact surface, temperature conditions, rolling speed, process lubricant, etc.

提高摩擦系数或摩擦角是较复杂的,因为在轧制条件下,摩擦系数决定于许多因素,如工具、变形金属的表面状态和化学成分,接触表面的单位压力,温度条件,轧制速度,工艺润滑剂等。

In the actual production, the biting conditions are mainly improved from the following two aspects:

实际生产中主要从以下两个方面改善咬入条件:

①Change the surface state of the rolled piece or roll to improve the friction angle. For example, remove the oxide scale on the billet (the oxide scale on the billet surface reduces the friction coefficient); or intentionally make the surface of the rolling groove rough during the pass turning or mark the groove hole before use to increase the friction coefficient.

①改变轧件或轧辊的表面状态,以提高摩擦角。如清除坯料上的氧化铁皮(钢坯表面的氧化铁皮使摩擦系数降低);也可在孔型车削时有意使得轧槽表面粗糙或使用前在槽孔上刻痕,以增大摩擦系数。

②Adjust rolling speed reasonably. The practice shows that the friction coefficient decreases with the increase of rolling speed. Therefore, it can achieve natural biting at low speed, and then improve the biting condition with the rolling piece entering the roll gap, and then gradually increase the rolling speed to a stable rolling state.

②合理调整轧制速度。实践表明,摩擦系数是随轧制速度的提高而降低的。因此,可以实现低速自然咬入,之后随着轧件进入辊缝使咬入条件好转,再逐渐提高轧制速度达到稳定的轧制状态。

1.3.1.2 Continuous Rolling Constant and Steel Tensile Coefficient
1.3.1.2 连轧常数及拉钢系数

(1) Continuous rolling constant.

(1) 连轧常数。

A rolling piece is rolled in two or more rolling mills at the same time, and the volume of the rolling piece passing through each rolling mill is equal in unit time, which is called continuous rolling.

一根轧件同时在两架以上轧机中进行轧制,并保持在单位时间内通过各架轧机的轧件体积相等,称为连轧。

Each stand of continuous rolling is arranged in sequence, the rolling pieces are rolled by several mills at the same time, and each stand is connected with each other through the rolling pieces, so that the deformation conditions, motion conditions and mechanical conditions of rolling have a series of characteristics.

连轧各机架依次顺序排列,轧件同时通过数架轧机进行轧制,各个机架通过轧件相互联系,从而使轧制的变形条件、运动条件和力学条件等都具有一系列特点。

As the rolling piece passes through each rolling mill in sequence, the rolling piece enters into the next rolling mill depending on the horizontal component of the force of the previous rolling mill (during this period, the inlet and outlet guide guards have a certain lateral restraint effect on it), so it is necessary to ensure that each rolling mill can bite into the rolling piece smoothly. This requires reasonable technology, pass design, and the correctness of the shape and size of the head of the rolled piece. Generally, two or three sets of flying shears are set in the continuous rolling production line, which are used to cut the head, tail and break in case of accident.

由于轧件依次顺序通过各道轧机,轧件依靠上一机架的作用力的水平分力进入下一机架(在此期间进、出口导卫对其有一定的侧向约束作用),因此要确保每一机架对进入该道次的轧件顺利咬入。这就要求合理的工艺、孔型设计,同时要保证轧件头部形状和尺寸的正确性。通常连轧生产线中都设有 2~3 台飞剪,用于轧件的切头、切尾和事故状态下的碎断。

In continuous rolling, with the reduction of the section and the increase of the rolling speed, it is necessary to keep the same second flow rate of each stand on the rolling line to maintain the normal rolling conditions. The relationship is as follows:

连续轧制时,随着轧件断面的缩小,其轧制速度递增,要保持正常的轧制条件就必须遵守轧件在轧制线上每一机架的秒流量保持相等的原则。其关系式为:

$$F_1v_1 = F_2v_2 = \cdots = F_nv_n = C$$

Where F_1, F_2, $\cdots$, F_n ——Cross section area of rolled piece passing through each frame, mm^2;

v_1, v_2, $\cdots$, v_n ——Rolling speed of rolling piece passing through each frame, m/s;

C ——Second flow rate of each stand;

Lower corner 1, 2, $\cdots$, n ——Frame serial number.

式中 F_1, F_2, …, F_n——分别为轧件通过各机架时的轧件断面面积，mm^2；
v_1, v_2, …, v_n——分别为轧件通过各机架时的轧制速度，m/s；
C——各机架轧件的秒流量；
下角 1, 2, …, n——机架序号。

This formula can also be simplified as:
此式还可简化为：

$$F_1 D_1 N_1 = F_2 D_2 N_2 = \cdots = F_n D_n N_n = C$$

Where D_1, D_2, …, D_n——The working diameter of each frame, mm;
N_1, N_2, …, N_n——Roll speed of each frame, r/min.

式中 D_1, D_2, …, D_n——分别为各机架的轧辊工作直径，mm；
N_1, N_2, …, N_n——分别为各机架的轧辊转速，r/min。

When rolling in each stand, the second flow rate is equal, which is a constant. This constant is called continuous rolling constant. When C stands for continuous rolling constant, there are:

轧件在各机架轧制时的秒流量相等，即为一个常数，这个常数称为连轧常数。以 C 代表连轧常数时有：

$$C_1 = C_2 = \cdots = C_n = C$$

The factors that affect the metal flow rate per second: one is the section area of the rolled piece, the other is the rolling speed.

影响金属秒流量的因素：一个是轧件断面面积，另一个是轧制速度。

Once the cross-section area of the rolled piece is adjusted, it will be fixed (in fact, there is wear due to friction, and the pass area tends to increase continuously). Only by adjusting the rolling speed can the balance relationship of metal flow per second be satisfied.

轧件断面面积一旦调整好就固定不变（实际上，由于有摩擦而存在磨损，孔型面积有不断变大的趋势），只有通过调整轧制速度来满足金属秒流量平衡关系。

The change of tension on rolling piece is caused by the difference of metal flow per second between adjacent stands, so adjusting the rolling speed of each stand can change the second flow of metal rolling piece, so as to achieve the purpose of controlling tension. However, in practical application, the area of rolled piece cannot give accurate value, so the concept of metal elongation coefficient is generally used to describe it.

轧件上的张力变化是由于轧件通过相邻机架的金属秒流量差引起的，所以调整各机架轧制速度就可以改变金属轧件的秒流量，从而达到控制张力的目的。但在实际应用中，轧件面积无法给出精确的数值，故一般采用金属延伸系数的概念来加以描述。

In a continuous small bar mill, the elongation coefficient R_n of the n stand shall be equal to the ratio of the speed of the n stand to the speed of the $n-1$ stand, that is:

在连续小型棒材轧机中，n 机架的延伸系数 R_n 应等于 n 机架的速度和 $n-1$ 机架速度之比，即：

$$R_n = \frac{v_n}{v_n - 1}$$

According to the above formula, as long as the rolling reference speed of the reference frame and the extension coefficient R_n of each frame are given, the rolling speed of each frame can be calculated, and the speed of each frame can be set accordingly. However, because the elongation coefficient given by the operator is experiential, and the specific conditions and conditions of rolling each billet, such as the change of boundary dimension and temperature, cannot be exactly the same, which results in the destruction of the above relationship. Therefore, in order to maintain the new balance of the above relationship in the continuous rolling process, micro tension control and looper adjustment functions are set in the control system.

根据上式，只要给出基准机架的轧制基准速度和各机架的延伸系数，就可求出各机架的轧制速度，据此进行各机架的速度设定。但是，因为操作者给出的延伸系数 R_n 带有经验性，加上轧制每根钢坯的具体条件和状况，如外形尺寸和温度变化等不可能完全一样，其结果导致上述关系遭到破坏，所以连续轧制过程中为了维持上述关系新的平衡，均在控制系统中设置了微张力控制和活套调节功能。

Micro tension control and looper adjustment belong to the scope of tension control. Micro tension control is generally used between the stands with large section and small stand spacing, which are not easy to form looper, such as rough rolling mill and medium rolling mill; while looper tension-free control is used between the stands with small section and easy to form looper, such as finishing mill.

微张力控制和活套调节都属于张力控制的范围。微张力控制一般用在轧件断面大、机架间距小、不易形成活套的机架之间，如粗轧机组和中轧机组中；而活套无张力调节则是用在轧件断面小、易于形成活套的机架之间，如精轧机组中。

The rolling speed can be divided into reverse and forward adjustment according to the control direction. For single line continuous rolling mill, it is more reasonable to use reverse adjustment, that is, the final finishing mill stand is selected as the reference stand, and the rolling speed of the upstream stand is adjusted against the direction of the rolling line, so as to control the rolling tension of the whole rolling line.

轧制速度按控制方向有逆调和顺调之分。对于单线连续轧机，采用逆调较为合理，即选用最后精轧机架为基准机架，逆轧制线方向调节上游机架的轧制速度，以此来控制全轧线的轧制张力。

Compared with prosodic modulation, reverse modulation has the following advantages:

与顺调相比，逆调有以下优点：

1) It can reduce the speed fluctuation of the auxiliary drive behind the reference frame of the finishing mill.

1) 可以减少精轧机基准机架后的辅助传动的速度波动。

2) The roller speed of the upstream stand is slower than that of the downstream stand, and the dynamic characteristics of the system can be improved.

2) 上游机架轧辊速度较下游机架慢些，与顺调相比系统动特性可以得到一些改善。

（2）Steel tensile coefficient.

（2）拉钢系数。

In continuous rolling, it is very difficult, even impossible, to keep the constant rolling constant by keeping the theoretical second flow equal. In order to make the rolling process go on smoothly, the operation technology of piling or pulling steel is often used consciously.

在连续轧制时，保持理论上的秒流量相等使连轧常数恒定是相当困难的，甚至是办不到的。为使轧制过程能够顺利进行，常有意识地采用堆钢或拉钢的操作技术。

The advantages and disadvantages of pull rolling are that there will be no accidents due to steel piling. The disadvantages are that the size of the head, middle and tail of the rolled piece is not uniform, especially in the finishing mill, which will directly affect the quality of the finished product and make the size of the head and tail of the rolled piece exceed the tolerance range.

拉钢轧制有利也有弊，有利是说不会出现因堆钢而产生的事故，有弊是指轧件头、中、尾尺寸不均匀，特别是在精轧机组，将直接影响到成品质量，使轧件的头尾尺寸超出公差范围。

In the process of small bar continuous rolling, looper rolling is generally adopted between the frames with looper, that is, tension-free rolling. In the other frame between the uses of mild tension rolling, that is, micro tension rolling.

小型棒材连轧的过程中，一般在设有活套器的机架间采用活套轧制，即无张力轧制。而在其他机架之间采用轻微拉钢轧制，即微张力轧制。

The drawing steel coefficient K_n is a representation of drawing steel or piling steel. To represent the second tensile steel coefficient, there are:

拉钢系数是拉钢或堆钢的一种表示方法。以 K_n 代表第 n 道次拉钢系数，则有：

$$K_n = F_n D_n N_n (1 + S_n) / F_{n-1} D_{n-1} N_{n-1} (1 + S_{n-1}) = C_n / C_{n-1}$$

When K_n is less than 1, it is piling steel rolling; when it is more than 1, it is pulling steel rolling.

当 $K_n<1$ 时，为堆钢轧制；当 $K_n>1$ 时，为拉钢轧制。

Drawing steel ratio is another representation of piling steel and drawing steel. The ε_n represent the drawing steel ratio of the n pass, there are:

拉钢率是堆钢与拉钢的另一种表示方法。以 ε_n 代表第 n 道次的拉钢率，则有：

$$\varepsilon_n = (C_n - C_{n-1}) / C_{n-1} \times 100\%$$
$$= (C_n / C_{n-1} - 1) \times 100\%$$
$$= (K_n - 1) \times 100\%$$

When ε_n is less than 0, it is piling steel rolling; when it is more than 0, it is pulling steel rolling.

当 $\varepsilon_n<0$ 时，为堆钢轧制；当 $\varepsilon>0$ 时，为拉钢轧制。

Theoretically, the second flow rate of each stand is equal and the continuous rolling constant is constant. After considering the influence of forward slip, the relationship still exists. But after considering the operation conditions of steel stacking and steel pulling, the second flow rate of each stand is not equal, and the continuous rolling constant does not exist, but is produced under the condition of establishing a new balance relationship.

从理论上讲，连续轧制时各机架的秒流量相等，连轧常数是恒定的。在考虑前滑影响后这种关系仍然存在。但当考虑了堆钢和拉钢的操作条件后，实际上各机架的秒流量已不相等，连轧常数已不存在，而是在建立了一种新的平衡关系情况下进行生产的。

1.3.2 Rolling Process and Equipment
1.3.2 轧制工艺及设备

There are 18 rolling mills in total, which are arranged alternately in the form of horizontal and vertical rolling mills (among them, 14, 16 and 18 are horizontal and vertical convertible rolling mills), and they are divided into three units of rough rolling, medium rolling and finish rolling. Each unit is composed of six rolling mills. The finishing mill is equipped with 6 vertical loopers. In order to improve the dimensional accuracy of products, the rolling piece is micro tension rolling in the roughing and middle rolling mill and no tension looping rolling in the finishing mill. During rolling, the rolling mill of the whole line is arranged in a flat and alternate way to avoid the torsion of the rolling piece, so as to improve the product quality; the pass system of the rolling mill adopts the ellipse round hole system. The whole line mill is a high rigidity short stress line mill, and the maximum rolling speed of finishing mill is 18m/s.

轧机共18架,呈平、立轧机交替布置(其中14架、16架、18架为平立可转换轧机),并分为粗轧、中轧、精轧三个机组,每个机组由6架轧机组成。精轧机组设6个立式活套。轧件在粗中轧机组中为微张力轧制,在精轧机组中为无张力活套轧制,以提高产品的尺寸精度。轧制时,全线轧机为平、立交替布置,避免轧件的扭转,以提高产品质量;轧机孔型系统采用椭圆-圆孔型系统。全线轧机为高刚度短应力线轧机,精轧机组最高轧制速度为18m/s。

In front of the rough rolling mill, there is a clip shear, which is used to protect the equipment in case of rolling line equipment failure or steel stacking accident, as well as during commissioning and trial rolling. After 6 and 12 stands, flying shears are set respectively, which are used for cutting head, cutting tail and accident breaking (low temperature of head and tail steel will bend and blossom). Cutting head and breaking part of rolling piece fall into the waste hopper under the platform, and the waste hopper will be transported out of the workshop by forklift; after finishing rolling unit, flying shear 3 is set, which are used for multiple length and section cutting of rolling piece, and flying shear 3 has optimized cutting function to ensure that the rolling piece in the tail section can be put into the cooling bed smoothly.

在粗轧机前设卡断剪,用于轧线设备故障或者堆钢事故时卡断轧件,以保护设备;以及在调试试轧期间使用。6架、12架后分别设飞剪,用于轧件切头、切尾和事故碎断(头尾钢温较低会弯头、开花)轧件切头及碎断部分落入平台下的废料斗内,由叉车将废料斗运出车间;精轧机组后设3号飞剪,用于轧件的倍尺分段剪切,3号飞剪具有优化剪切功能以保证尾段轧件顺利上冷床。

1.3.2.1 Rough Rolling Process and Equipment
1.3.2.1 粗轧工艺及设备

The main function of rough rolling is to compress and extend the billets preliminarily, to obtain the rolled pieces with proper temperature, correct section shape, qualified size, good surface, regular end and suitable length. The rough rolling mill usually adopts flat interchange rolling instead of rolling without torsion, and the number of stands is usually 6. Generally, flat box vertical box rolling is adopted, and the average pass extension coefficient is 1.28~1.32.

粗轧的主要功能是使坯料得到初步压缩和延伸,得到温度合适、断面形状正确、尺寸合格、表面良好、端头规矩、长度适合工艺要求的轧件。粗轧机多采用平—立交替轧机无扭轧制,机架数通常为6架,一般采用平箱—立箱,平均道次延伸系数为1.28~1.32。

In rough rolling stage, micro tension or low tension rolling is widely used, because the section size of the rolled piece is large and it is not sensitive to tension, so it is very difficult and uneconomical to set looper to realize tension free rolling.

在粗轧阶段普遍采用微张力或低张力轧制,因为此时轧件断面尺寸较大,对张力不敏感,设置活套实现无张力轧制十分困难也极不经济。

It is necessary to cut the head and tail after rough rolling. The heat dissipation condition of both ends of the head and tail of the rolling piece is different from that of the middle part. The temperature of both ends of the head and tail of the rolling piece is low and the plasticity is poor. At the same time, the shape of the head of the rolling piece is irregular due to the low temperature, wide spread and uneven deformation of the end of the rolling piece during rolling deformation, which will cause the entrance guide to be blocked or unable to bite during continuous rolling. Therefore, the end must be cut off after 6 passes of rough rolling. Generally, the length of cutting head and tail is 70~200mm.

粗轧后的切头、切尾工序是必要的。轧件头尾两端的散热条件不同于中间部位,轧件头尾两端温度较低,塑性较差;同时轧件端部在轧制变形时由于温度较低,宽展较大,变形不均造成轧件头部形状不规则,这些在继续轧制时都会导致堵塞入口导卫或不能咬入。为此在经过6道次粗轧后必须将端部切去。通常切头、切尾长度在70~200mm。

The rough rolling mill area includes six cantilever rough rolling mills, descaling machines, conveying roller tables, removal devices, heat preservation covers, billet pinch rolls, etc. arranged alternately in horizontal and vertical direction.

粗轧机区域包括的机械设备有:六架平立交替布置的悬臂式粗轧机、除鳞机、输送辊道、剔除装置、保温罩、钢坯夹送辊等。

(1) Rough rolling mills.

(1) 粗轧机组。

There are six rolling mills in the rough rolling mill of the bar mill, which are arranged in the form of flat interchange. In this way, the rolling operation without torsion can be realized and the accident links in the rolling process can be reduced. All six rough rolling mills are two high short stress line mills, which have the advantages of large capacity, compact structure, convenient operation and maintenance, good sealing performance and long service life.

棒材厂的粗轧机组共有六架轧机,呈平立交替布置,这样可以实现无扭转轧制作业,减少轧钢生产过程中的事故环节。六架粗轧机组都是二辊短应力线轧机,具有能力大、结构紧凑、操作维护方便、密封性好、寿命长等优点。

The rough rolling mill has the following characteristics:
粗轧机组具有以下几个特征:

1) The horizontal and vertical forms are alternately arranged, no twist rolling, and the rolling line is the same.

1) 水平与垂直两种形式交替布置,无扭轧制,轧制线同定不变。

2) The roller ring is directly installed on the cantilever end of the roller shaft. It is convenient and quick to replace the roller ring with a crane and a special spreader.

2) 辊环直接安装在轧辊轴的悬臂端头上,用天车配合专用吊具更换辊环,方便快捷。

3) The radial load of the roller shaft is borne by the oil film bearing, and the axial load is borne by the axial thrust tapered roller bearing. The bearing capacity is large, the strength of the shaft is high, and the working life is long.

3) 轧辊轴的径向负荷由油膜轴承承受,轴向负荷由轴向推力圆锥滚子轴承承受,承载能力大,轴的强度高,工作寿命长。

4) When the failure of oil film bearing, thrust bearing, roll shaft and other key parts affects the production, the roll shaft components can be replaced as a whole to shorten the time of accident shutdown and reduce the labor intensity of emergency repair.

4) 在油膜轴承、止推轴承及轧辊轴等关键零部件发生故障影响生产时,可整体更换轧辊轴组件,缩短事故停产时间,降低抢修的劳动强度。

5) According to the different working speed of each rolling mill, different types of reducers are equipped. Among them, No.1 rolling mill is equipped with double-stage planetary reducer, No.2, No.3 and No.5 rolling mill is equipped with single-stage planetary reducer, and No.4 and No.6 rolling mill is equipped with single-stage ordinary helical gear reducer.

5) 根据每架轧机工作转速的不同,配备了不同类型的减速机。其中1号轧机配备的为双级行星减速机,2号、3号和5号轧机配备的是单级行星减速机,4号和6号轧机配备的是单级普通斜齿轮减速机。

(2) Descaling machine.

(2) 除鳞机。

1) Installation position of descaling machine.

1) 除鳞机的安装位置。

The descaling machine is arranged on the outlet side of the heating furnace, about 8.5m away from the outlet of the heating furnace.

除鳞机被安置在加热炉出口侧,距离加热炉出口约为8.5m。

2) Functions of descaling machine.

2) 除鳞机的功能。

The function of the descaling machine is to remove the oxide scale from the billet just coming out of the heating furnace, which is beneficial to ensure that the billet can smoothly bite into the first roughing mill, protect the first rolling mill and guide equipment, and ensure the quality of the final product.

除鳞机的功能是把从加热炉中刚出来的钢坯上的氧化铁皮除掉,这对于保证钢坯顺利咬入第一架粗轧机,保护第一架轧机及导卫设备,以及保证最终产品的质量都有一定的好处。

3) Working principle of descaling machine.

3) 除鳞机的工作原理。

The working principle of the descaling machine is that when the billet passes through the descaling machine, each high-pressure water nozzle on the annular pipe in the descaling machine sprays a certain shape of high-pressure water to the billet at a certain angle (15°), the pressure of the water is 160~200MPa, and the iron oxide scale is peeled off by the strong impact force of high-pressure water. In order to prevent the excessive temperature drop in the descaling process of billet, the operation speed of billet is required to be faster in the descaling process. Therefore, this descaling method can also be called rapid descaling method.

除鳞机的工作原理是在钢坯经过除鳞机时,除鳞机内的环形管上的各高压水喷嘴向钢坯按一定的角度(15°)喷射一定形状的高压水,水的压力为160~200MPa,利用高压水的强大冲击力将氧化铁皮剥落。为了防止钢坯在除鳞过程中的温降过大,在除鳞过程中要求钢坯的运行速度较快,因此,这种除鳞方式又可称为快速除鳞法。

4) Structure and main components of descaling machine.

4) 除鳞机的结构与主要组成部分。

①Main structure of descaling machine.

①除鳞机的主体结构。

The external part of the descaling machine is a box welded with steel plates, and the inside of the box is equipped with a circular pipe nozzle and a roller table. In addition, in order to prevent the high-pressure water sprayed on the billet from moving to the furnace along the billet, a curtain composed of a chain is installed at the billet entrance of the descaling machine. The main function of curtain is to block water.

除鳞机外部是用钢板焊接的箱体,箱体内装有环形管喷嘴和辊道。另外,为了防止喷到钢坯上的高压水顺钢坯向加热炉运动到炉内,在除鳞机上钢坯入口处,安装有由链条组成的幕帘。幕帘的主要作用是阻水。

In addition to the main equipment, it is also equipped with booster pump station, valve station, pipeline and other auxiliary equipment.

除主体设备外,还配备有增压泵站、阀台、管线等辅助设备。

②Nozzle.

②喷嘴。

The core component of the descaling machine is the nozzle, which is made of cemented carbide steel and has a general working life of 5~6 months. When the water quality is not good, the service life is relatively low, only 2~3 months. In the production process, it can be judged whether the nozzle is in good condition according to the descaling effect, or it can be judged whether the nozzle is in good condition by observing the water flow shape of the nozzle from the peephole of the descaling box. If it needs to be replaced, it shall be replaced in time to ensure the descaling effect.

除鳞机的核心部件是喷嘴,其制造材料为硬质合金钢,一般工作寿命为5~6个月,在水质不好时寿命较低,只有 2~3 个月,在生产过程中,可根据除鳞的效果来判断喷嘴是否完好,也可从除鳞箱的窥视孔观察喷嘴喷出的水流形状来判断喷嘴是否完好。需要更换时应及时更换,以确保除鳞效果。

(3) Transfer roller table.

(3) 传输辊道。

1) Position and function of transfer roller table.

1) 传输辊道的位置与功能。

The transfer roller table is installed between the outlet of the heating furnace and the billet pinch roller. Its main function is to transfer the steel from the heating furnace to the first roughing mill, and to ensure the realization of the fast descaling mode. It also plays a certain role in soaking and heat preservation for the billet to be rolled after descaling.

传输辊道安装在加热炉出口和钢坯夹送辊之间,其主要功能是把从加热炉出来的钢坯传送到第一架粗轧机,并能保证快速除鳞方式的实现,还对除鳞以后的待轧钢坯起一定的均热保温作用。

2) Structure and main composition of transmission roller table.

2) 传输辊道的结构及主要组成。

The main components of the transmission roller table are roller, transmission device, guide groove, heat preservation cover and other parts. The roller is driven by AC variable frequency motor separately, and both ends are supported by rolling bearings. The lubrication mode of the bearing is dry oil automatic lubrication (high temperature resistant grease is recommended). The middle bearing shaft of the roller is cooled by internal circulating water, which can extend the service life of the bearing and the roller.

传输辊道的主要组成部分有辊子、传动装置、导槽、保温罩等部件。辊子由交流变频电机单独驱动,两端采用滚动轴承支承,轴承的润滑方式为干油自动润滑(建议采用耐高温的润滑脂),辊子的中间支承轴采用内部循环水冷却,这样可延长轴承和辊子的使用寿命。

(4) Insulation cover.

(4) 保温罩。

The heat preservation cover is located at the position of the removal device, and its main function is to keep the billet on the roller table warm, which is immovable in normal steel rolling production. When repairing or replacing the roller, use the crown block to lift the heat preservation cover away. After the maintenance, use the crown block to lift the heat preservation cover.

保温罩处于剔除装置位置,其主要作用是对辊道上的钢坯保温,在正常轧钢生产中是不动的,在检修或更换辊子时用天车把保温罩吊走,检修完后再用天车把保温罩吊装好。

(5) Knockout device.

(5) 剔除装置。

The removal device is located between the descaling machine and the baffle, and its main function is to remove the unqualified billet.

剔除装置位于除鳞机与挡板之间,主要作用是剔除不合格的钢坯。

(6) Billet pinch roll.

(6) 钢坯夹送辊。

1) Installation position and main functions of pinch roll.

1) 夹送辊安装位置及主要功能。

Pinch roll is installed between No. 1 roughing mill and billet conveying roller table, its main function is to assist No. 1 roughing mill in biting when necessary, in this case, biting is called forced biting.

夹送辊安装在1号粗轧机和钢坯传送辊道之间,其主要功能是在必要时协助1号粗轧机咬入,在这种情况下的咬入又称为强迫咬入。

2) Working principle of pinch roll.

2) 夹送辊的工作原理。

The lower roll is driven by an electric motor through a reducer and rotates at a constant speed. When the billet passes through the pinch roll, if it needs to be pinch, the upper roll is pulled down through the hydraulic cylinder to make the upper roll and the lower roll tightly grip the billet, because the lower roll is the driving roll and the upper roll is the free roll, which has a certain pinch force on the billet.

下辊由电动机通过一个减速机驱动并按一定的速度匀速转动。当钢坯通过夹送辊时,如需要夹送,就通过液压缸把上辊向下拉,使上辊和下辊紧夹住钢坯,因下辊是主动辊,而上辊是自由辊,这样就对钢坯产生了一定的夹送力。

3) Structure of pinch roll.

3) 夹送辊的结构。

The main parts of pinch roll are motor, reducer, lower pinch roll, upper pinch roll, two hydraulic cylinders and coupling, etc. because the upper and lower pinch roll often contact with high-temperature billet during operation, in order to prevent the roll body and roll shaft from thermal deformation and extend the service life of the roll, a water spray cooling device is installed above the side of the upper roll to spray water outside the upper roll for cooling, and the lower roll is not provided with a single one. The independent external cooling device, whose external cooling is mainly realized by the water flowing down from the upper roller. In addition, the upper roll and the lower roll have the function of internal water cooling, and the upper roll and the lower roll can be cooled by the rotary joint on the operation side. In order to improve the working life of the equipment and reduce the variety of spare parts, the interchangeability of the upper and lower pinch rolls is considered in the design and manufacturing process of the equipment. In this way, the upper and lower rolls can be exchanged continuously, so that the wear degree of the upper and lower rolls is basically the same.

夹送辊主要部件有电机、减速机、下夹送辊、上夹送辊、两个液压缸及联轴器等,由于上、下夹送辊在工作时经常与高温钢坯接触,为防止辊身和辊轴的热变形,延长辊子的使用寿命,在上辊的侧上方安装有喷水冷却装置,对上辊进行外部喷水冷却,下辊没设置单独的外部冷却装置,其外部冷却主要靠从上辊流下来的水来实现。另外,上辊及下辊都具有内水冷功能，可通过操作侧的旋转接头对上、下辊进行水冷。因为在实际工作中上辊和下辊的磨损程度不同，为了提高设备的工作寿命，减少备件的品种，在设备的设计制造过程中考虑了使上、下夹送辊具有互换性，这样可经过不断调换上、下辊，使上、下辊的磨损程度基本保持一致。

1.3.2.2 Medium and Finishing Rolling Process and Equipment
1.3.2.2 中、精轧工艺及设备

(1) Medium and finishing rolling process.
(1) 中、精轧工艺。

The function of the middle and finishing rolling unit is to make the rolled products extend by rolling from frame to frame in the rolling groove, and finally roll out the finished product with correct shape, qualified size and good surface. The middle and finishing rolling mills each have 6 rolling mills, and the stand numbers are 7~18. Among them, No. 14, No. 16, and No. 18 are flat-to-stand convertible racks. There are eight horizontal loopers in the rolling mill between No. 10~18, and No. 2 flying shear between the middle rolling mill and the finishing rolling mill, which have the functions of cutting head, tail cutting and breaking.

中、精轧机组的功能是使轧件在轧槽中经逐架压缩而延伸，最终轧出形状正确、尺寸合格、表面良好的成品。中、精轧机组各有6架轧机，机架号为7~18号。其中，14号、16号、18号为平-立可转换机架。10~18号轧机间设有八个水平活套装置，中轧机组与精轧机组间设2号飞剪，具有切头、切尾和碎断功能。

The round rolled product rolled from the rough rolling mill passes through the No. 1 shear head and enters the middle rolling mill. Then, the rolled product passes the No. 2 flying shear head, and the finished steel is rolled through the finishing mill. Among them, rolling mills No. 7 to No. 10 use current memory micro-tension rolling, and rolling mill No. 10 to No. 18 use looper tension-free rolling.

从粗轧机轧出的圆轧件，经1号剪切头，进入中轧机组，然后轧件经2号飞剪切头，通过精轧机轧出成品钢材。其中，7~10号轧机采用电流记忆法微张力轧制，10~18号轧机采用活套无张力轧制。

At present, the hole-type systems used in bar rolling mills are mostly elliptical-square hole-type systems and elliptical-round hole-type systems. Bar mills use different hole-type systems depending on the grouping of the finished product. For smaller $\phi14$ round steel, an integrated type combining an ellipse-square hole system and an ellipse-round hole system is adopted, that is, an ellipse-square hole system is used between the 8th to 10th racks. It not only solves the problem of too large a total extension, but also it is necessary to use a torsion guide at the exit of the square rolling product, while also ensuring the product quality. Try to use the common hole pattern, so that the rolling groove can give full play to its superior commonality, and the time for changing the variety can be shortened.

目前，棒材轧机采用的孔型系统多为椭圆—方孔型系统和椭圆—圆孔型系统。棒材厂根据成品分组不同而采用不同的孔型系统。对于尺寸较小的 $\phi14$ 圆钢，采用了椭圆—方孔型系统与椭圆—圆孔型系统相结合的综合型式,即在 8~10 号机架间采用了椭圆—方孔型系统,这样,既解决了总延伸过大的问题,也有必要在方轧件出口使用扭转导卫,同时也保证了产品质量。尽量采用共用孔型,使轧槽充分发挥其优越的共用性,且可缩短更换品种的时间。

From the perspective of the rolling process, twist-free and tension-free rolling is the most ideal state in the rolling process. Torsion-free creates conditions for increasing rolling speed and tension-free rolling, and it also simplifies the guidance device. Tension-free rolling improves the uniformity of cross-sectional dimensions in the length direction of the rolled product. Based on this goal, the rolling lines of the middle and finishing rolling mills of the bar mill are fixed, and all rolling mills are arranged in an upright position, so there is no twist in the rolled products. Each stand is driven by a DC motor separately, which makes the rolling mill very flexible. The micro-tension rolling is implemented between the 7th to 10th stands, and the loopless tensionless rolling is implemented between the 10th to 18th stands. This rolling process is closest to the ideal rolling state, thereby ensuring the production line. Quality and high yield.

从轧制工艺角度来说,无扭无张力轧制是轧制过程中的最理想状态。无扭为提高轧制速度、进行无张力轧制创造了条件,同时也简化了导卫装置。无张力轧制改善了轧件长度方向上断面尺寸的均匀性。基于这一目标,棒材厂中、精轧机组的轧制线固定,所有轧机均为平立布置,因而轧件无扭转。各机架均为直流电机单独驱动,使轧机具有很大的灵活性。在 7~10 号机架间实现微张力轧制,在 10~18 号机架间实现活套无张力轧制,这种轧制工艺最接近于理想的轧制状态,从而保证了生产线的优质高产。

(2) Medium rolling mill.

(2) 中轧机组。

The medium rolling mill is located between the roughing and finishing mills and consists of six chuck mills. The unit layout is shown in Figure 1-3. This type of horizontal and vertical alternate arrangement of the unit can realize continuous rolling of bars in several working stands at the same time. The rolling speed of the working stand conforms to the principle of "equal flow per second". The speed matching between rolling mills is realized by cascade speed regulation of electrical system. For the convenience of use and adjustment, each rolling mill is driven by a water-cooled DC motor.

中轧机组位于粗、精轧机之间,由六架卡盘式轧机组成。机组布置如图 1-3 所示。机组这种平立交替布置形式,可以实现棒材同时在几个工作机座中连续无扭转轧制。工作机座的轧制速度符合"秒流量相等"的原则。轧机间的速度匹配由电气系统的级联调速实现。为了便于使用和调整,每台轧机由一台水冷直流电机单独驱动。

The six chuck mill of the medium rolling mill is designed and manufactured by Danieli company of Italy. This is a short stress line mill. The structure of the mill is compact, the volume is small and the rigidity is improved because of the chuck type stand and the four pull rod mechanism. Four bearing blocks of the rolling mill are sleeved on four pull rods which are supported by chuck bracket. Because the pull rod is close to the rolling center line, the size and installation position of the pull rod are considered in the design, so the mill obtains high rigidity. During rolling,

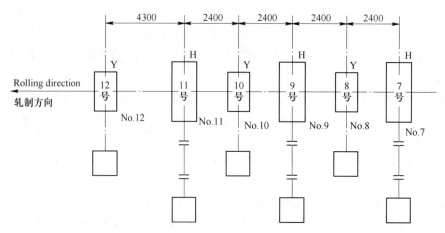

Figure 1-3　Layout plan of medium rolling mill
图 1-3　中轧机组平面布置图

the gap size is adjusted by the gap device. During adjustment, a set of transmission mechanism is driven by hydraulic motor (or wrench manually) to obtain the required roll gap, and one side bearing seat can also be adjusted independently by the device. In order to roll different products, the corresponding groove can be replaced. The movement is driven by the hydraulic motor screw system on the sliding support to move the rolling mill along the roll axis, so that the rolling groove required for rolling is aligned with the rolling line. The rolling torque is transferred from DC motor→main reducer→cardan shaft to roll, and the rolling speed is regulated by DC motor.

中轧机组的六架卡盘式轧机由意大利达涅利公司设计制造。这是一种短应力线轧机。由于采用了卡盘式机架和四拉杆机构，使得轧机结构紧凑、体积小，提高了刚度。该种轧机的四个轴承座套装在四个拉杆上，拉杆由卡盘支架支撑。由于拉杆靠近轧制中心线，设计时考虑了拉杆的尺寸和安装位置，由此，轧机获得了高刚度。轧制时辊缝大小，是通过辊缝装置调节的。调整时，由液压马达（或扳手手动）驱动一套传动机构，得到要求的辊缝，用该装置还可以单独对一侧轴承座进行调整。为了轧制不同的产品，可以更换相应的轧槽。该动作由滑动支座上的液压马达丝杠系统带动轧机沿轧辊轴线方向移动，使得轧制需要的轧槽对准轧制线。轧制力矩由直流电机→主减速机→万向轴传递到轧辊，轧制速度由直流电机调速调节。

The radial rolling force of the chuck mill is borne by four rows of cylindrical roller bearings installed in the bearing pedestal, and the axial force is borne by the thrust bearing installed in the bearing pedestal at the end of the roller. The cardan shaft bracket is self balancing type, which can adapt to the roll distance and keep in the corresponding position. The rolling mill is also equipped with cooling water pipe, hydraulic pipe and lubricating pipe.

卡盘式轧机的径向轧制力由装在轴承座内的四列圆柱滚子轴承承受，轴向力由装在轧辊尾部的轴承座内的推力轴承承受。其万向轴托架是自平衡型，能与辊距相适位而保持在相应位置。在该轧机上，还配有冷却水管、液压管路、润滑管路。

（3）Finish rolling mill.

(3) 精轧机组。

1) Three frame two roller chuck type horizontal unit.

1) 三架两辊卡盘式水平机组。

Model: "DOMS3400".

型号:"DOMS3400"。

Location: units 13, 15 and 17.

位置:第13号、15号和第17号机组。

Characteristic:

特点:

①In the rolling part, the chuck type compact frame is used.

①轧制部分使用的是卡盘式紧凑型机架。

②The rigidity of the rolling mill is ensured by the body size of the four pull rod bolts and the location of the four pull rod bolts close to the rolling line.

②轧机的刚度是靠四条拉杆螺栓本身的形体尺寸和四个拉杆螺栓的坐落位置紧靠轧线来保证的。

③This type of mill has the characteristics of compact overall layout and high stand rigidity.

③采用此种形式的轧机有整体布局紧凑和机架刚度极高的特点。

④It can bear large load.

④能承受较大的载荷。

⑤The rolling line is fixed.

⑤轧线固定。

⑥Quick rack replacement.

⑥可快速更换机架。

⑦The rolling mill adopts four row short cylindrical roller bearings to bear the radial load of the shaft, and two-way thrust ball bearings are used at the end of the roller shaft to bear the axial load.

⑦轧机采用四列短圆柱滚子轴承,承受轴的径向载荷,并在轧辊轴的尾端采用双向止推球轴承用于承受轴向载荷。

⑧The frame balance adopts the mechanical balance form.

⑧机架平衡采用的是机械平衡形式。

⑨The frame and reducer are connected by universal retractable transmission shaft.

⑨机架和减速机之间由万向可伸缩传动轴联接。

⑩The lubricating pressure of thin oil in finishing mill area is 0.2MPa.

⑩精轧机区稀油润滑压力为0.2MPa。

⑪The 13, 14 and 15 chuck stands and 16, 17 and 18 chuck stands in finishing mill area are interchangeable.

⑪精轧机区13、14、15卡盘机架及16、17、18卡盘机架是可以互换的。

2) Three stand two high chuck type horizontal vertical conversion mill.

2) 三架两辊卡盘式可平立转换的轧机。

Model: DCRS3400.

型号：DCRS3400。

Location: frame 14, 16, 18.

位置：第14、16、18号机架。

Features: in addition to the compact layout of the whole rotating part of the rotating system driven by hydraulic pressure, the other features are the same as the horizontal frame.

特点：除旋转系统由液压驱动整个旋转部分布局十分紧凑外，其余与水平式机架特点相同。

1.3.2.3 Cooling and Lubrication
1.3.2.3 冷却与润滑

（1）Roll cooling.

（1）轧辊冷却。

In the rolling process, the roll pass surface first contacts with hot steel, and then is cooled by water. Under the alternate action of heating and cooling, the roll surface appears network crack, which is called craze phenomenon. The strong oxidation of the metal at the crack causes the crack to expand, leading to the increase of friction coefficient and the acceleration of wear. In this way, the surface of the roll will be worn and damaged due to improper cooling of the roll. If the cooling water at the exit of the rolled piece is sufficient, the roll can be sufficiently cooled and the cooling water can be prevented from entering the pass along with the rolled piece. This cooling method can prolong the service life of the pass. On the contrary, if the cooling water enters into the roll with the rolled piece, the water will vaporize rapidly under the high temperature and high pressure in the deformation area, which will not only cause the slip and steel piling accident of the rolled piece, but also cause the wear and damage of the roll surface. The amount of roll cooling water should be sufficient, but the larger the amount is, the higher the cost will be, so as long as it can meet the requirements of roll cooling. The pressure of cooling water in the bar mill shall not be less than 4Pa.

轧辊孔型表面在轧制过程中先与热钢接触，继而又被水冷却，在加热与冷却的交替作用下，轧辊表面出现网状裂纹，即所谓龟裂现象。裂纹处金属发生强烈氧化，使裂纹扩展，导致摩擦系数增大，磨损加快。这样，由于轧辊冷却不当会造成轧辊表面磨损和破坏。如果轧件出口处冷却水充足，那么既能充分冷却轧辊，又能防止冷却水随同轧件进入孔型。此种冷却方法可延长孔型使用寿命。反之，若冷却水随轧件进入轧辊，水在变形区的高温高压下迅速汽化，不但容易造成轧件打滑堆钢事故，也使轧辊表面磨损和破坏。轧辊冷却水用量要充足，但用量越大，成本就会随之提高，所以只要能够满足轧辊冷却要求即可。棒材厂冷却水压力不低于4Pa。

（2）Cooling of rolling guide.

（2）滚动导卫的冷却。

If the guide wheel of the rolling guide rail contacts with the high-temperature rolled piece for a long time, if it is not cooled, it will heat up rapidly until it is red hot, resulting in serious wear of the guide wheel, serious damage of the guide wheel bearing, etc., which makes the guide rail lose the function of supporting the rolled piece. Therefore, it is necessary to ensure the sufficient cooling of the guide wheel. The cooling water enters the guide box from the system, reaches the support arm of the guide wheel through the pipeline, and then sprays out from the three opening guide wheels to achieve the purpose of cooling. It is worth noting that the rolling guide and roll share the same cooling water system, and this cooling water impurity is the highest, which is easy to block the cooling pipeline of the guide. Therefore, the operator needs to check the cooling water supply status frequently to prevent the accident of steel piling due to the burning of the guide wheel.

滚动导卫的导轮长时间与高温轧件接触，如不进行冷却就会迅速升温直至红热，造成导轮磨损严重，导轮轴承严重破坏等现象，使导卫失去扶持轧件的作用。因此，必须保证导轮的充分冷却。冷却水从系统进入导卫盒，经管路到达导轮支撑臂，然后从三个开孔向导轮喷出，达到冷却目的。值得注意的是，滚动导卫与轧辊共用同一系统的冷却水，而这种冷却水杂质含量较高，易堵塞导卫的冷却管路，因此操作人员需经常检查冷却水供位状况，防止发生因导轮烧毁而堆钢的事故。

(3) Lubrication of roller bearing.

(3) 轧辊轴承的润滑。

The lubrication mode of roller bearing is oil air lubrication.

轧辊轴承的润滑方式为油气润滑。

(4) Lubrication of rolling guide.

(4) 滚动导卫的润滑。

The lubrication mode of guide wheel bearing of rolling guide is oil air lubrication. Oil air lubrication is a kind of lubrication method in which the mixture of oil mist and air is forced to be input into the bearing through a special channel by using compressed air which has fully removed moisture, dust and other foreign matters to atomize the lubricating oil. Compared with dry oil lubrication, it has the advantages of good lubrication effect. However, in the actual production, if the oil and gas supply is not normal, the stator bearing will be damaged. Operators should pay special attention to the supply of oil and gas. Oil air lubrication is adopted for roller guide in bar mill.

滚动导卫的导轮轴承润滑方式为油气润滑。油气润滑是使用充分清除了水分、灰尘等异物的压缩空气将润滑油雾化，然后使油雾与空气的混合物强制性地通过专门通路输入轴承内部的润滑方式。与干油润滑相比，具有润滑效果好的优点。但是，在实际生产中，如油气供应不正常，就会破坏导轮轴承。操作人员要特别注意油气的供应情况。棒材厂辊式导卫采用的是油气润滑方式。

1.3.2.4　Looper

1.3.2.4　活套

(1) Purpose of looper setting.

(1) 设置活套的目的。

Looper is a kind of guiding device set between rolling mills in order to keep a good shape and size of rolling piece and carry out tension-free rolling. The looper produced in rolling mill often changes due to the wear of pass and the change of rolling temperature. For the adjustment of looper change, the looper scanner is generally used to detect the looper amount, and then the looper amount is adjusted by adjusting the motor speed.

活套是为了保持良好的轧件形状、尺寸、进行无张力轧制而在轧机之间所设置的一种导向装置。轧机间产生的活套常常由于孔型的磨损、轧材温度的变化等而产生变动。活套变动的调整，一般用活套扫描器来检测活套量，再通过调整电机转速来调整活套量。

(2) Type of looper.
(2) 活套的种类。

There are two kinds of loopers: vertical looper and side looper. The vertical looper has small capacity and is mainly used in medium and finishing mills, while the side looper has large capacity and is mainly used in medium and finishing mills. There are 8 horizontal loopers in the middle and finishing mill of bar mill.

活套有立活套和侧活套两种。立活套容量小，多用于中、精轧机，侧活套容量大，多用于中、精轧机组间。棒材厂中、精轧机组共有 8 个水平活套。

(3) Loop up and down process.
(3) 活套起落过程。

The looper scanner is in the starting state. When the head of the rolled piece is bited into the A-frame, a signal is generated. The looper scanner detects the existence of the rolled piece. When the rolled piece is bite into the B-frame again, an impact speed drop is generated, which makes the rolled piece between the two frames produce a certain looper, and the signal of the rolled piece biting into the B-frame excites the looper support roller to rise to help form a horizontal looper. According to the preset height of the looper, the looper scanner can speed up the upstream carriage in a certain proportion. In the interactive process of continuously detecting looper height and rolling mill speed-up, looper is formed and the height reaches the preset value. When the end of the rolled piece leaves the A-frame, a signal is generated, and the looper support roll falls down, and the looper disappears.

活套扫描器处于启动状态，当轧件头部咬入 A 机架时，产生一个信号，活套扫描器检测出轧件的存在，轧件再咬入 B 机架时，产生冲击速降，使两机架之间的轧件产生一定的活套，且轧件咬入 B 机架的信号激发活套支承辊升起帮助形成水平活套。活套扫描器根据预设定的活套高度使上游机架入按一定比例升速。在连续检测活套高度与轧机升速的交互过程中，活套形成，高度达到预设定值。当轧件尾部离开 A 机架时产生一个信号，活套支承辊落下，活套随之消失。

1.3.3 Rolling Pass System
1.3.3 轧制孔型系统

1.3.3.1 Pass System for Rolling Round Steel
1.3.3.1 轧制圆钢的孔型系统

The pass system of round steel here refers to the last 3~5 passes of rolling round steel,

i. e. finish rolling pass system. The common round steel pass systems are as follows.

圆钢的孔型系统在这里是指轧制圆钢的最后 3~5 个孔型，即精轧孔型系统。常见的圆钢孔型系统有如下四种。

(1) Square-ellipse-circular hole system (Shown in Figure 1-4).

(1) 方—椭圆—圆孔型系统（见图 1-4）。

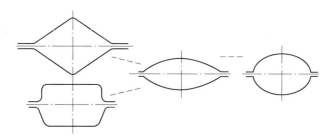

Figure 1-4   Square-ellipse-circular hole system
图 1-4   方—椭圆—圆孔型系统

The advantages of this pass system are: large extension system; square rolling can be automatically aligned in the elliptical pass, stable rolling; it can be well connected with other extended pass systems, and its disadvantages are: uneven deformation of square rolling in the elliptical pass; groove depth of square pass; poor pass interoperability. Because of the large elongation coefficient of this pass system, it is widely used in small and wire mill to roll round steel under 32mm.

这种孔型系统的优点是：延伸系统较大；方轧件在椭圆孔型中可以自动找正，轧制稳定；能与其他延伸孔型系统很好衔接，其缺点是：方轧件在椭圆孔型中变形不均匀；方孔型切槽深；孔型的共用性差。由于这种孔型系统的延伸系数大，所以被广泛应用于小型和线材轧机轧制 32mm 以下的圆钢。

(2) Circle-ellipse-circle hole system (Shown in Figure 1-5).

(2) 圆—椭圆—圆孔型系统（见图 1-5）。

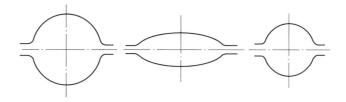

Figure 1-5   Circle-ellipse-circle hole system
图 1-5   圆—椭圆—圆孔型系统

Compared with the square ellipse round pass system, the advantages of this pass system are: uniform deformation and cooling of the rolled piece; easy to remove the oxide scale on the surface of the rolled piece, good surface quality of the finished product; easy to use the apron; the size of the finished product is more accurate; it can roll out a variety of round bars from the middle round hole, so it has a large commonality. Its disadvantages are: small elongation coefficient; unstable rolling of ellipse parts in the round hole, which needs to be clamped by accurately adjusted clamping plate, otherwise it is easy to produce "ears" in the round hole of the pass. This pass system is widely used in small-scale and wire mill to roll round steel below 40mm. In the finishing mill of high-speed wire rod mill, this pass system can be used to produce various specifications of wire rod.

与方—椭圆—圆孔型系统相比，这种孔型系统的优点是：轧件变形和冷却均匀；易于去除轧件表面的氧化铁皮，成品表面质量好；便于使用围盘；成品尺寸比较精确；可以从中间圆孔轧出多种规格的圆钢，故共用性较大。其缺点是：延伸系数较小；椭圆件在圆孔中轧制不稳定，需要使用经过精确调整的夹板夹持，否则在孔型圆孔型中容易出"耳子"，这种孔型系统被广泛应用于小型和线材轧机轧制 40mm 以下的圆钢。在高速线材轧机的精轧机组，采用这种孔型系统可以生产多种规格的线材。

(3) Ellipse-vertical ellipse-ellipse-circular hole system (Shown in Figure 1-6).

(3) 椭圆—立椭圆—椭圆—圆孔型系统（见图 1-6）。

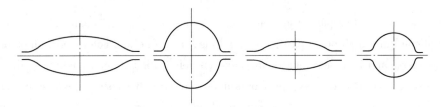

Figure 1-6　Ellipse—vertical ellipse—ellipse—circular hole system

图 1-6　椭圆—立椭圆—椭圆—圆孔型系统

The advantages of this pass system are: the deformation of the rolled piece is uniform; the oxide scale on the surface of the rolled piece is easy to be removed, and the surface quality of the finished product is good; the elliptical piece can be automatically aligned in the vertical elliptical pass, and the rolling is stable. Its disadvantages are: the elongation coefficient is small; because of the repeated stress of the rolled piece, it is easy to loose the central part, even when the steel quality is poor, it will appear axial cracks. This pass system is generally used for rolling low plasticity alloy steel or small and wire continuous rolling mill.

这种孔型系统的优点是：轧件变形均匀；易于去除轧件表面氧化铁皮，成品表面质量好；椭圆件在立椭圆孔型中能自动找正，轧制稳定。其缺点是：延伸系数较小；由于轧件产生反复应力，容易出现中心部分疏松，甚至当钢质不良时会出现轴心裂纹。这种孔型系统一般用于轧制塑性较低的合金钢或小型和线材连轧机上。

(4) Universal pass system (Shown in Figure 1-7).

(4) 万能孔型系统（见图 1-7）。

The advantages of this pass system are: strong commonality, a set of pass can be used to produce several adjacent specifications of round steel by adjusting the roll; the deformation of the rolled piece is uniform; it is easy to remove the oxide scale on the surface of the rolled piece, and the surface quality of the finished product is good. Its disadvantages are: the extension system is small; it is not easy to use the shroud; when the vertical rolling hole is not designed properly, the rolling piece is easy to twist. This pass system is suitable for rolling 18~200mm round steel.

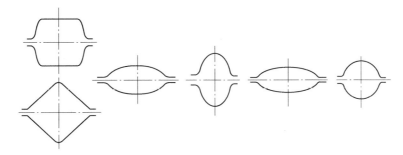

Figure 1-7 Universal pass system
图 1-7 万能孔型系统

这种孔型系统的优点是：共用性强，可以用一套孔型通过调整轧辊的方法，轧出几种相邻规格的圆钢；轧件变形均匀；易于去除轧件表面氧化铁皮，成品表面质量好。其缺点是：延伸系统较小；不易于使用围盘；立轧孔设计不当时，轧件容易扭转。这种孔型系统适用于轧制 18~200mm 的圆钢。

1.3.3.2 Pass Design of Finished Round Steel
1.3.3.2 圆钢成品孔型设计

The round shape of the round steel product is the last one. The design of the round shape of the round steel product directly affects the dimensional accuracy of the finished product, the adjustment of the rolling mill and the service life of the hole shape. When designing the round shape of the round steel product, it should generally be considered to minimize the change in ovality and to make full use of the allowable deviation range, that is, to ensure the maximum adjustment range. In order to reduce overfilling and facilitate adjustment, the shape of the round steel hole is a round hole with an expansion angle. At present, the widely used hole formation method is shown in Figure 1-8 (a).

圆钢成品孔型是轧制圆钢的最后一个孔型，圆钢成品孔型设计的好坏直接影响到成品的尺寸精度、轧机调整和孔型寿命。设计圆钢成品孔型时，一般应考虑到使椭圆度变化最小，并且能充分利用所允许的偏差范围，即能保证调整范围最大。为了减少过充满和便于调整，圆钢成品孔的形状采用带有扩张角的圆形孔。目前广泛使用的成品孔构成方法如图 1-8 (a) 所示。

The radius of the base circle of the finished hole is $R = 0.5[d-(0\sim1.0)\Delta_-](1.007\sim1.02)$, where d is the nominal diameter of the round steel or the standard diameter; Δ_- is the allowable negative deviation; 1.007~1.02 is the expansion coefficient, and the specific value depends on the final rolling temperature and steel type. Each steel type can be taken as:

成品孔的基圆半径 $R = 0.5[d-(0\sim1.0)\Delta_-](1.007\sim1.02)$，其中 d 为圆钢的公称直径或称之为标准直径；Δ_- 为允许负偏差；1.007~1.02 为膨胀系数，其具体数值根据终轧

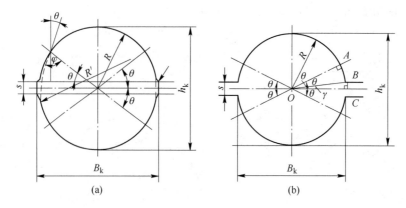

Figure 1-8 Composition of round steel finished hole
图 1-8 圆钢成品孔的构成

温度和钢种而定，各钢种可取为：

Ordinary steel	Carbon tool steel	Ball bearing	High speed steel
普通钢	碳素工具钢	滚珠轴承	高速钢
1.011~1.015	1.015~1.018	1.018~1.02	1.007~1.009

The width B_k of the finished product blank is: $B_k = [d+(0.5 \sim 1.0)\Delta_+] (1.007 \sim 1.02)$, where Δ_+ is the allowable positive deviation.

成品空的宽度 B_k 为：$B_k = [d+(0.5 \sim 1.0)\Delta_+] (1.007 \sim 1.02)$，其中的 Δ_+ 为允许正偏差。

The expansion half angle θ of the finished hole can be generally taken as $\theta = 20° \sim 30°$, and commonly used $\theta = 30°$.

成品孔的扩张半角 θ，一般可取为 $\theta = 20° \sim 30°$，常用 $\theta = 30°$。

The expansion radius R' of the finished hole shall be determined as follows, i.e. the side angle β shall be determined first, and its value is:

成品孔的扩张半径 R' 应按如下步骤确定，即先确定出侧角 β，其值为：

$$\beta = \arctan \frac{B_k - 2R\cos\theta}{2R\sin\theta - s}$$

When β is less than θ, the expansion radius R' can be calculated. If $\beta = \theta$, then only two sides of the pass can be expanded by tangent line; if $\beta < \theta$, then the value of R' can be determined according to the following formula

当按上式求出 β 值小于 θ 才能求扩张半径 R'。若 $\beta = \theta$ 时，则只能在孔型的两侧用切线扩张；当 $\beta < \theta$ 时，则可按下式确定 R' 值

$$R' = \frac{2R\sin\theta - s}{4\cos\beta\sin(\theta - \beta)}$$

If $\beta > \theta$, there are two adjustment methods, one is to adjust the values of B_k, R and s to make $\beta \leq \theta$; the other is to adjust the angle of θ to make $\beta = \theta$ and expand it with tangent. At this time [Shown in Figure 1-8 (b)]

若 $\beta > \theta$ 时，有两种调整方法，其一，调整 B_k、R 和 s 值，使 $\beta \leq \theta$；其二，调整 θ 角，

使 $\beta=\theta$，并用切线扩张。这时 [见图 1-8（b）]

$$\gamma = \arctan \frac{s}{B_k}$$

$$\alpha = \arccos \frac{R}{OB} = \arccos \frac{2R}{\sqrt{B_k^2 + s^2}}$$

$$\theta = \alpha + \gamma = \arccos \frac{2R}{\sqrt{B_k^2 + s^2}} + \arctan \frac{s}{B_k}$$

For bar rolling mill, because the range of positive and negative deviation of the finished product is very small, the finished round hole is usually expanded by tangent line, and the value range of expansion angle is as follows (Shown in Table 1-8).

棒材轧机，由于成品正负偏差范围很小，成品圆孔通常用切线扩张，扩张角取值范围见表 1-8。

Table 1-8　The value range of expansion angle

表 1-8　扩张角取值范围

Finished product diameter/mm 成品直径/mm	φ5.5	φ6.5	φ7	φ8	φ9	φ10	φ12	φ14
Expansion angle $\theta/(°)$ 扩张角 $\theta/(°)$	30	30	25	25	20	20	20	20

Then, $B_k = 2R/\cos\theta - s\tan\theta$.

这时，$B_k = 2R/\cos\theta - s\tan\theta$。

The roll gap s can be selected according to the diameter d of the rolled round steel according to Table 1-9, and the radius of the outer fillet $r = 0.5 \sim 1$mm.

辊缝 s 可根据所轧圆钢直径 d 按表 1-9 选取，外圆角半径 $r=0.5 \sim 1$mm。

Table 1-9　The relationship between the gap s and d of the finished round steel

表 1-9　圆钢成品孔辊缝 s 与 d 的关系

d/mm	6~9	10~19	20~28	30~70	70~200
s/mm	1~1.5	1.5~2	2~3	3~4	4~8

It should be pointed out that the above dimensional relationship is applicable to the finished pass of rolling general round steel. For rolling some alloy steel, it is necessary to design the finished pass according to the production process and other requirements, sometimes not only without negative deviation, but also for positive deviation. However, the above pass composition method is still applicable to the latter.

应指出，上述尺寸关系适用于轧制一般圆钢的成品孔型，对于轧制某些合金钢，则应根据生产工艺和其他要求，有时不但不用负偏差，而且还要用于正偏差设计成品孔，但上述的孔型构成方法仍适用于后者。

1.3.3.3 Design of Other Finishing Passes
1.3.3.3 其他精轧孔型设计

So far, other finishing passes have been determined based on empirical data. At this time, the pass size is determined instead of the rolled piece size, which is different from the extended pass design. In order to be reliable, three methods of calculating metal deformation introduced in the extended pass system can be used after determining each finishing pass. Of course, other methods can also be used to check the fullness of the rolled piece in the pass. When the hole filling degree is not appropriate, the hole size shall be modified. Next, according to the round steel pass system to introduce the design of other finishing holes.

到目前为止，其他精轧孔型都是根据经验数据确定的，此时确定的是孔型尺寸，而不是轧件尺寸，这一点与延伸孔型设计不同。为了可靠，在确定各精轧孔型之后，可以利用延伸孔型系统中介绍的三种计算金属变形的方法，当然也可以利用其他方法，验算轧件在孔型中的充满程度。当孔型充满程度不合适时，应修改孔型尺寸。下面按圆钢孔型系统介绍其他精轧孔的设计。

(1) Square-ellipse-circular hole system.
(1) 方—椭圆—圆孔型系统。

The composition of oval square finish rolling pass is shown in Figure 1-9, and the relationship between its size and finished round steel is shown in Table 1-10.

椭圆-方精轧孔型的构成如图 1-9 所示，其尺寸与成品圆钢的关系见表 1-10。

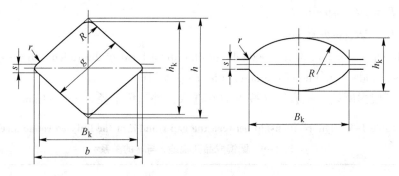

Figure 1-9 Oval square pass size
图 1-9 椭圆—方孔型尺寸

The internal and external radius of the elliptical pass is
椭圆孔型的内外半径为

$$r = 1.0 \sim 1.5 \text{mm}$$

$$R = \frac{(h_k - s)^2 + B_k^2}{4(h_k - s)}$$

The formation height h, width b, radius R and R of inner and outer fillet of square pass are respectively: $H = (1.4 \sim 1.41)a$; $b = (1.41 \sim 1.42)a$; $r = (0.19 \sim 0.2)a$; $r = (0.1 \sim 0.15)a$; roll gap s is taken as 1.5~4mm when rolling round steel with diameter less than 34mm, and $S=$

4~6mm when $D>34$mm, but it should be noted that s value and R value should correspond, even if $s<\left(\dfrac{1}{0.707}R-\dfrac{0.414}{0.707}r\right)$, to ensure the condition of obtaining correct square section.

方孔型的构成高度 h，宽度 b 以及内外圆角半径 R 和 r 分别为：$h=(1.4\sim1.41)a$；$b=(1.41\sim1.42)a$；$R=(0.19\sim0.2)a$；$r=(0.1\sim0.15)a$；辊缝 s 当轧制直径小于 34mm 的圆钢时取 $s=1.5\sim4$mm，$d>34$mm 时可取 $s=4\sim6$mm，但要注意 s 值和 R 值应相对应，即使 $s<\left(\dfrac{1}{0.707}R-\dfrac{0.414}{0.707}r\right)$，以保证获得正确方形断面的条件。

Table 1-10 The relationship between the size of ellipse and square pass and the diameter D of finished round steel

表 1-10 椭圆和方孔型构成尺寸与成品圆钢直径 d 的关系

Finished product specification d/mm 成品规格 d/mm	The relationship between the size of ellipse pass and d 成品前椭圆孔型尺寸与 d 的关系		The relationship between the side length a and d of the front hole of the finished product 成品前方孔边长 a 与 d 的关系
	$\dfrac{h_k}{d}$	$\dfrac{B_k}{d}$	
6~9	0.70~0.78	1.64~1.96	$(1.0\sim1.08)d$
9~11	0.74~0.82	1.56~1.84	$(1\sim1.08)d$
12~19	0.78~0.86	1.42~1.70	$(1\sim1.14)d$
20~28	0.82~0.83	1.34~1.64	$(1\sim1.14)d$
30~40	0.86~0.90	1.32~1.60	$d+(3\sim7)$
40~50	约0.91	约1.4	$d+(8\sim12)$
50~60	约0.92	约1.4	$d+(12\sim15)$
60~80	约0.92	约1.4	$d+(12\sim15)$

In order to determine the size of the rolled piece in the finishing pass, the width coefficient β of the rolled piece in each finishing pass can be selected according to the side length a of the square section or the diameter $2R$ of the finished product. Refer to the data in Table 1-10. In this way or by other methods, the width b of the rolled piece shall be smaller than the width B_k of the rolling groove, and it is friendly to make $B_k/b\leqslant0.95$ or $B_k/b=0.95\sim0.85$, otherwise the pass size shall be modified accordingly.

为了确定轧件在精轧孔型中的尺寸，根据方断面边长 a 或成品直径 $2R$ 来选择轧件在各精轧孔型中的宽展系数 β 可参考表 1-10 中的数据。这样或用其他方法求出的轧件宽度 b 应小于轧槽宽度 B_k，并使 $B_k/b\leqslant0.95$ 或 $B_k/b=0.95\sim0.85$ 为宜，否则应对孔型尺寸作相应的修改。

When the side length a of square piece and the spread coefficient β of rolled piece in the finished pass and elliptical pass are determined according to Table 1-10 and Table 1-11, the height and width of elliptical piece can be determined according to the relationship between reduction and spread coefficient according to the aforementioned design method of two square clamp one flat, and then the size of elliptical pass can be determined according to the fullness of pass.

当根据表 1-10 及表 1-11 确定出方件边长 a 和确定轧件在成品孔型和椭圆孔型中的宽展

系数 β 后，也可按类似两方夹一扁的前述延伸孔型设计方法，根据压下量和宽展系数的关系来确定椭圆件的高度和宽度，再根据轧件尺寸考虑孔型的充满度来确定椭圆孔型的尺寸。

Table 1-11　Data of spread coefficient β of rolled piece in elliptical square pass

表 1-11　轧件在椭圆—方孔型中的宽展系数 β 的数据

d/mm	β		
	Finished pass 成品孔型	Elliptical pass 椭圆孔型	Square pass 方孔型
6~9	0.4~0.6	1.0~2.0	0.4~0.8
10~32	0.3~0.5	0.9~1.3	0.4~0.75

(2) Circle-ellipse-circle hole system.

(2) 圆—椭圆—圆孔型系统。

The diameter D of the base circle of the elliptical front pass is,

椭圆前孔型的基圆直径 D 为，

When the diameter D of round steel is 8~12mm：

当圆钢的直径 D 为 8~12mm 时：

$D = h_k = (1.18 \sim 1.22)d$

When the diameter D of round steel is 13~30mm：

当圆钢直径 D 为 13~30mm 时：

$D = h_k = (1.21 \sim 1.26)d$

Its shape is the same as the finished hole, and also has a 30° expansion angle.

其形状同成品孔，也带有30°的扩张角。

The spread coefficient β of the rolled piece in the round ellipse finish rolling pass can be selected according to Table 1-12.

轧件在圆—椭圆精轧孔型中的宽展系数 β 可按表 1-12 选取。

Table 1-12　The spread coefficient β of rolling piece in round ellipse finish rolling pass

表 1-12　轧件在圆—椭圆精轧孔型中的宽展系数 β

pass 孔型	Finished pass 成品孔型	Elliptical pass 椭圆孔型	Circular hole pass 圆孔型	Elliptical pass 椭圆孔型	
				$d = 15 \sim 20$mm	$d \geq 20 \sim 25$mm
β	0.3~0.5	0.8~1.2	0.4~0.5	0.85~1.2	0.50~0.85

In the oval finish rolling pass, the pass size can also be determined first according to the spread coefficient of the rolled piece in the finished pass and oval pass, and then according to the required fullness.

椭圆精轧孔型时，同样也可按两圆夹一扁的方法，根据轧件在成品孔型和椭圆孔型中的宽展系数先确定轧件尺寸，然后根据所要求充满度确定孔型尺寸。

(3) Universal pass system.

(3) 万能（通用）孔型系统。

1) Commonality of universal pass system.

1) 万能孔型系统孔型的共用性。

The utility degree of a group of universal finishing passes varies according to the diameter of round steel, as shown in Table 1-13. D and d in the table are the maximum and minimum diameter of round steel in rolling a group of round steel. $D-d$ is better not to exceed the data in the table. This is because the larger the difference of $D-d$ is, the smaller the height width ratio of designed vertical pressure will be, and the more unstable the rolled piece is in the vertical pressure hole.

一组万能精轧孔型的公用程度依圆钢的直径而异，见表 1-13。表中的 D 和 d 分别为轧制一组圆钢中最大和最小圆钢直径，$D-d$ 最好不超过表中的数据，这是因为 $D-d$ 的差值越大，设计出的立压力高宽比将越小，轧件在立压孔中越不稳定。

Table 1-13 Sharing degree of a group of universal finish rolling passes

表 1-13 一组万能精轧孔型的共用程度

Round steel diameter/mm 圆钢直径/mm	14~16	16~30	30~50	50~80	>80
Diameter difference of adjacent round steel $D-d$/mm 相邻圆钢直径差 $D-d$/mm	2	3	4~5	5	10

2) Design of oval pass before finished product.

2) 成品前椭圆孔型的设计。

Refer to Table 1-14 for the composition dimension of elliptical pass. The height H_k of the pass is determined by the minimum round steel diameter d, and the width B_k is determined by the maximum round steel diameter D. the relationship between B_k and H_k, D and d is shown in Table 1-14. In the initial design, it is better to make the value of H_k smaller so as to facilitate adjustment and necessary modification.

椭圆孔型的构成尺寸参见表 1-14，其孔型的高度 H_k 是按最小圆钢直径 d 确定，其宽度尺寸 B_k 是按最大圆钢直径 D 确定，其 B_k 和 H_k 与 D 和 d 的关系见表 1-14。在初设计时，最好使 H_k 值小些，以便于调整和做必要的修改。

Table 1-14 The relationship between the dimensions of elliptical pass H_k and B_k and D and d

表 1-14 椭圆孔型尺寸 H_k 和 B_k 与 D 和 d 的关系

Round steel diameter/mm 圆钢直径/mm	14~18	18~32	40~100	100~180
H_k/d	0.75~0.88	0.80~0.9	0.88~0.94	0.85~0.95
B_k/D	1.5~1.8	1.38~1.78	1.26~1.60	1.22~1.40

The roll gap s can be $s \leq 0.01D_0$, D_0 is the roll diameter. The internal and external arc radii R and r of the pass are the same as before.

辊缝 s 可取 $s \leq 0.01D_0$，D_0 为轧辊直径。孔型的内外圆弧半径 R 和 r 的取法同前所述。

1.3.3.4 Round Steel Pass Design Example
1.3.3.4 圆钢孔型设计实例

Example: A $\phi400/\phi250\times5$ small steel rolling mill is driven by two AC motors respectively, and the speed of the finished frame is 6.5mm/s. Trial design of finish-rolled hole pattern for rolled 20mm round steel.

例：某 $\phi400/\phi250\times5$ 小型轧钢车间分别由两台交流电机传动，成品机架速度为 6.5mm/s。试设计轧制 20mm 圆钢的精轧孔型。

Solution:
解：

For rolling 20 mm round steel on $\phi250$ rolling mill, either square ellipse element pass system or circle ellipse circle pass system can be adopted. Considering the uniform deformation and the use of disc, the circle ellipse circle pass system is adopted.

According to the national standard GB 702—86, the allowable deviation of 20mm round steel is 3 groups ±0.5mm, then the size of the finished pass is:

在 $\phi250$ 轧机上轧制 20mm 圆钢可采用方—椭圆—圆孔型系统，也可以采用圆—椭圆—圆孔型系统，考虑到变形均匀、使用圆盘，确定采用圆—椭圆—圆孔型系统。

按国家标准 GB702—86，20mm 圆钢的允许偏差 3 组±0.5mm，则成品孔型的尺寸为：

$$B_k = [d + (0.5 \sim 0.1)\Delta_+] \times (0.007 \sim 1.02) = [20 + 0.7 \times 0.5] \times 1.011 = 20.6\text{mm}$$

Because $s=2$mm; $\theta=30°$
因为 $s=2$mm; $\theta=30°$
So
则

$$\beta = \arctan\frac{B_k - 2R\cos\theta}{2R\sin\theta - s} = \arctan\frac{20.6 - 19.8\cos30°}{19.8\sin30° - 2} = 23.6°$$

Since $\beta<\theta$, R' can be obtained as
因为 $\beta<\theta$, 故可求出 R' 为

$$R' = \frac{2R\sin\theta - s}{4\cos\beta\sin(\theta - \beta)} = \frac{19.8\sin30° - 2}{4\cos23.6°\sin(30° - 23.6°)} = 19.3\text{mm}$$

Determine the size of the oval hole before the finished product:
确定成品前椭圆孔型尺寸为：

$$h_k = (0.80 \sim 0.83)d = 0.80 \times 20 = 16\text{mm}$$
$$B_k = (1.34 \sim 1.64)d = 1.6 \times 20 = 32\text{mm}$$

Taking the roll gap $s=3$mm, the ellipse radius R is:
取辊缝 $s=3$mm，则椭圆半径 R 为：

$$R = \frac{(h_k - s)^2 + B_k^2}{4(h_k - s)} = \frac{(16-3)^2 + 32^2}{4(16-3)}$$

The diameter D of the base circle radius of the ellipse front hole is:
椭圆前圆孔型的基圆半径直径 D 为：

$$D = h_k = (1.21 \sim 1.26)d = 1.25 \times 2 = 25 \text{mm}$$

The determination of other dimensions is similar to that of the finished hole.

其他尺寸的确定与成品孔类同。

Draw each hole pattern according to the design size as shown in Figure 1-10.

按设计尺寸画出各个孔型如图 1-10 所示。

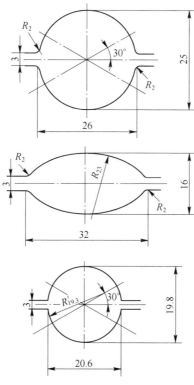

Figure 1-10　Finishedpass

图 1-10　精轧孔型

Check the fullness of the finished rolled pattern. Table 1-12 shows the range of the value of the broadening coefficient of each finish rolling hole. According to the knowledge of rolling principles, in general, when the diameter of the finished round steel is large, the lower limit of the broadening coefficient is taken, and vice versa. In this example, a 20mm round steel is designed. The expansion coefficient should be a small value. The expansion coefficient $\beta_t = 0.7$ in the elliptical hole pattern and the $\beta_y = 0.35$ in the finished hole pattern.

验算精轧孔型的充满情况。表 1-12 出了各精轧孔型宽展系数的取值范围，由轧制原理的知识可知，在一般情况下，当成品圆钢直径大时，宽展系数取下限，反之则相反，本例是设计 20mm 圆钢，则宽展系数应该取偏小值，取椭圆孔型中的宽展系数 $\beta_t = 0.7$，成品孔型中的宽展系数 $\beta_y = 0.35$。

The size of the oval rolled product is：

椭圆轧件尺寸为：

$$h = 16 \text{mm}, \ b = 25 + (25-16) \times 0.7 = 31.3 \text{mm}$$

The dimensions of the rolled product in the finished product are:

成品孔型中轧件的尺寸为:

$$h = 19.8mm, \ b = 16+(31.3-19.8) \times 0.35 = 20mm$$

The calculation results show that the filling degree of the elliptical hole pattern is $31.3/32 = 0.98$, which is too large. Taking the filling degree of the oval hole pattern as 0.9, then $B_k = 31.3/0.9 = 34.8mm$ for the oval hole pattern.

计算结果表明,椭圆孔型的充满程度为 $31.3/32 = 0.98$,充满程度太大。取椭圆孔型的充满程度为 0.9,则椭圆孔型的 $B_k = 31.3/0.9 = 34.8mm$。

According to $h_k = 16mm$ and $B_k = 34.8mm$, calculate the ellipse arc radius $R = (16-3)^2 + 34.8^2/4 \times (16-3) = 26.54$.

再依 $h_k = 16mm$, $B_k = 34.8mm$,计算椭圆圆弧半径 $R = (16-3)^2 + 34.8^2/4 \times (16-3) = 26.54$。

1.3.4 Rollers and Guides
1.3.4 轧辊与导卫

1.3.4.1 Roller
1.3.4.1 轧辊

(1) The structure of roller.

(1) 轧辊结构。

The basic structure of the roller is divided into three parts: roller body, roller neck and roller head, as shown in Figure 1-11.

轧辊的基本结构分为三个部分:辊身、辊颈和辊头,如图 1-11 所示。

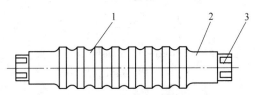

Figure 1-11 The structure of roll
1—Roller body; 2—Roller neck; 3—Roller head
图 1-11 轧辊的结构
1—辊身; 2—辊颈; 3—辊头

The roller body is the part where the roller is in contact with the rolled product and plastically deforms the rolled product. The roller body of the bar mill is a cylinder, and a hole pattern is arranged on it by processing means such as turning and grinding. When non-porous rolling is used, the roller body part does not open the rolling groove, and it is still a standard cylinder.

辊身是轧辊与轧件接触并使轧件产生塑性变形的部分。棒材轧机的轧辊辊身是圆柱体,通过车、磨等加工手段在其上配置孔型。采用无孔型轧制时,其辊身部分不开轧槽,仍为标准的圆柱体。

1) roller size.

1）轧辊尺寸。

The diameter of the roller body and the length of the roller body are the basic parameters of the size of the marking roller, and are also important characteristics of the marking mill.

辊身直径与辊身长度是标志轧辊尺寸的基本参数,也是标志轧机的重要特性参数。

①Roller body diameter.

①轧辊的辊身直径。

The roller body diameter includes four types: maximum diameter, minimum diameter, nominal diameter, and working diameter.

轧辊的辊身直径包括最大直径、最小直径、公称直径、工作直径四种。

Nominal diameter. Also called the nominal diameter, it usually refers to the center moment of the rolling mill herringbone gear. In China, the former Soviet Union and the United States usually use the nominal diameter to indicate the size of the rolling mill. However, some countries also define the rolling mill size by the maximum roll diameter.

公称直径。也称名义直径,通常是指轧机人字齿轮中心矩。在我国,国外通常用名义直径表示轧机大小。但有的国家也以最大辊径来定义轧机大小。

Work roller diameter. The work roller diameter refers to the roller diameter at a certain point of the rolling groove corresponding to the actual exit speed of the rolled product, also known as the rolling diameter.

工作辊径。工作辊径是指与轧件实际出口速度相对应的轧槽某一点的轧辊直径,亦称轧制直径。

Maximum diameter. Working diameter of new roller.

最大直径。新辊时的工作直径。

Minimum diameter. Working diameter of the roller after the last turning.

最小直径。最后一次车削后的轧辊工作直径。

②Roller length.

②辊身长度。

The ratio of the length L of the roller body to the nominal diameter D of the roller has certain requirements for different rolling mills. For section steel rolling mills, it is usually 1.5 to 2.5, and for finishing rolling mills, it is 1.5 to 2.0. L/D signifies the roller bending stiffness. The smaller the ratio, the higher the roller stiffness and the smaller the bending moment.

轧辊辊身长度 L 与轧辊名义直径 D 之比,对于不同轧机都有一定的要求。对于型钢轧机通常为 1.5~2.5,精轧机则为 1.5~2.0。L/D 标志着轧辊抗弯刚度,比值越小,轧辊刚度越高,随之弯矩越小。

2）returning of roller.

2）轧辊的再车削。

In the process of rolling, the surface of roller is constantly worn. After a period of time, the roller surface wear will seriously affect the dimensional accuracy and surface quality of the product. The total allowable weight of the roll varies according to the groove type, pass design and roller material. Because the greater the amount of repeated turning, the longer the service life of the roller, it is always hoped that the amount of repeated turning is larger. However, due to the limitation of contact angle, roller surface state (especially surface crack) and surface cold hard layer, as well as the influence of factors such as bite condition, bending strength and bending deflection, the amount of heavy vehicle is limited.

在轧制过程中,轧辊表面不断被磨损。经过一段时间后,辊面磨损较严重将影响产品尺寸精度和表面质量,此时则需重车。轧辊总的允许重车量根据轧槽型式、孔型设计和轧辊材质等不同而不同。因为轧辊的重车量越大,轧辊的使用寿命则越长,所以总是希望重车量大些。但是由于接触倾角、轧辊表面状态(特别是表面裂纹)与表面冷硬层的限制,以及咬入条件,弯曲强度与弯曲挠度等因素的影响,使其重车量受到一定限制。

The turning coefficient of roller can be expressed as follows:

轧辊的车削系数可用下式表示:

$$K = (D_{max} - D_{min})/D_o$$

Where K——Turning coefficient;
　　D_{max}——Maximum roller diameter, mm;
　　D_{min}——Minimum roller diameter, mm;
　　D_o——Nominal roller diameter, mm.

式中 K——车削系数;
　　D_{max}——最大辊径, mm;
　　D_{min}——最小辊径, mm;
　　D_o——名义辊径, mm。

3) Roller neck.

3) 辊颈。

The roller neck, also known as the roller neck, is the supporting part of the roller. It is in contact with the roller bearing to support the roller on the bearing. The roller neck is cylindrical and tapered, and most of the roller neck of bar mill is cylindrical.

辊颈亦称辊脖子,是轧辊的支承部分。它与轧辊轴承接触,使轧辊支撑在轴承上。辊颈有圆筒形和带锥形的,棒材轧机的轧辊辊颈大多数为圆筒形。

The roller neck size is diameter d and length l, which is related to bearing form and working load. In view of the strength, it is necessary to take a larger value of the roller neck to strengthen the roller safety and prevent the phenomenon of the roller neck breaking. But structurally it is limited by bearing size. In addition, the transition angle r between the roller neck and the roller body should be larger to prevent the roller from breaking easily due to stress concentration.

辊颈尺寸为直径 d 和长度 l,它与轴承形式及工作载荷有关。从强度考虑,将辊颈取较大值,对加强轧辊安全防止出现断辊颈现象是必要的。但结构上它受到轴承尺寸限制。另外,辊颈与辊身的过渡角 r 应选大些,以防止应力集中容易断辊。

4) Roller head.

4) 辊头。

The roller head is the connecting part of the roller and the driving device, which is used to transmit the torque and drive the roller to rotate. Its shape usually has two kinds of flat head and plum head.

辊头是轧辊与驱动装置的联接部分，用来传递扭矩，带动轧辊转动。其形状通常有扁头和梅花头两种。

The plum head has simple structure and low machining accuracy. It is generally used in the rolling mill with low speed and small roll spacing. The roll head of the advanced continuous bar mill is generally flat head, mainly considering the convenience of machining and the relatively easy import and export of the rack.

梅花头结构简单，加工精度要求不高。一般用在速度不高，轧辊间距较小的轧机处。先进的连续棒材轧机的辊头一般都采用扁头，主要是考虑机加工方便及机架进出口比较容易。

（2）Working conditions and characteristics of roller.

（2）轧辊的工作条件及特点。

1）Working conditions of roller.

1）轧辊的工作条件。

The roller body often works under the most heavy load conditions such as high temperature, high pressure and impact. Such as:

轧辊辊身经常在高温、高压、受冲击等最繁重的负荷条件下工作。如：

①The roller of section mill has to bear a lot of rolling pressure, up to several hundred tons or even thousands of tons.

①型钢轧机的轧辊要承受很大的轧制压力，高达几百吨乃至上千吨。

②It is often subjected to shock loads and alternating fatigue loads.

②经常承受着冲击负荷与交变的疲劳负荷。

③In the high temperature state, internal stress is generated due to water spray cooling.

③在高温状态下由于喷水冷却而产生内应力等。

Therefore, the working environment of roller is very bad.

因此，轧辊的工作环境相当恶劣。

2）Working characteristics of roller.

2）轧辊的工作特点。

The roller shall be selected with the following characteristics:

选择轧辊时要具有以下特点：

①It has high strength to withstand strong bending moment and torque.

①有很高的强度，以承受强大的弯矩和扭矩。

②It has enough rigidity to reduce the elastic deformation of the roll to ensure the high precision of the workpiece size.

②有足够的刚度，以减少轧辊的弹性变形来保证轧件尺寸的高精度。

③It has high surface hardness and wear resistance to ensure the highest quality of the rolled piece and improve the single groove life. This is especially true for finished pass rolls.

③有较高的表面硬度和耐磨性，以保证轧件的高质量并提高单槽寿命。对于成品道次轧辊尤其如此。

④It has good structure stability to resist the influence of high temperature.

④有良好的组织稳定性，以抵抗轧件的高温影响。

(3) Main performance index of roller.

(3) 轧辊的主要性能指标。

With the development of composite roller technology, the concept of what is the main technical index of roller is changing. The traditional two major technical indexes are wear resistance and strength, that is, to pursue the roller with less wear and not easy to break. The overall strength of the roller mainly depends on the core. To ensure the core strength of the roller, there is no technical obstacle at present, so the focus is gradually shifted to the working layer. There are two phenomena in the consumption of the working layer of the roller, one is the gradual wear of the roller surface, the other is the consumption caused by various cracking phenomena (including hot cracking, peeling, etc.). In view of these two consumption phenomena, there are two requirements for the working layer of the roller, namely, wear resistance and accident resistance. The level of accident resistance mainly depends on the toughness index of the material. Therefore, the main performance index of the roller should be the core strength and the toughness and wear resistance of the working layer.

随着复合轧辊技术的发展，什么是轧辊的主要技术指标，其观念也在变化。传统的两大技术指标是耐磨性和强度，即追求磨损少而不易折断的轧辊。轧辊的整体强度主要取决于芯部，要保障轧辊的芯部强度，目前在技术上已不存在障碍，因此侧重点逐渐移向工作层。轧辊工作层的消耗有两种现象，一是辊面的逐渐磨损，一是因各种开裂现象引起的消耗（包括热裂、剥落等）。针对这两种消耗现象，对轧辊的工作层就产生了两方面的要求，即耐磨性和抗事故性。抗事故性的高低主要取决于材料的韧性指标。因此，轧辊的主要性能指标应该是芯部强度和工作层的韧性及耐磨性。

1) Wear resistance.

1) 耐磨性。

Wear is the most common form of roller damage, which will endanger the surface accuracy and dimensional accuracy of the rolled piece.

磨损是轧辊最常见的损伤形式，它会危及轧件的表面精度和尺寸精度。

Wear originates from the action of friction, friction body, temperature and medium. The friction system is composed of the relative movement of the workpiece and the roller under the rolling load. The scale or chips of the rolled piece will form the abrasive grains between the roller gaps. In addition to abrasive wear, the roller surface also suffers periodic hot and cold fatigue, which promotes the fatigue wear of the roller surface. In addition, the roller surface may be corroded and worn due to the action of medium, such as oxidation wear.

磨损起源于摩擦力的作用，摩擦体的作用，以及温度和介质的作用。轧件和轧辊在轧制载荷下做相对运动构成摩擦体系。轧件的氧化铁皮或磨屑本身将构成辊缝间的磨粒。除磨粒磨损外，热轧条件下辊面还承受周期性的冷热疲劳，助长了辊面的疲劳磨损。此外，辊面上还可能因介质的作用而发生腐蚀磨损，如氧化磨损。

Because of the complex wear mechanism between rollers, there is no wear resistance index which can be used to characterize the wear resistance of rollers. Therefore, the wear resistance of roller can only be judged indirectly according to the composition and hardness.

正是因为辊缝间的复杂磨损机制,至今还没有任何一种耐磨性指标能适合于表征轧辊的磨损抗力。因此,轧辊的耐磨性只能间接地根据成分和硬度来判断。

The selection of roller hardness is based on the rolling piece, rolling mill and rolling conditions, and sometimes more importantly, the operation habits and roller experience, as well as the consideration of cost. If we pursue high hardness, we should not only increase the purchase cost, but also be restricted by the anti accident. The harder the roller is, the worse its toughness is, the more likely it is to have all kinds of fracture damage, including hot crack. This is because the wear resistance and toughness of materials often contradict each other, so it is not easy to pursue single high hardness.

选择轧辊硬度的依据是轧件、轧机和轧制条件,有时更重要的是操作习惯和用辊经验,也包括对成本的考虑。如果一味追求高硬度,不仅要增加购置成本,而且还要受抗事故性的制约。越硬的轧辊往往韧性越差,越容易发生各式各样的断裂损伤,包括热裂在内。这是因为材料的耐磨性和韧性经常是相互抵触的,所以,不能简单地追求单项高硬度。

The hardness of the roller is generally measured by the shore hardness, because the shore hardness tester can be made into a portable one, which can be used on the roller, and only leaves a small indentation on the roller surface. There are several types of shore hardness tester, D type is generally used in China. Recently, the Leeb hardness tester has been paid more attention by the roll industry. The hardness is indicated by the bounce height of the hammer head by the shore hardness tester, while the hardness is indicated by the rebound speed by the Leeb hardness tester. Of course, Vickers, Rockwell or other hardness are not excluded, as long as it is convenient to use and there is a legal comparison table with shore hardness.

轧辊的硬度一般用肖氏硬度来度量,因为肖氏硬度计可做成便携式的,可放在轧辊上使用,而且只在辊面上留下很小的压痕。肖氏硬度计有若干类型,我国一般采用 D 型。近来里氏硬度计已受到轧辊行业的重视。肖氏硬度计根据锤头弹跳高度来标示硬度,而里氏硬度计则根据回弹速度来标示硬度。当然也不排除维氏、洛氏或其他硬度,只要使用方便,并与肖氏硬度有法定对照表即可。

2) Resilience.

2) 韧性。

The toughness index will be selected according to the main damage mechanism of the roller surface. The toughness indexes that can be selected include yield strength, impact toughness, cold and hot fatigue, fracture toughness, contact fatigue strength, etc. According to the specific service conditions and the main damage mechanism of the roller, generally only one of them needs to be selected as the index of the roller toughness.

韧性指标将根据辊面的主要损伤机制来选择,可以选择的韧性指标有屈服强度、冲击韧性、冷热疲劳、断裂韧性、接触疲劳强度等。根据轧辊的具体服役条件和主要损伤机制,一般只需选择其中一项作为轧辊韧性的指标。

In the process of casting, heat treatment, hardfacing and cutting, the roller may leave residual stress due to incongruous deformation. The residual stress and the working stress will be superposed, which will cause different damage phenomena. Therefore, the influence of residual stress, especially the residual stress in the working layer, should be taken into account when analyzing the anti accident performance of roller.

轧辊在铸造、热处理、堆焊以至切削过程中都有可能因变形不协调而留下残余应力。使用时残余应力将与工作应力叠加,共同引发不同的损伤现象。因此,分析轧辊的抗事故性能时,要同时顾及其残余应力,尤其是工作层内的残余应力的影响。

3) Core strength.

3) 芯部强度。

The strength requirements of the working layer and the core (including the roll neck and the driving part) of the roller are not concerned. Generally speaking, roller strength refers to the core strength to resist bending and torsional stress and avoid roller breakage. Because the mechanical stress produced by roughing stand is greater than that produced by finishing stand, the higher the front frame number, the higher the requirement of roller strength.

轧辊的工作层和芯部(包括辊颈和传动部分)对强度的要求是不同的。泛指的轧辊强度一般应为芯部强度,用以抵抗弯扭应力,避免断辊。由于粗轧机座所产生的机械应力大于精轧机座所产生的机械应力,因此越靠前的架次对轧辊强度的要求越高。

Sometimes, the working layer of roller also requires strength directly, for example, the compressive strength can represent the ability of the roller surface to resist the pressure crack. In addition, the testing cost of some toughness indexes is very high, so sometimes the strength is roughly used to express the ability of roll surface to resist various cracks.

轧辊工作层有时也直接要求强度,如抗压强度可代表辊面的抗压裂的能力。此外,有些韧性指标的测试费用很高,故有时也会粗略地用强度来表示辊面抗各种裂纹的能力。

(4) Material of roller.

(4) 轧辊的材料。

1) Cast iron roller.

1) 铸铁轧辊。

In the current standard of our country, cast iron roller is divided into four categories: chilled cast iron roller, infinite chilled cast iron roller, nodular cast iron roller and high chromium cast iron roller. Due to the differences in composition, inoculation, cold speed, heat treatment and other aspects, the cast iron roller has formed a huge system, which can be used in different rolling mills and rolling conditions due to the differences in structure and performance. The composition and properties of various cast iron rollers can be found in the national standard GB/T 1504—91.

在我国的现行标准中,铸铁轧辊又分为冷硬铸铁轧辊、无限冷硬铸铁轧辊、球墨铸铁轧辊和高铬铸铁轧辊四大类。由于成分、孕育、冷速、热处理等方面的细致差异,铸铁轧辊已形成了一个庞大的体系,彼此间组织结构和性能的差异,使之可用于不同的轧机和轧制条件。各种铸铁轧辊的成分和性能,都可以在国家标准 GB/T 1504—91 中查到。

①Chilled cast iron roller.

①冷硬铸铁轧辊。

The structure of working layer is matrix plus carbide, and the core is gray iron. The transition position (i. e. transition layer) between them can be clearly identified on the fracture surface. The cold hard cast iron roller is famous for its wear resistance. According to the grades, the hardness of chilled cast iron roller can reach 55~85HSD (D-type shore hardness value), which is widely used as medium and finishing roller in small and wire rod mill.

工作层组织为基体加碳化物,芯部为灰铸铁组织。其间的过渡位置(即过渡层)在断口上明确可辨。冷硬铸铁轧辊以耐磨著称。根据牌号的不同,冷硬铸铁轧辊的硬度可达 55~85HSD(D 型肖氏硬度值),在小型和线棒材轧机上被广泛用作中轧和精轧辊。

②Infinite chilled cast iron roller.

②无限冷硬铸铁轧辊。

The structure of the working layer is matrix with carbide and flake graphite, and the core is gray iron. There is a gradual transition (or a wide transition layer), and there is no clear transition position on the fracture surface. Due to the appearance of graphite, the lower limit of hardness can be lower than that of chilled cast iron, and high hardness can be obtained after adding alloy, so the hardness range is 50~85HSD. The hardness transition of infinite cold hard cast iron roller is smooth, and it can resist hot crack. It is also widely used in medium and finishing mills of small and line bars.

工作层组织为基体加碳化物和细片状石墨,芯部为灰铸铁组织。其间为逐渐过渡(或过渡层很宽),断口上无明确可辨的过渡位置。由于石墨的出现,硬度下限可低于冷硬铸铁,加入合金后也能获得很高的硬度,故硬度范围达 50~85HSD。无限冷硬铸铁轧辊硬度过渡平缓,抗热裂,同样被广泛用于小型和线棒材的中轧机和精轧机上。

③Nodular cast iron roller.

③球墨铸铁轧辊。

Due to the spheroidizing effect of magnesium or rare earth elements on graphite, the structure of nodular cast iron roller is matrix plus spheroidal graphite, and there may be different amount of carbides in the working layer. The hardness range of nodular cast iron roller is very wide, up to 42~80HSD. Depending on the effect of alloy elements and heat treatment, the nodular cast iron roller with needle like structure can be obtained. Therefore, sometimes the hardness can be higher under the condition of less carbides. The hardness of nodular cast iron roller with needle structure can reach 50~85HSD. In addition, the semi cold hard nodular cast iron roller can be made by using the method of metal mold sand lining, which can reduce the hardness to 35~55HSD and has no obvious drop to the core, so as to meet the needs of some deep hole rollers. Nodular cast iron roller is the most widely used roller, from the rough rolling, medium rolling to finish rolling of small and line bars are widely used.

由于镁或稀土元素对石墨的球化作用,球墨铸铁轧辊的组织为基体加球状石墨,工作层内可有数量不等的碳化物。球墨铸铁轧辊的硬度范围很宽,达 42~80HSD。依靠合金元素和热处理的作用,可获得基体为针状组织的球墨铸铁轧辊,因此,有时可在较少碳化物的条件下获得较高的硬度。具有针状组织的球墨铸铁轧辊,硬度可达 50~85HSD。此外,还可以采用金属型衬砂的方法制成半冷硬球墨铸铁轧辊，使硬度降至 35~55HSD，到芯部无明显落差，以适应某些深孔型轧辊的需要。球墨铸铁轧辊是应用量最多的轧辊，从小型和线棒材的粗轧、中轧到精轧均被大量采用。

④High chromium cast iron roller.

④高铬铸铁轧辊。

In general, M_3C carbide in cast iron roller structure forms continuous phase in eutectic, so it is brittle. The eutectic carbide in high chromium cast iron is M_7C_3 type, which forms discontinuous phase and reduces brittleness. Therefore, it is paid more attention by roller industry. With the difference of chromium content and heat treatment process, the hardness range of high chromium cast iron roller is very wide, 55~95HSD. In small and wire rod mill, only finishing mill uses high chromium cast iron roller.

一般铸铁轧辊组织中的 M_3C 型碳化物在共晶时形成连续相，因此较脆。高铬铸铁中的共晶碳化物是 M_7C_3 型的，构成断续相，减小了脆性，因此很受轧辊行业的重视。随着铬含量和热处理工艺的不同，高铬铸铁轧辊的硬度范围也很宽，为 55~95HSD。在小型和线棒材轧机上一般只有精轧机采用高铬铸铁轧辊。

2) Cast steel roller.

2) 铸钢轧辊。

In China's current standards, cast steel rollers are divided into four categories: common cast steel roller, alloy cast steel roller, semi steel roller and graphite steel roller.

在我国的现行标准中，铸钢轧辊又分为普通铸钢轧辊、合金铸钢轧辊、半钢轧辊和石墨钢轧辊四大类。

①Common cast steel roller. The common cast steel roller is made of high quality carbon steel, containing 0.6%~0.8% carbon (mass fraction) and sometimes 1% manganese.

①普通铸钢轧辊。普通铸钢轧辊是由优质碳素钢铸成，含碳（质量分数）0.6%~0.8%，有时含 1%的锰。

②Alloy cast steel roller. The alloy cast steel roller can contain less than 1.8% manganese, less than 1.2% chromium, less than 0.5% nickel and less than 0.45% molybdenum.

②合金铸钢轧辊。合金铸钢轧辊可含 1.8%以下的锰，1.2%以下的铬，0.5%以下的镍，0.45%以下的钼。

③Semi steel roller. Semi steel roller contains 1.3%~1.7% carbon to improve hardness and wear resistance.

③半钢轧辊。半钢轧辊含 1.3%~1.7%的碳，以提高硬度，改善耐磨性。

④Graphite steel roller. Due to the inoculation of silicon calcium and silicon iron, free graphite is formed in the structure of the graphite steel roller, which improves the hot cracking resistance of the roller.

④石墨钢轧辊。石墨钢轧辊因经过硅钙和硅铁等的孕育作用，组织中有游离石墨生成，从而改善了轧辊的抗热裂性。

The upper limit of the hardness of the cast steel roller can only reach the lower limit of the hardness of the cast iron roller, but it has high strength and can be used to open deep pass. The roughing mill of some small rolling mills adopts cast steel roller. The composition and hardness of various cast steel rollers can be found in the national standard GB 1503—89.

铸钢轧辊的硬度上限只能达到铸铁轧辊的硬度下限,但是强度高,可开深孔型。部分小型轧机的粗轧机采用铸钢轧辊。各种铸钢轧辊的成分和硬度,都可以在我国国家标准GB 1503—89 中查到。

3) Forged steel roller.

3) 锻钢轧辊。

Forged steel roller is mainly used for supporting roller and working roller of cold rolled strip. In the field of hot rolling, forged steel hot rolling rollers are mainly used in non-ferrous metal industry. Forged steel rollers are occasionally used in roughing mills of small mills.

锻钢轧辊主要用于支承辊和冷轧带钢的工作辊。在热轧领域,主要是有色金属业采用锻钢热轧辊。在小型轧机的粗轧机上偶尔也采用锻钢轧辊。

4) Special roller.

4) 特种轧辊。

There are some kinds of rollers, most of which are new technology and new material rollers developed in recent years, which have not been included in the national standard, are all classified as special rollers.

有一些轧辊品种,大部分是近年来开发的新工艺和新材质轧辊,尚未纳入国家标准的一律归入特种轧辊。

①Cemented carbide roller.

①硬质合金轧辊。

The cemented carbide roller is made by pressing and sintering tungsten carbide powder and cobalt powder. The cemented carbide has good heat conduction performance, high hardness and low drop at high temperature, good wear resistance, good heat-resistant fatigue performance and high strength, which is suitable for high-speed rolling. Therefore, the cemented carbide roller ring has been widely used in high-speed wire rod finishing mill. Although the price of cemented carbide roller is expensive, because of its excellent wear resistance, some bar mills have also used cemented carbide roller in finishing mill, especially for finished products. The high-speed wire roller rings are all made of solid carbide. However, it is not practical for small and bar mills to use integral cemented carbide rollers. In order to reduce the consumption of cemented carbide, the roller ring of composite (cast in method composite cemented carbide) is generally made, and then the finishing roller of combined small rolling mill is made by mechanical combination.

硬质合金轧辊是用碳化钨粉和钴粉压制和烧结而成,碳化钨硬质合金具有良好的热传导性能,在高温下硬度高且下降小,耐磨性好,耐热疲劳性能好,强度高,适合于高速轧制,因此高速线材精轧机组已普遍采用硬质合金辊环。尽管硬质合金轧辊的价格昂贵,但因其优异的耐磨性,部分棒材轧机也已经在精轧机组,尤其是成品架次,采用了硬质合金轧辊。高线辊环都是由整体硬质合金制造。但小型和棒材轧机采用整体硬质合金轧辊是不现实的。为减少硬质合金的消耗,一般制成复合(铸入法复合硬质合金)辊环,然后用机械组合的方法制造成组合式小型轧机精轧辊。

The compound roller ring is generally made by casting. The pass part is hard alloy, and the inner layer is ductile metal, such as nodular cast iron. The cemented carbide and nodular cast iron are metallurgical bonded. The nodular cast iron and the mandrel are connected by key.

复合辊环一般用镶铸方法制成，其孔型部分为硬质合金，内层为韧性金属，如球墨铸铁。硬质合金与球墨铸铁之间为冶金结合。球墨铸铁与芯轴之间则靠键联接。

②High speed steel roller.

②高速钢轧辊。

The application of high-speed steel roller in hot strip mill has achieved remarkable results. Now, it has been used in small and wire rod mill.

高速钢轧辊在热带钢轧机上的应用已经获得了引人注目的效果，现在，也已经在小型和线棒材轧机上开始被采用。

The alloy content (mass fraction) of high-speed steel roller has exceeded 15%. Its composition range is: $w(C)=1\%\sim2\%$, $w(Cr)=3\%\sim10\%$, $w(Mo)=2\%\sim7\%$, $w(V)=2\%\sim7\%$, $w(W)=1\%\sim5\%$, $w(Nb)=0\%\sim5\%$, $w(Co)=0\%\sim5\%$. The hardness after heat treatment is 75~85HSD.

高速钢轧辊的合金含量（质量分数）已超过15%，其成分范围为：$w(C)=1\%\sim2\%$，$w(Cr)=3\%\sim10\%$，$w(Mo)=2\%\sim7\%$，$w(V)=2\%\sim7\%$，$w(W)=1\%\sim5\%$，$w(Nb)=0\%\sim5\%$，$w(Co)=0\%\sim5\%$，热处理后的硬度为75~85HSD。

The manufacturing methods of high speed steel roller are as follows:

高速钢轧辊的制造方法有如下几种：

Centrifugal casting. Centrifugal casting is a powerful way to produce composite roller. Only high-speed steel is used in the working layer, and ductile iron is generally used in the core and neck of the roller.

离心铸造法。离心铸造是生产复合轧辊的有力手段，只需要在工作层部位采用高速钢，芯部和辊颈部均采用韧性较好的材料，一般是球墨铸铁。

Continuous casting compound method (CPC method). High speed steel is cast into the gap between the steel mandrel and the annular crystallizer, and the coating process is gradually completed by pulling ingot.

连续铸造复合法（CPC方法）。在钢制芯轴和环形结晶器件形成的缝隙间铸入高速钢，同时以拉锭方式逐渐完成涂敷过程。

Electroslag casting. Electroslag casting is a technology of baton Electric Welding Research Institute in Ukraine. It is based on the principle of direct electroslag casting or the principle of conversion casting. It is used to produce casting roller. The mold form is similar to CPC method.

电渣铸造法。电渣铸造法是乌克兰巴顿电焊研究院的一项技术，基于电渣冶金的直接电渣熔铸原理或转铸原理的工艺方法，用于生产铸造轧辊，铸型类似CPC方法。

Integral high speed steel roller (powder high speed steel hot isostatic pressing). The whole forging high-speed steel roller has been advanced to powder hot isostatic pressing (hip) forming, which is used in much smaller cold rolling mill, such as the work roller of Sendzimir mill. However, it does not exclude the use of this method to produce high-speed steel roller rings for composite wire and bar mills.

整体高速钢轧辊（粉末高速钢热等静压成型法）。整锻高速钢轧辊已进步到粉末热等静压（HIP）成型，用在小得多辊冷轧机上，如森吉米尔轧机工作辊。但并不排除用此法生产复合的线棒材轧机用的高速钢辊环。

(5) Selection of roller.

（5）轧辊的选择。

Generally speaking, from the first roughing mill to the finishing mill, the reduction is from large to small, but the rolling speed is higher and higher. Therefore, the general principle of roller selection is: in the roughing mill, the roller is subject to large impact load, large rolling force, and high temperature and alternating thermal stress. Therefore, the roughing roller is mainly based on strength and thermal cracking resistance, taking into account wear resistance. In the finishing mill, the section of the rolled piece has become very small, and the pass deformation is the smallest. Therefore, the requirement of roller strength is no longer the main condition, and the emphasis should be placed on ensuring the dimensional accuracy and surface quality of the products. Therefore, the finishing roller is mainly wear-resistant, taking into account the hot cracking resistance and strength. In the medium rolling mill, the working environment of the roller is between the rough and finishing mill rollers. Therefore, the requirements of the roller are between the two.

总的说来，从粗轧第一机架到精轧的成品机架，压下量从大到小，但轧制速度越来越高。因此，一般的选辊原则是：在粗轧机组，轧辊受冲击载荷大，轧制力也较大，而且受高温和交变热应力作用，所以，粗轧辊以强度和抗热裂性为主，兼顾耐磨性，在精轧机组，轧件的断面已变得很小，道次变形量也较小。所以对轧辊强度要求不再是主要条件，重点要放在保证产品尺寸精度和表面质量上，所以，精轧辊以耐磨为主，兼顾抗热裂性和强度，在中轧机组，轧辊的工作环境介于粗、精轧机的轧辊之间。因此，对所用轧辊的要求也介于二者之间。

1) Selection of rollers for rough rolling mill.

1）粗轧机组轧辊的选择。

For heavy load roughing mill, if the size of continuous casting billet is large, the pass is too deep, and the stress of groove root is highly concentrated, resulting in excessive bending and torsion stress of roller, and the cast steel roller is also broken, the forged steel roller has to be used, such as Cr-Mn-Mo or Cr-Ni-Mo series Forged steel roller with carbon content (mass fraction) of 0.5%~0.6%. This kind of roller has a strong biting ability at the same time.

大负荷粗轧机架，如连铸方坯尺寸大，孔型太深，槽根应力高度集中，致使轧辊弯扭应力过大，铸钢轧辊也有断辊现象时，只好采用锻钢轧辊，如碳含量（质量分数）为0.5%~0.6%的Cr-Mn-Mo或Cr-Ni-Mo系列锻钢轧辊。这种轧辊同时具有较强的咬钢能力。

For the roughing mill with large heat load, if the hot crack is produced and propagated too fast, which causes the hot crack phenomenon to be too serious, the graphite steel roll can be used to improve it.

热负荷大的粗轧机架，如热裂纹的产生和扩展过快，致使热裂现象过于严重时，采用石墨钢轧辊即可得到改善。

In the roughing mill with small heat load, the steel rolling quantity of semi steel roller is higher. When the hot crack of semi steel roller is serious, alloy cast steel roller with carbon content of about 0.7% can be used.

在热负荷较小的粗轧机架上,半钢轧辊的轧钢量较高。半钢轧辊热裂严重时,可改用含碳0.7%左右的合金铸钢轧辊。

In addition, the rolling speed of the roughing mill is relatively low, sometimes lower than 0.2m/s, which is easy to cause serious hot cracking when using ordinary cast steel roller. Therefore, as long as the strength allows, the use of nodular cast iron roller more occasions, such as shore hardness of about 45 pearlite (and sometimes ferrite) nodular cast iron roller. Some roughing mills have many passes, and the pearlitic nodular cast iron roller with high hardness (about 55 shore hardness) can also be used in the back several passes.

另外,粗轧机架的轧制速度相对较低,有时低于0.2m/s,使用普通铸钢轧辊时容易发生严重热裂。因此,只要强度允许,采用球墨铸铁轧辊的场合更多,如肖氏硬度45左右的珠光体(有时还含铁素体)球墨铸铁轧辊。有的轧机粗轧机组的架次较多,也可在后面几架采用较高硬度(肖氏硬度55左右)的珠光体球墨铸铁轧辊。

2) Selection of rollers for medium rolling mill.

2) 中轧机组轧辊的选择。

Almost no pearlitic nodular cast iron roller with shore hardness of about 55~65 was used in the medium rolling mill. In the middle rolling mill with more sorties, it is necessary to select them in sections. For example, the hardness of the first few stands is about 55, and the latter stands is about 65.

中轧机组几乎无例外地采用肖氏硬度55~65的珠光体球墨铸铁轧辊。在架次较多的中轧机组中须分段选用,如前几架的肖氏硬度选在55左右,而后几架选在肖氏硬度65左右。

3) Selection of rollers for finish rolling mill.

3) 精轧机组轧辊的选择。

There are many kinds of wear-resistant products to be selected because of the high requirements of the finishing mill. Most of the finishing mills of small rolling mill still choose the nodular cast iron roller, of course, the Pearlite Nodular Cast iron roller with higher hardness or the acicular structure (bainite+martensite) nodular cast iron roller with higher hardness (70~75HSD) and the infinite chilled cast iron roller with higher alloy content (70~80HSD) are selected. High hardness cast iron roller generally needs to be made into composite roller, and centrifugal casting process is adopted. If you want to further improve the wear resistance of precision roller, you can choose special roller. Among them, high speed steel roller and cemented carbide roller have achieved excellent rolling effect. The high-speed steel roller for small-scale finish rolling is generally made by centrifugal casting method. The working layer hardness of high-speed steel can reach 75~80HSD, and the core can still be made of nodular cast iron. With reference to the experience of using carbide roller ring in high speed wire rod finishing mill, carbide roller also came into being and was used in the finishing mill stand of bar mill, especially in the finished mill stand. This kind of cemented carbide roller is composed of cemented carbide roller ring and steel mandrel. Generally, the cemented carbide roller ring is a composite cemented carbide roller ring with nodular cast iron in the inner layer.

精轧机架对耐磨性的要求较高，因此，有许多耐磨品种可供选择。大多数小型轧机的精轧机组至今仍首选球墨铸铁轧辊，当然是选用其中硬度较高的珠光体球墨铸铁轧辊或硬度更高的针状组织（贝氏体+马氏体）球墨铸铁轧辊（70~75HSD）和合金含量较高的无限冷硬铸铁轧辊（70~80HSD）。高硬度铸铁轧辊一般需制成复合轧辊，并采用离心铸造工艺。如果想进一步提高精轧辊的耐磨性，可以选择特种轧辊。其中，高速钢轧辊和硬质合金轧辊已经取得了优异的轧制效果。小型精轧用的高速钢轧辊一般也用离心铸造方法制成，高速钢的工作层硬度做到75~80HSD即可，芯部仍可用球墨铸铁。借鉴高速线材精轧机采用硬质合金辊环的经验，硬质合金轧辊也应运而生，并用到了棒材轧机的精轧机架上，尤其是其成品机架。这种硬质合金轧辊由硬质合金辊环和钢制芯轴机械组合而成，硬质合金辊环部分一般是内层铸有球墨铸铁的镶铸式复合硬质合金辊环。

(6) Use and maintenance of roller.

(6) 轧辊的使用与维护。

Roller is the most important production tool in bar production. Therefore, the correct use of roller and careful maintenance is directly related to the improvement of product quality and mill production capacity. At present, the effective operation rate of the normal production workshop is generally lower than 80%, and a considerable part of the stop time is used to change the roller and groove, so any measure to prolong the life of the groove will produce obvious economic benefits. On the premise that the mill form and roller material have been determined, how to correctly use and maintain the roller to promote the high yield, high quality and low consumption in the rolling production is particularly important.

轧辊是棒材生产中最重要的生产工具，因此，正确使用轧辊并进行精心维护直接关系到产品质量的改善和轧机生产能力的提高。目前，正常生产的车间有效作业率一般都低于80%，而相当一部分的停车时间是用来换辊和换槽，所以任何一项使轧槽寿命得以延长的措施都会产生明显的经济效益。在轧机形式、轧辊材质已经确定的前提下，如何正确进行轧辊的使用与维护，以促进轧钢生产中的高产、优质、低消耗就显得尤为重要。

1) Use of roller.

1) 轧辊的使用。

In the process of using the roller, we should pay attention to the problems of incoming inspection, roller turning, roller transportation, roller disassembly and so on.

轧辊的使用过程中，应注意来料检查、轧辊车削、轧辊运输及轧辊拆装等方面问题。

①Incoming inspection.

①来料检查。

After the roller is transported from the production plant to the use plant, the acceptance shall be carried out in strict accordance with the technical requirements of the drawings, such as the dimensional accuracy, roller hardness, surface quality, etc. of all parts. The unqualified products shall be resolutely removed to prevent leaving more hidden dangers for the next step of putting into use.

轧辊从生产厂运到使用厂后,应严格按照图样的技术要求进行验收,如各处的尺寸精度、轧辊硬度、表面质量等,不合格产品应坚决剔除,以防止给下一步的投入使用留下更多的隐患。

Common casting defects of roller are as follows:

常见的轧辊铸造缺陷有:

Shrinkage cavity and looseness; discontinuity caused by repair casting; bright band and bright spot; hot crack caused by blocked shrinkage; shallow, too thick or uneven depth of chill layer of composite casting roller; poor combination of outer layer and core of composite casting roller; severe electric wear or poor service performance of roller caused by chemical composition and structure. Including: too few carbides in cast iron roller, too much graphite or insufficient graphite in cast iron roller, and too many pearlites in cast iron roller matrix; pinholes on roller surface or sub surface caused by poor gas or deoxidation or "rain" in horizontal centrifugal casting.

缩孔和疏松;补浇引起的不连续性;亮带和亮斑;收缩受阻引起的热裂;复合浇铸的轧辊激冷层深度过浅,过厚或不匀;复合浇铸轧辊外层和芯部结合不良;由化学成分和结构引起的轧辊严重磨损或使用性能不良。包括:铸钢辊中碳化物的数量过少,铸铁轧辊中石墨量不足或石墨过多,铸铁轧辊基体中珠光体数量过多等几个因素;由气体或脱氧不良或卧式离心铸造时的"下雨"造成的轧辊表面或亚表层针孔。

②Roll turning.

②轧辊车削。

The correct turning of roller pass is closely related to the production of bar, because it directly affects the dimensional tolerance of products. In the production of rough and poor precision roller, it will cause additional difficulties to the adjustment work and produce more waste products.

轧辊孔型车削的正确与棒材生产关系很大,因其直接影响产品尺寸公差。加工粗糙、精度差的轧辊在生产中会给调整工作造成额外困难,产生更多的废品。

When the roller is in service under the hot rolling condition, there will inevitably be hot cracks on the roll surface, so each time the roller is repaired, not only the worn pass should be repaired, but also the hot cracks should be repaired as much as possible. The remaining cracks will propagate quickly.

轧辊在热轧条件下服役时,辊面上不可避免地会产生稀疏不等的热裂纹,故每次修磨时,不仅要把磨损的孔型修复,还应尽量把热裂纹修尽。残留的裂纹会很快地扩展。

When turning the roller, it must be carried out in strict accordance with the technical requirements. After that, the pass template shall be used for careful measurement. As the pattern plate will wear in repeated use, it will inevitably produce errors when it is used to measure the hole pattern for a long time, so it is necessary to check the pattern plate frequently. In the process of roller processing, pay attention to avoid the stress concentration caused by poor processing, such as leaving car marks. Due to the change of cross-section, the roller body and roller neck are the places of stress concentration, which are easy to break in use. During the processing, there should be transition fillets, which should not be too small.

在车削轧辊时一定要严格按技术要求进行。之后要用孔型样板进行认真测量。由于样板在反复使用中也会产生磨损，长期用来测量孔型，必然产生误差，所以要经常校验样板。轧辊加工过程中要注意避免加工不良造成的应力集中，如留有车痕。辊身和辊颈处由于有断面变化，所以是应力集中之处，在使用过程中容易断裂。加工过程中应有过渡圆角，且不宜过小。

After the finished roller is inspected and confirmed to be qualified, the roller code shall be marked on the obvious position, and the card shall be registered, indicating the date, roller diameter value, applicable varieties, sorties, etc.

车完的轧辊经过检查，确认合格后，要在明显位置上标注轧辊代号，并登记卡片，注明日期、辊径值、适用的品种、架次等内容。

③Roller transportation.

③轧辊运输。

A series of transportation problems are involved in the process of roller from good rolling to installation to rolling mill or removal from rolling mill to processing workshop and storage. At this time, pay attention to protect the roller surface and rolling groove. It can not only prevent the collision between rollers, but also prevent the collision between rollers and other objects. Each process should be handled with care to prevent the roll from being damaged or the rolling groove from falling. For the roller transported to the destination, the surface quality of the roller shall be checked carefully to avoid the phenomenon of using the damaged roller.

轧辊从车好到安装到轧机上或从轧机上拆下运到加工车间及存放等一系列过程中都涉及运输问题。此时，要注意保护轧辊的表面和轧槽。既防止轧辊之间的互相磕碰与撞击，还要防止轧辊与其他物体的撞击。每个过程都要轻起轻放，防止撞坏轧辊或轧槽掉块。运到目的地的轧辊还要认真检查轧辊表面质量，避免出现使用已经损坏的轧辊的现象。

④Roller disassembly.

④轧辊拆装。

Before the roller is installed, it is necessary to carefully check the roller number, measure the roller diameter, and check the surface quality and processing quality. Ensure that qualified rollers are used.

轧辊在安装前一定要认真核对辊号、测量辊径、检查表面质虽和各处加工质量。确保使用合格的轧辊。

The oil and dirt on the roller surface shall be cleaned carefully. All contact surfaces shall be kept clean, and attention shall be paid to the coordination of contact surfaces during installation. Such as roll flat head and joint, roll neck and bearing inner ring, dynamic labyrinth, etc.

要认真清理轧辊表面的油污、脏污。各接触面要保持清洁，安装时要注意接触面的配合情况。如轧辊扁头与接轴处，轧辊辊颈与轴承内圈、动态迷宫等处。

The removed roller shall be carefully inspected for wear, placed in the specified place, and recorded for the next use after turning.

拆下来的轧辊要仔细检查磨损情况，放到规定地点，并做好记录，以便经过车削后下次使用。

2) Roller maintenance.

2) 轧辊的维护。

In steel rolling production, proper maintenance of the roller can effectively reduce the wear of the roller, prolong the service life of the roller, improve the operation rate of the rolling mill, reduce the production cost, improve the surface quality of the product, improve the dimensional accuracy, reduce the accident rate, and reduce the labor intensity of the operators. Therefore, it is necessary to strengthen roller maintenance. In the process of producing and using roller, we should pay attention to strengthen the maintenance of roller from the following aspects.

在轧钢生产中，正确地维护轧辊可以有效地减少轧辊的磨损，延长轧辊使用寿命，提高轧机作业率，降低生产成本，同时还有利于改善产品表面质量和提高尺寸精度，而且还可以减少事故率，减少操作人员的劳动强度。所以，加强轧辊维护是十分必要的。在生产和使用轧辊的过程中，要注意从以下几个方面加强轧辊的维护。

①Pre start up inspection.

①开车前检查。

Before start-up, the roller shall be inspected comprehensively to prevent using the roller whose groove has been excessively worn or whose surface has been partially peeled off. In the direction of the roller axis, the rolling groove shall be free of relative dislocation to ensure the correct shape and qualified dimensional accuracy of the rolled piece. In order to prevent the relative staggered movement of the roller, the axial direction of the roller shall be tightened so that the roller does not change its position under the action of force, and the radial direction shall be checked to ensure that the axis of the two rollers is parallel and the gap values of each roller are the same.

开车前，应对轧辊进行全面检查，防止使用轧槽已经过度磨损或轧辊表面已有局部剥落的轧辊。在轧辊轴线方向上，轧槽应无相对错位，以保证轧件的形状正确和合格的尺寸精度。为防止轧辊相对错动，应使轧辊轴向紧固，使轧辊在受力作用时不改变应有位置，同时检查径向应保证两辊轴线平行，各处辊缝值相同。

②Inspection during production.

②生产过程中检查。

During production, pay attention to the use of rollers at any time. Whether the cooling water is sufficient and the steel temperature is appropriate. Avoid steel temperature too low or rolling blackhead steel, prevent roller break accident. Once problems are found, they should be solved in time. The groove that has reached the rolling tonnage shall be replaced in time. If the rated tonnage is not reached, but the surface of the rolling groove has cracks, falling blocks and other defects or the rolling groove has been excessively worn, it is also necessary to replace it in time, fill in the record truthfully, and analyze the reasons when asking. When dealing with the accident, it is necessary to prevent the gas cutting flame from spraying to the surface of the rolling groove.

生产中要随时注意观察轧辊使用情况。冷却水是否充足，钢温是否合适。避免钢温过低或轧制黑头钢，防止出现断辊事故。一旦发现问题要及时解决。已到轧制吨位的轧槽，要及时更换。未到额定吨位，但轧槽表面已有裂纹、掉块等缺陷或轧槽已过度磨损，也要及时更换，并如实填写记录，问时要分析原因。处理事故时要防止气割火焰喷射到轧槽表面。

③Prevent fracture.

③防止出现断裂。

Broken roller is the enemy of production, which seriously affects the operation rate and increases the production cost, so we should try to avoid it. There are many reasons for roller breaking, but they can be roughly divided into two categories: one is the material or manufacturing quality of the roller, that is, the internal causes of the roller. The other is the process conditions and use conditions, i.e. external reasons. The following are several common forms of roller break and possible causes:

断辊是生产的大敌，严重影响作业率和提高作业生产成本，所以要力求避免。断辊原因很多，但大致可分为两类：一类是属于轧辊材质或制造质量，即轧辊内在原因。另一类则是工艺条件和使用情况，即外部原因。以下是几种常见的断辊形式和可能的原因：

The middle part of the roller body is broken and the fracture is perpendicular to the roller axis, as shown in Figure 1-12 (a). The reason is that the bending stress of the roller exceeds the tensile strength or fatigue limit of the roller material. The causes include excessive reduction, low temperature steel rolling, improper heating of new roller or severe hot cracking of roller.

辊身中部折断、断口与轧辊轴向垂直，如图 1-12（a）所示。其原因是轧辊承受的弯曲应力超过轧辊材料的抗拉强度或疲劳极限。导致的因素包括压下量过大、轧制低温钢、新辊使用时受热不当或轧辊有严重热裂。

The roller body is not broken in the middle. The possible causes include local overload or uneven load on the roller, thermal stress caused by excessive temperature gradient of the roller body, residual force inside the roller body, casting defects, etc.

辊身非中部折断。其可能的原因包括轧辊局部过载或不均匀受载、由辊身温度梯度过大引起的热应力、辊身内部的残余位力、铸造缺陷等。

The hole part is broken, as shown in Figure 1-12 (b). Here, the roller section is the smallest and the bending stress is the largest. In addition, the serious ring crack in the pass often becomes the source of its fatigue crack.

孔型部位折断，如图 1-12（b）所示。此处轧辊截面最小、承受弯曲应力最大。另外，孔型内的严重环裂常常成为其疲劳裂纹源。

The junction of roller body and roller neck is broken, as shown in Figure 1-12 (c). Generally, it belongs to fatigue fracture. The interface between roller body and roller neck is a stress concentration area, which is very sensitive to cracks and notches. Such cracks and notches include car marks, casting and metallurgical defects, small hot cracks and repair welding areas. The causes of this kind of broken roller are hot crack caused by bearing failure, axial force, shrinkage cavity, abrasion and crack at the end of roller body, or excessive load on roller, etc. Improper operation of the rolling mill often causes such roller breakage.

辊身与辊颈交界处折断，如图 1-12（c）所示。一般属于疲劳断裂。辊身与辊颈的交界处是应力集中区域,对裂纹和缺口非常敏感。这种裂纹和缺口包括车痕、铸造和冶金缺陷、小的热裂纹和补焊区。导致此类断辊的起因是由轴承故障引起的热裂,辊身端部的轴向力、缩孔、磨蚀和裂纹,或轧辊承受过大载荷等。轧机的操作不当经常引起此类断辊。

The roller body and the roller neck are broken at the same time. It is usually caused by severe overload.

辊身和辊颈同时断裂。通常是由于严重过载造成。

The roller body is broken. High hardness roller, especially after surface quenching treatment and containing high residual internal stress, is easy to break.

辊身碎裂。高硬度轧辊,尤其是经表面淬火处理和含有高的残余内应力的轧辊易发生此类断辊。

The roller neck or head is broken, as shown in Figure 1-12 (d). If the roller head torque is too large due to the maximum screwdown and bearing seizure, such kind of roller break may occur.

辊颈或辊头扭断,如图 1-12 (d) 所示。压下最过大,轴承咬死等导致辊头扭矩过大时有可能造成此类断辊。

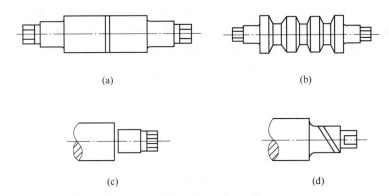

Figure 1-12 Several forms of broken roller
(a) The middle part of the roller body is broken; (b) The pass part is broken;
(c) The junction between the roller body and the roller neck is broken; (d) The roller neck is twisted

图 1-12 断辊的几种形式
(a) 辊身中部断;(b) 孔型部位断;(c) 辊身与辊颈交界处断;(d) 辊颈扭断

Generally speaking, the increase of rolling pressure, such as too much pass reduction, especially low-temperature steel rolling is the important reason for roll breakage. Sometimes, because of the bad cooling, it will cause the hot crack and the huge thermal stress, which will reduce the strength of the roller significantly, and it is also easy to break the roller.

一般来说,轧制压力的增大,如道次压下量过大,特别是低温钢轧制是断辊的重要原因。有时因冷却不良造成热裂并产生巨大的热应力,使轧辊强度显著减少,也极易造成断辊。

Sometimes the cause of fracture can be found out from the fracture, such as sand hole, inclusion, crack, etc. Once there is a roller break accident, it should be recorded in the roller record and the cause of the accident should be analyzed.

造成断裂的原因有时可从断口处检查出来，如砂眼、夹杂、裂纹等。一旦出现断辊事故，要如实记载到轧辊记录中，并分析事故原因。

④Pay attention to cooling quality.

④注意冷却质量。

Cooling water is very important for the service life of rollers. Poor cooling or inadequate cooling will cause a huge thermal gradient in the surface layer of the roller, which will accelerate the generation of thermal stress of roller spalling. In addition, the relatively high temperature reduces the strength and wear resistance of the roller. Therefore, attention should be paid to:

冷却水对轧辊寿命相当重要。冷却不好或冷却不充分会在轧辊表层内引起巨大的热梯度，导致加速产生轧辊剥落的热应力。此外，相对较高的温度降低了轧辊的强度和耐磨性。因此在生产中要注意：

Regularly check the cooling water quantity, water pressure and water quality. Prevent small water volume, low water pressure, poor water quality and water pipe blockage.

坚持定期检查冷却水量、水压和水质。防止水量小、水压低、水质差、水管堵的现象发生。

Turn on the cooling water before driving, and turn off the cooling water after stopping.

开车前先开冷却水，停车后再关冷却水。

It is one of the factors that cooling water is cut off during rolling and then water is supplied suddenly. Therefore, sudden water supply is not allowed after water supply is cut off.

轧制时中断使用冷却水，然后突然给水，是导致断辊事故的因素之一。因此，断水后绝对不允许骤然给水。

The water-cooled surface of roller should be as large as possible to improve the cooling efficiency.

轧辊水冷面应尽可能大些，以提高冷却效率。

The cooling water shall be as close as possible to the exit side of the rolled piece.

冷却水应尽可能靠近轧件出口侧。

The flow direction of water should be sprayed on the rolling groove along the tangent direction. At this time, the relative speed of cooling water and rolling groove is large and the cooling effect is good. When the water directly washes the groove (the spray angle is close to the vertical), the splash will spread, so the cooling effect is not good.

水的流向应沿切线方向喷射到轧槽上，此时冷却水与轧槽相对速度大，冷却效果好。当水直冲轧槽（喷射角度接近垂直）时将飞溅散开，这样冷却的效果并不好。

The entire width of the groove shall be covered with water.

轧槽整个宽度应被水覆盖。

⑤Roller storage.

⑤轧辊存贮。

The quality of roller management and roller consumption is an important part of the assessment of steel rolling production. The roller shall be managed by specially assigned person, and the roller management card and roller account shall be established.

轧辊管理的好坏和轧辊消耗量是考核轧钢生产的重要内容。轧辊要设专人管理，并建立轧辊管理卡和轧辊台账。

The contents of roller management card shall include roller number, original roller diameter, roller material, manufacturer and other original conditions. In the roller management card, the usage shall be recorded in detail, such as the number of usage, the amount of each grinding, the amount of rolling in each groove, etc., as well as the position in the roller turnover process, the analysis and identification of the abnormal usage and the broken roller. Roller management card is shown in Figure 1-13.

轧辊管理卡的内容应该包括轧辊编号，原始辊径、轧辊材质、生产厂家等原始情况。在轧辊管理卡中，应详细记录其使用情况，如使用次数、每次修磨量、每槽轧制量等，以及轧辊周转过程中的位置以及使用异常情况和断辊情况的分析鉴定。轧辊管理卡如图1-13所示。

Rolling mill number 轧机架次					Roll product 轧制产品				
Configuration pass hole model 配置孔型号					Roller diameter 轧辊直径				
Roller material 轧辊材质					Chemical composition 化学成分				
Roller number 轧辊编号					Depth of hard layer 冷硬层深度				
Manufacturer 生产厂家					Arrival date 到货日期				
Date 日期		Rolling tons 轧制吨数	Number of grooves used 使用轧槽数	Total rolling tons 轧制总吨数	Diameter after returning/mm 重车后直径/mm	Observation while turning 车削时的观察			Remarks 备注
Get into 入	Get out 出					Hardness 硬度	Pockmark 麻点	Crack 裂纹	Surface 表面	

Figure 1-13 Roller record card
图1-13 轧辊记录卡片

The following problems should be paid attention to in roller storage:
轧辊贮存中需要注意的问题有:

The appearance quality and dimension accuracy shall be strictly checked according to the drawings and technical requirements before the roller is put into the process. For those who fail to meet the requirements, the cause analysis record shall be made and the objection shall be handled. It is strictly forbidden to put the unqualified products into the warehouse.

轧辊入序前要按图纸及技术要求严格检查外观质量和尺寸精度。不合格者，要做好原因分析记录，待异议处理，严禁不合格品入库。

Fill in the warehousing account and inventory ledger carefully.
认真填写入库台账和库存分类账。

The rollers shall be classified and hoisted on the roller storage rack one by one. Don't leave it on wet ground. It is not allowed to lift more than two at the same time to prevent collision, extrusion and friction. The roller shall be placed on a special shelf to prevent extrusion deformation in piles, and cardboard or rubber shall be placed between the two rollers.

将轧辊一根根地分类吊放于轧辊存放架上。不能随意放在潮湿地上。吊运时不得两根以上同时吊运，以防互相碰撞、挤压和摩擦。轧辊应放在专用架子上，以防止成堆堆放挤压变形，在两辊间隔以硬纸板或橡胶。

Roller shall be stored and registered according to rolling specification and frame number. And corresponding to warehouse location number.

轧辊按轧制规格、机架号存放、登记。并与库房货位号对应。

Check and report the consumption of roller on a regular basis every month, and make economic benefit accounting.

每月定期核算并报轧辊消耗量，做出经济效益核算。

1.3.4.2　Guide Guard
1.3.4.2　导卫

Guide and guard device plays an important role in bar production, and its working environment is very bad in use, which is mainly manifested as irregular stress, high temperature and chilling alternating heat shock, friction at high speed, and poor lubrication conditions. According to production statistics of some manufacturers, more than 50% of rolling accidents are caused by guide and guard devices. In terms of product quality, more than 60% of the waste products (referring to the unqualified geometry and surface state) are caused by the untimely replacement, improper adjustment or loose fixation of the guide and guard device. Therefore, while paying attention to the design of the guide and guard device, we should pay close attention to the use of the guide and guard device in production, and often maintain, adjust and replace the guide and guard device. The best way is to accumulate the use data of the guide and guard device for a long period of time for different varieties and specifications, so as to achieve the planned and timely replacement or adjustment after inspection when replacing the roll or rolling groove.

导卫装置在棒材生产中占有相当重要的地位，并且在使用中其工作环境相当恶劣，主要表现为受力不规则、受高温和激冷交变热冲击、在高速下摩擦、润滑条件十分恶劣。据一些厂家生产统计表明，约有50%以上的轧制事故是由导卫装置造成的。在产品质量方面，其成品废品（指几何形状和表面状态不合格）约有60%以上是由导卫装置不及时更换、调整不当或固定松动而造成的。因此，在重视导卫装置设计的同时，在生产中要密切注意导卫装置的使用情况，经常对导卫装置进行维护、调整和更换，最好是通过对不同的品种、规格进行长时期的导卫使用数据的积累，从而在更换轧辊或轧槽时做到有计划地及时更换或检查后及时调整。

(1) Concept of guide guard.

(1) 导卫的概念。

Guide guard refers to a device that is installed before and after the roll pass in the production of section steel, and the auxiliary rolled piece enters and exports the roll pass according to the required direction and state.

导卫是指在型钢生产中，安装在轧辊孔型前后，辅助轧件按所需方向和状态进入、导出轧辊孔型的装置。

(2) The role of guide guard.

(2) 导卫的作用。

In order to produce section steel smoothly and safely, guide and guard device is an indispensable guide device in section steel rolling mill. Its function is to make the rolling piece enter and exit the pass according to the specified direction and position, so as to avoid the rolling piece from winding the roll or deviating from the rolling line. Improve the working conditions of roll rolling and ensure the safety of personnel and equipment.

为了顺利而安全地进行型钢生产,导卫装置是型钢轧机中必不可少的诱导装置。其作用是使轧件按规定的方向和位置进出孔型,避免轧件缠绕轧辊或偏离轧制线。改善轧辊轧制的工作条件,同时保证人身和设备的安全。

In the production of steel rolling, only the correct pass design, without the correct design and installation of guide device, can not produce qualified products. The guide device has a great influence on the product quality, sometimes it can make up the defect of the pass properly, so we should pay attention to the guide device.

在轧钢生产中,仅有正确的孔型设计,而无正确的导卫装置的设计与安装是不能生产出合格产品的。导卫装置对产品质量有很大影响,有时还可适当弥补孔型的缺陷,故应对导卫给予重视。

(3) Types and components of guide and guard devices.

(3) 导卫装置的类型及部件。

The guide and guard device can be divided into inlet guide and outlet guide according to its use, and can be divided into sliding guide and rolling guide according to its type.

导卫装置按其用处可分为入口导卫和出口导卫;按其类型可分为滑动导卫和滚动导卫。

The main components of the guide device for bar rolling mill are guide beam, guide plate, guide plate box, guide tube, guide roller, rolling guide plate, torsion roller and other devices which can make the workpiece deform and twist outside the pass.

构成棒材轧机用导卫装置的主要部件有导板梁、导板、导板盒、导管、导管盒、导辊、滚动导卫板、扭转辊等能使轧件在孔型以外产生变形和扭转的装置。

(4) Maintenance of guide guard.

(4) 导卫的维护。

Attention shall be paid to the following aspects during use and maintenance:

在使用和维护时应注意以下几个方面:

1) Ensure the machining quality of all parts. High quality dimensional accuracy and surface machining quality of components are the prerequisite for their function.

1) 确保各部件加工质量。部件高质量的尺寸精度及表面加工质量是其发挥作用的前提保证。

2) Assembly quality and adjustment quality shall be ensured. After the quality of the components of the guide guard is guaranteed, it is also required to be installed carefully. Bolts shall be of proper length and tightened in place. The thickness of gasket shall be appropriate. Water cooling pipeline and oil lubrication channel shall be unblocked to prevent blockage. The parts shall be used after inspection to ensure there is no damage or deformation. The surface oil stain, iron sheet and other dirt shall be cleaned. The rolling guide shall ensure the flexible rolling of the roll, and

the roll gap shall correspond to the thickness of the rolled piece. The twist angle of the twist guide is suitable. The clearance of the cutting wheel is moderate. The passage of rolling piece is smooth.

2）装配质量与调整质量要确保。导卫的部件质量得到保证后，还要求在装配时要认真仔细，安装到位。螺栓要长度恰当，拧紧到位。垫片厚度要合适。水冷管路，油润滑通道确保畅通，防止堵塞。部件要检查后才使用，确认无损坏，变形。表面油污、铁皮等脏污要清理干净。滚动导卫要确保辊子滚动灵活，辊缝与轧件厚度对应。扭转导卫的扭转角度合适。切分轮间隙大小适中。轧件通路畅通。

3）Before use, check whether the appearance quality, size and mark number of the guide parts conform to the drawing, frame number, pass code and rolling specification.

3）在使用前要检查导卫部件的外观质量、尺寸和标记号是否符合图样、机架号、孔型代号和轧制规格。

4）Adhere to frequent inspection and adjustment in production. If any problem is found, it shall be solved in time, and the damaged guide or excessively worn parts shall be replaced with new ones.

4）坚持生产中的勤检查、勤调整。发现问题及时解决，已损坏的导卫或过度磨损的部件要更换新件。

5）The guide should be aligned with the rolling line by adjusting the crossbeam or changing the thickness of the gasket.

5）导卫要与轧制线对中，通过调整横梁或改变垫片厚度来实现。

6）The replaced guide guard shall be checked and maintained in a timely manner. The replaced parts shall be replaced, the worn parts shall be reprocessed, and the irreparable parts shall be scrapped seriously. Ensure that the guide and guard put into use next time are free of defects and hidden dangers.

6）换下来的导卫要及时进行全面检查和维护。该换的部件要更换，磨损的部件要再加工，无法修复的应认真执行报废标准。保证下次投入使用的导卫无缺陷、无隐患。

7）Hoisting of guide guard. Cranes shall be used during handling and installation. In the process of lifting, it should be handled with care to prevent impact damage.

7）导卫的吊装。搬运和安装过程中要使用吊车。在吊运过程中要轻起轻放，防止撞击损坏。

8）Ensure the quality of cooling water. The cooling water plays a role in the normal operation of the guide, prolongs the service life and improves the quality of the rolled piece. Before use, it is necessary to check whether the water pipeline is smooth or not and whether the water nozzle parts are in good condition. In use, it is necessary to observe whether there is cooling water and the amount of water to ensure the best effect.

8）保证冷却水的质量。冷却水对于导卫正常发挥作用、延长使用寿命、提高轧件质量关系很大，使用前应检查水管线畅通与否，水嘴零件是否完好。在使用中应注意观察有无冷却水，水量大小，保证达到最佳效果。

9）Lubrication is mainly for rotating parts. If the oil is not supplied on time or the oil quantity is insufficient, the components will not rotate, resulting in accidents. Therefore, check the oil sign before using the guide guard. Pay attention to observe the lubrication during the use. If there is a problem, it should be solved in time.

9）润滑主要针对转动部件而言。如果不按时供油或油量不足，将使部件无法转动，导致事故发生。所以使用导卫前要检查油符，使用过程中要注意观察润滑情况，如出现问题，要及时解决。

10) Storage of guide guard. The quality of guide management and guide consumption is an important content to evaluate and reflect the quality of rolling production. A special person shall be assigned to manage the storage of the guide, and a guide account and guide management card shall be established. The content of the guide guard management card shall include the guide guard number, applicable varieties, which sorties to use and the position of the guide guard in the turn-over process towel.

10）导卫的存贮。导卫管理的好坏和导卫消耗量是考核和反映轧钢生产好坏的重要内容。导卫的存贮要设有专人管理，并建立导卫台账和导卫管理卡。导卫管理卡的内容应该包括导卫编号、适用品种、用于哪一架次以及导卫在周转过程中的位置。

All parts and the whole of the guide guard shall be carefully inspected for size and processing quality when entering the warehouse, and shall be strictly accepted according to the drawings. Records shall be made for the unqualified ones, and the unqualified ones shall not be put into storage.

导卫的各部件及整体在入库时要认真检查尺寸及加工质量，严格按图样验收。不合格者，要做好记录，严禁不合格者入库。

The assembled guide guard and all parts in stock shall be placed on special shelves according to the specifications, varieties and rack positions used, and shall not be placed on the ground at will to avoid damage and rust.

组装好的导卫和库存的各部件应分别按照所用规格、品种、机架位置放于专用架子上，不能随意放于地上，以免损坏和生锈。

Task 1.4　Finishing of Steel Bars
任务1.4　棒材的精整

Mission objectives
任务目标

(1) Familiar with the composition, structure and principle of the finishing line equipment for bar and round steel;

（1）熟悉棒材圆钢精整线设备的组成、构造及原理；

(2) It can operate the cooling bed, shearing, transportation, weighing, bundling, packaging and other equipment in the finishing area of bar and round steel.

（2）能够对棒材圆钢精整区的冷床、剪切、运输、称重、打捆、包装等设备进行操作。

1.4.1 Process of Finishing Area

1.4.1 精整区工艺流程

No. 3 double length flying shear→front roller table of cooling bed→12°cooling bed input roller table→lifting skirt plate→straightening plate→step cooling bed (dynamic and static rack) →alignment roller table→steel chain and unloading trolley→cooling bed output roller table→No. 4 fixed length cold shear →surface inspection of collection area and short length removal →collection and packing→inspection and warehousing

3号倍尺飞剪→冷床前段辊道→12°冷床输入辊道→升降裙板→矫直板→步进冷床（动静齿条）→对齐辊道→排钢链及卸钢小车→冷床输出辊道→4号定尺冷剪→收集区表面检查及短尺剔除→收集打包→检验入库

1.4.2 Multiple Shear of Flying Shear

1.4.2 飞剪倍尺剪切

（1）Overview.

（1）概述。

After the billet is rolled into bar by rolling mill, the length of the finished rolled piece is much longer than that of the cold bed. Therefore, it must be cut into the length that the cooling bed can receive. After cutting, the bar cooled by the cooling bed is cut into the length of the finished product. In order to avoid producing short length steel and improve the yield, the length of the upper cooling bed steel is generally cut into the integral multiple of the fixed length. This shearing process is called multiple length shearing, which is completed by multiple length shearing (No. 3 shearing) downstream of the rolling mill.

钢坯经轧机轧制成棒材之后，成品轧件的长度远大于冷床所能接收的长度。因此，必须经剪切成为冷床所能接收的长度，剪切后经冷床冷却的棒材再由定尺剪剪切为成品定尺长度。为了避免在定尺剪切过程中产生短尺钢，提高成材率，一般都把上冷床钢的长度剪切成为定尺的整倍数，这个剪切过程称为倍尺剪切，由轧机下游的倍尺剪（3号剪）剪切完成。

In the actual production, the length of billet is not absolutely constant, but changes randomly within an allowable range, which leads to the continuous change of the length of the steel rolled out by the finishing mill, which may make the length of the last section of steel after multiple length shearing less than the minimum length that the cold bed can receive. In this case, the steel whose length is less than the minimum length that can be received by the cooling bed is usually cut and broken after the shear of multiple length steel. If the length of the last section of steel at this time is only slightly less than the minimum length that the cooling bed can receive, but greater than the fixed length, in this case, breaking the tail steel will cause great waste and reduce the yield.

在实际生产中，钢坯的长度不是绝对不变的，而是在一个可允许的范围内随机变化，这就导致了精轧机轧出钢材的长度不断变化，可能使经倍尺剪切后所得到的最后一段钢的长度小于冷床所能接收的最小长度。这种情况下，通常是在倍尺钢剪切结束后，把长度小于冷床所能接收最小长度的钢由碎断剪碎断。如果这时最后一段钢的长度只是稍小于冷床所能接收的最小长度，而大于定尺长度，这种情况下碎断尾钢，会造成很大浪费，降低成材率。

In order to solve the above problems, the optimal cutting process of multiple length steel is produced. The purpose of this process is to obtain the maximum number of finished bars from a given steel, reduce the length of short bars, and improve the yield.

为了解决以上问题，就产生了倍尺钢优化剪切工艺。该工艺的目的是从给定的钢中得到最大数量的成品长度的棒材，减少短尺，以提高成材率。

The basic idea of optimizing the shearing process is that when the length of tail steel is less than the minimum length that can be received by the cooling bed, it is greater than that of the finished product When the length of the tail steel is less than one foot, a part of the length of the double foot steel before the tail steel, that is to say, the length of the double foot steel of the penultimate upper cooling bed is reserved for the tail steel, so that the length of the tail steel can reach the length that the cooling bed can receive, while the length of the penultimate second one is reduced, but the conditions for the upper cooling bed can still be met. In the other case, if the tail steel length is too small, after optimization with the penultimate steel, both of themcan not reach the length of the upper cooling bed, the optimization starts with the penultimate steel. The final optimization result is that the last three steels can reach the minimum length of the upper cooling bed, and the tail steel whose length is less than the fixed length is received by the short length collecting bed. If its length is less than the minimum length that can be received by the short bed, it will be cut by crushing.

优化剪切工艺的基本思想就是当尾钢长度小于冷床所能接收的最小长度而大于成品定尺长度时，就把尾钢之前的一段倍尺钢，即倒数第二段上冷床的倍尺钢的长度留一部分给尾钢，使尾钢长度达到冷床所能接收的长度，而倒数第二根长度有所减少，但仍能满足上冷床条件。另一种情况下，如果尾钢长度太小，在与倒数第二根钢优化后，二者都达不到上冷床长度，则优化从倒数第三根倍尺钢开始。最终优化结果是使最后三根钢都能达到最小上冷床长度，最后长度小于定尺长的尾钢由短尺收集床接收。如果其长度还小于短尺床所能接收的最小长度则由碎断剪碎断。

The equipment needed to realize the optimal cutting process is mainly the double length shear (No. 3 shear), breaking shear and a series of hot metal detectors distributed in the main rolling line and its downstream. There is a pinch roll in front of the No. 3 shear, and there is a guide between the No. 3 shear and the broken shear. The equipment layout is shown in Figure 1-14.

实现倍尺优化剪切工艺所需要的设备主要是倍尺剪（3号剪）、碎断剪以及一系列分布在主轧线以及其下游的热金属探测器。在3号剪前有夹送辊，3号剪与碎断剪之间是导向器。设备布置如图1-14所示。

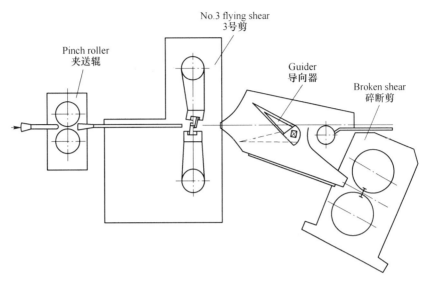

Figure 1-14　Equipment in double length shear area
图 1-14　倍尺剪区设备

(2) Double length shear (No. 3 shear).

(2) 倍尺剪（3 号剪）。

No. 3 shear is located at the downstream of the rolling mill and after the water quenching line. It is mainly used for cutting head, tail, multiple length cutting, accident cutting, optimization cutting, etc.

3 号剪位于轧机下游，淬水线之后，主要用于轧件的切头、切尾、倍尺剪切、事故剪切，优化剪切等。

The shear is a combination shear (crank/convolution). According to the production needs, the shear crank or convolution mode is selected, which is driven and transformed by the hydraulic cylinder. Two blades for crank and two blades for flying shear.

该剪为组合剪（曲柄/回旋），根据生产需要选择剪子曲柄或回旋方式，由液压缸驱动变换。两个刀片用于曲柄，两个刀片用于飞剪。

(3) Scissor.

(3) 碎断剪。

The crushing shear is located behind the No. 3 shear. It is a continuous rotary shear. During the rolling process, the crushing shear is in continuous operation. It is used for short steel breaking and accidental cutting. The broken shear structure is shown in Figure 1-15.

碎断剪位于 3 号剪后，是一台连续回转剪，在轧制过程中，碎断剪处于连续运转状态。用于短尺钢的碎断，事故剪切。碎断剪结构如图 1-15 所示。

(4) Pinch roller.

(4) 夹送辊。

The pinch roller is located on the entrance side of No. 3 shear. Its role is to pull the rolling stock when the tail of the rolling stock leaves the final rolling stand. The structure of pinching silver

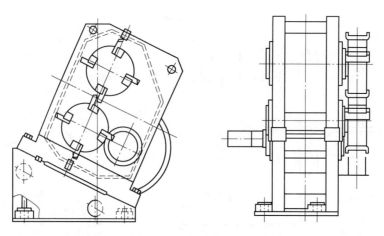

Figure 1-15　Broken shear structure
图 1-15　碎断剪结构

is shown in Figure 1-16.

夹送辊位于 3 号剪入口侧。它的作用是当轧件尾部脱离终轧机架后，拉送轧件。夹送辊的结构如图 1-16 所示。

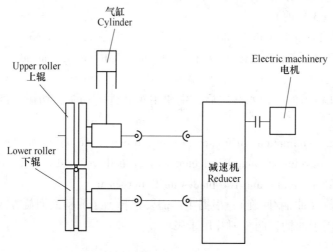

Figure 1-16　Pinch roller
图 1-16　夹送辊

The pinch roller is a horizontal pinch roller, mainly including motor drive device, universal joint shaft, frame, roll group, cylinder, etc. The upper roller is movable and driven by the air cylinder. When the air cylinder starts, the upper roller is pressed down so that the rolled piece can be clamped.

该夹送辊为水平式夹送辊，主要包括电机传动装置、万向接轴、机架、辊组、气缸等。上辊可动，由气缸驱动，当气缸开动时，上辊压下，以便能夹持轧件。

To ensure alignment with the center of rolling line, the position of lower roll can be adjusted. The stroke of the upper roll driving cylinder can also be adjusted.

为保证与轧制线中心对正，下辊位置可进行调整。上辊驱动气缸的行程也可调整。

A guide device is also arranged at the inlet and outlet of the pinch roller. The guide device is mainly a bell mouth conduit, the conduit is fixed on the bracket, and the height of the conduit bracket can be adjusted to align with the rolling line. According to different rolling products, different specifications of conduit are selected.

在夹送辊入口和出口还装有导向装置。导向装置主要是一个喇叭口导管，导管固定在支架上，导管支架高度可以调整，以便对准轧制线。根据不同的轧制产品，选用不同规格的导管。

There are grooves on the pinch roller, which shall be replaced according to the cross section of different products.

夹送辊上有槽，根据不同产品的断面进行更换。

The installation and adjustment of pinch roller is to prepare the used roll according to the product specifications, add the correct thickness of gasket on the shaft, then cover the roll ring on the shaft, install the flange, and screw the flange on the shaft end with bolts.

夹送辊的安装、调整是根据生产产品规格，准备好所使用的辊子，在轴上加上正确厚度的垫片，再把辊环套在轴上，安装法兰盘，并用螺栓把法兰拧紧在轴端上。

Control the opening and closing of pinch roller at the ground station, check the alignment degree of the two roll center line and the rolled piece, if there is deviation, adjust the upper and lower rolls respectively. The pinch roll running test shall be carried out at the ground station before production.

在地面站控制夹送辊的打开、关闭，检查两辊中线与轧件的对准程度，如有偏差，可对上、下辊分别调整。生产前在地面站进行夹送辊运转测试。

1.4.3 Fixed Length Shear
1.4.3 定尺剪

（1）Overview.

（1）概述。

The bar is cooled to below 100℃ by the cold bed, and then sent to the cold flying shear according to the predetermined number of shear pieces for fixed length cutting, and then sent to the collection area by the sheared roller table.

棒材经冷床冷却到100℃以下，按预定剪切根数送入冷飞剪进行定尺剪切，由剪后辊道输入收集区。

（2）Sizing and optimization.

（2）定尺和优化。

No. 2 and No. 3 flying shears can be cut to a certain length and optimized.

2号、3号飞剪可以进行定尺剪切和优化剪切。

The operation sequence of the fixed length shear is similar to that of the head shear, and the steel divider is always at the lowest position. The minimum shear length is determined by rolling speed and shear period.

定尺剪切的操作顺序与头部剪切相类似,分钢器一直处于最低位。最小的剪切长度依据轧制速度、剪切周期来限定。

The optimized cut is performed by shear 2. The automatic control system calculates the length of the bar, and optimizes the combination according to the required fixed length, so that the shearing machine can cut the bar according to the multiple of the fixed length, so that the bar can be placed on the cold bed, and the fixed length can be cut on the cold pendulum shear, so as to save time.

优化剪切由 2 号剪执行。自动控制系统计算出棒材的长度,同时根据所需要的定尺长度进行优化组合,使剪切机按定尺长度的倍数进行剪切,以便将棒材放在冷床上,并在冷摆剪上定尺剪切,以节约时间。

Shear 3 is used as the fixed length shear. If the calculation is wrong or shear 2 is wrong, shear 3 can also carry out tail shear.

3 号剪当作定尺剪,如果计算错误或 2 号剪剪切错误,3 号剪也可进行尾部剪切。

When the tail stock is less than 3.5m, the tail stock shall be sheared and transported to the scrap shear; if the tail stock is more than 3.5m and less than 6m, it can be connected to the last bar, cut by cold shear and transported to the short scale collection system; if the tail stock is between 6~7.9m, the tail stock shall be added to the last bar for 6m, and the part with the tail stock more than 6m shall be broken; if the tail stock is between 8~12m, then It is divided into two parts and added to the last two bars, which are cut by cold shear and transported to the short scale collection system.

当尾料少于 3.5m 时进行尾部剪切并把尾料输送到碎断剪碎断;如果尾料超过 3.5m 少于 6m 时,可连在最后一根棒材上,由冷剪剪切并输送到短尺收集系统;如果尾料长度在 6~7.9m 之间,则尾部 6m 加到最后一根棒材上,尾部超过 6m 的部分进行碎断;如果尾料长度在 8~12m 之间,则将其一分为二,加在最后两根棒材上,由冷剪剪切并输送到短尺收集系统。

1.4.4 Process and Equipment in Cooling Bed Area
1.4.4 冷床区工艺及设备

(1) Equipment in cooling bed area.
(1) 冷床区设备。

The cooling bed area is the main area of the finishing section, which is responsible for the cooling and transportation of multiple steel. In order to realize the timely and accurate transportation and cooling of bars, the following equipment is equipped in the cooling bed area: transport roller table with brake apron, pneumatic steel puller, electromagnetic apron, step-by-step rack cooling bed, alignment roller table on the cooling bed, etc. The composition of cooling bed equipment is shown in Figure 1-17.

冷床区是精整工段的主要区域,它担负倍尺钢材的冷却和运输任务。为实现棒材及时准确的运输和冷却,在冷床区装备了以下设备:带制动裙板的运输辊道、气动拨钢器、电磁裙板、步进式齿条冷床、冷床上对齐辊道等。冷床设备组成如图 1-17 所示。

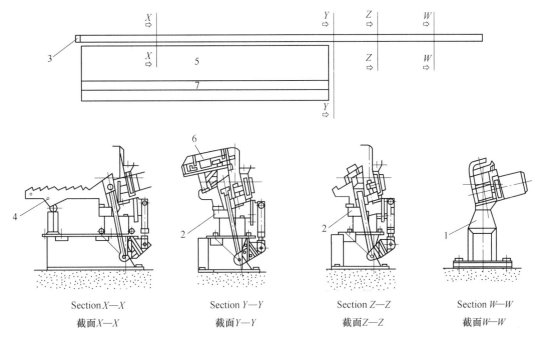

Figure 1-17　Composition diagram of cooling bed equipment
1—Input roller table; 2—Apron roller table; 3—Fixed baffle plate; 4—Straightening plate;
5—Cooling bed; 6—Pneumatic steel transfer; 7—Alignment roller table
图 1-17　冷床设备组成示意图
1—输入辊道；2—裙板辊道；3—固定挡板；4—矫直板；5—冷床；6—气动分钢器；7—对齐辊道

(2) Cooling bed inlet equipment.
(2) 冷床入口设备。

The cooling bed inlet equipment includes input roller table, lifting apron roller table, steel separator, safety baffle, etc. The roller table with brake apron is located after No. 3 shear and before the cooling bed. Its function is to transport and brake the multiple length steel rolled by the rolling mill, and transport the steel to the cooling bed, as shown in Figure 1-18.

冷床入口设备包括输入辊道、升降裙板辊道、分钢器、安全挡板等。带制动裙板的辊道位于 3 号剪后，冷床之前。它的作用是对轧机轧出的倍尺钢进行运输、制动，并把钢输送到冷床上，如图 1-18 所示。

(3) Cooling bed process.
(3) 冷床区工艺。

The cooling bed is a kind of start stop walking beam rack cooling bed, which is the main equipment in the bar finishing area, located between the apron roller table and the layering equipment in the cold shear area. The function of the cold bed is to air cool the hot bar after rolling, and then straighten, straighten, and transport it to the cold shear area after cooling for shear.

冷床是一种启停式步进梁齿条冷床，是棒材精整区的主要设备，位于裙板辊道与冷剪区成层设备之间。冷床的作用是对轧后热状态的棒材进行空冷，并矫直、齐头、冷却后输

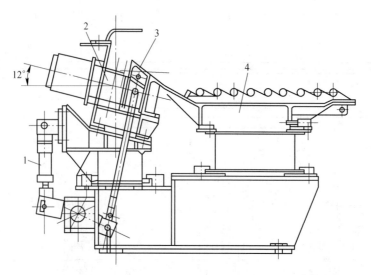

Figure 1-18　Cooling bed inlet equipment

1—Hydraulic cylinder；2—Cooling bed input roller table；3—Brake apron；4—Straightening plate

图 1-18　冷床输入设备

1—液压缸；2—冷床输入辊道；3—制动裙板；4—矫直板

送到冷剪区进行定尺剪切。

The cooling bed body is composed of step rack beam and fixed rack beam. The transmission of the moving rack is driven by two sets of DC motors and reducers. The two sets of transmission mechanisms are rigidly connected on the low-speed output shaft to ensure the synchronous operation of the two parts.

冷床本体由步进齿条梁和固定齿条梁组成。动齿条的传动由两套直流电机、减速机传动，两套传动机构在低速输出轴上刚性联接，保证两部分同步运行。

The movable beam is composed of 22 sections, each section is driven by a set of eccentric wheels fixed on the low-speed shaft, the eccentric wheel rotates for one circle, and the walking beam takes one step. The movable step beam rack drives the bar to move forward one step and positions the bar on the teeth of the next fixed rack. The brake plate delivers a piece of steel to the cooling bed every time, and the moving rack of the cooling bed enters once. A proximity switch for detecting the position of the step beam of the cooling bed is arranged on the low-speed axis of the cooling bed. Two of them are used to detect the deceleration position before the step beam stops. Two are used to detect that the step beam is in the parking position.

活动梁由 22 段组成，每段由一套固定在低速轴上的偏心轮驱动，偏心轮转动一周，步进梁走一步。活动步进梁齿条带动棒材向前移动一步把棒材定位在下一个定齿条的齿上。制动板每次向冷床输送一根钢，冷床动齿条进一次。在冷床低速轴上装有用于探测冷床步进梁位置的接近开关。其中两个用于探测步进梁停止前的减速位置。两个用于探测步进梁位于停车位置。

A microswitch is arranged on the last tooth of the output side of the cooling bed to detect the bar.

在冷床输出侧最后一个齿上装有一个微动开关,用于探测棒材。

Two high-speed transmission shafts of the cooling bed are respectively equipped with an air brake. The function of the two holding brakes is to keep the position of the step beam when the motor stops running, and to open the holding brake before the step beam moves. The band brake is only used for positioning the step beam, not for braking.

在冷床的两个高速传动轴上,各装有一个气动抱闸。两个抱闸的作用是当电机停转时,保持步进梁位置,当步进梁运动之前,抱闸打开。抱闸只用于步进梁定位,不用于制动。

On the inlet side of the cooling bed, there is a row of straightening plates connected with the fixed rack in the full length direction, and the straightening plates have the same size of teeth with the rack. As the multiple length steel rolled from the rolling mill is still at a high temperature when it is just put into the cooling bed, it is relatively soft and cannot be directly placed on the rack. Therefore, it should be straightened on the straightening plate first and cooled to a lower temperature, and then it can enter the rack after it has sufficient strength.

在冷床入口侧,在通长方向上有一列与固定齿条相联接的矫直板,矫直板上有与齿条尺寸相同的齿槽。由于从轧机轧出的倍尺钢,刚上冷床时仍处于较高温度,比较软,不能直接放到齿条上,所以先在矫直板上矫直,并冷却到较低温度,有足够强度后再进入齿条上。

In the second half of the cooling bed, there is a roller table with a fixed baffle plate at the end to align the steel on the cooling bed. The flush roller table is composed of 96 grooved rollers, each roller is driven by an AC motor separately, and the control of the roller table is divided into 8 sections.

在冷床后半部,有一条齐头辊道,端部有一个固定的齐头挡板,用来使冷床上的钢材齐头。齐头辊道由96个带槽的辊组成,每个辊由一台交流电机单独驱动,辊道的控制分为8段。

Project 2　Production of HRB
项目 2　螺纹钢棒材生产

Task 2.1　Preparation of the Raw Material
任务 2.1　坯料准备

Mission objectives
任务目标

（1）Master the types, defects and inspection methods of raw materials;
（1）掌握原料的种类、缺陷及检查方法；
（2）Be able to select the type and technical parameters of raw materials according to the actual production; be able to identify the defects of raw materials.
（2）能够根据实际生产选择原料种类及技术参数；能够识别原料缺陷。

2.1.1　Selection of the Billet
2.1.1　坯料的选择

With the change of the whole steel production process and the improvement of the level of small rolling mill, the billet of small rolling mill is also changing. Before the emergence and maturity of continuous casting technology, the billets used in small rolling mill were made by the continuous rolling of billets. At that time, the principle of selecting blank was: the cross-sectional area of the selected blank was the cross-sectional area of the minimum specification multiplied by the total elongation coefficient in the product plan.

随着整个钢铁生产工艺的变化和小型轧机水平的提高，小型轧机的坯料也在不断地变化之中。在连铸技术出现和成熟以前，小型轧机所用的坯料是钢锭经初轧—钢坯连轧机开坯而成的。当时选择坯料的原则是：以产品方案中最小规格的断面面积乘以总延伸系数即为所选择的坯料断面面积。

After the emergence of continuous casting technology, the earliest benefit is the small rolling mill, which directly uses the continuous casting slab as the raw material to heat and roll into the material at one time, and cancels the blooming, which can improve the metal yield rate by 8% ~ 12%, save energy consumption by 35% ~ 45%, and improve the surface quality and internal quality of the product, which is welcomed by iron and steel manufacturers and users. Therefore, the development of small rolling mill itself must consider the development of continuous casting and closely cooperate with continuous casting. In this way, after continuous casting, the principle of billet selection for small rolling mill has fundamental changes. It is not only the rationality of small rolling mill itself, but also the overall rationality of continuous casting small rolling mill, and even put the rationality of continuous casting in a more important position.

连铸技术出现后,最早受益的就是小型轧机,直接以连铸坯为原料一次加热轧制成材,取消了初轧开坯,可提高金属收得率8%~12%,节约能耗35%~45%,并可提高产品的表面质量和内在质量,深受钢铁制造厂和钢铁用户的欢迎。因此,小型轧机本身的发展一定要考虑连铸发展,并与连铸密切配合。这样,有了连铸后,小型轧机选择坯料的原则有了根本的变化,不再是只考虑小型轧机本身的合理性,而是要考虑连铸—小型轧机整体的合理性,甚至要将连铸的合理性放在更主要的地位。

From the point of view of rolling, the section of continuous casting slab should be as small as possible, which can reduce the number of passes, the number of stands of rolling mill, the investment and operation cost. Therefore, at the beginning of continuous casting, people hope to provide small section continuous casting slab for small rolling mill. After a period of exploration, it has been proved that it is impossible for the continuous caster to operate normally and the slab quality is not guaranteed to produce small cross section. After increasing the section size of continuous casting slab, a series of improvements have been made on other supporting technologies of continuous casting, and the production of billet caster has been stabilized and really entered the practical stage.

从轧钢的观点看连铸坯的断面要尽可能小一些,这样可以减少轧制道次,轧机的架数可以减少,投资和运行费用均可降低。因此,开始出现连铸时,人们希望能为小型轧机提供小断面的连铸坯。经过一段时间的摸索证明,要生产小规格的断面,连铸机不可能正常操作,铸坯质量也没有保证。后来把连铸坯的断面尺寸加大后,对连铸的其他配套技术进行了一系列改进,小方坯连铸机的生产才稳定下来,真正进入实用阶段。

2.1.2 Continuous Casting Billet
2.1.2 连铸坯

It is necessary for the small bar mill to use the continuous casting slab as the raw material in the market competition. At present, common carbon steel and low alloy steel small bar mills, most of the alloy steel small bar mills are all based on continuous casting billets, and the alloy steel grades and varieties based on continuous casting billets are further expanded. Because of the cost gap between the product quality and the secondary heating rolling, the small bar mill with small ingot as the raw material or bloom as the raw material is gradually eliminated in the market competition.

直接以连铸坯为原料,已是小型棒材轧机可能在市场竞争中存在的必要条件。目前,普通碳素钢和低合金钢小型棒材轧机、大部分合金钢小型棒材轧机都以连铸坯为原料,并且以连铸坯为原料的合金钢钢种和品种还在进一步扩大。以小钢锭为原料或以初轧开坯为原料的小型棒材轧机，由于产品质量和二次加热轧制导致成本上的差距，在市场竞争中正在逐渐被自然淘汰。

The reasonable selection of slab section has a great influence on the investment and operation of caster and small bar mill. The section of billets used in ordinary small bar mill should be about (130mm×130mm)~(150mm×150mm), the single weight of billets should be 1.5~2.0t, even 2.5t. With the increase of single weight, the amount of cutting head and tail is relatively reduced, and the sizing rate is increased, which is conducive to the improvement of metal yield. The progress of continuous casting technology is the main driving force to promote the development of metallurgical technology, including small bar mill. High speed continuous casting technology has been successfully used to produce 130mm×130mm continuous casting slab at a pulling speed of 4.3m/min, that is, the single flow output of the caster has reached 33t/h. For the caster itself, no matter from the quality or output point of view, there is no need for larger section billet. The small bar mill should make full use of the results of continuous casting to reduce the number of stands and deformation work in the rolling process.

合理选择连铸坯断面，对连铸机和小型棒材轧机的投资与操作都有很大的影响。普通小型棒材轧机使用的坯料断面应在（130mm×130mm）~（150mm×150mm）左右，坯料单重1.5~2.0t，甚至达2.5t。单重增加，切头、切尾量相对减少，定尺率提高，有利于提高金属的收得率。连铸技术的进步是推动包括小型棒材轧机在内的整个冶金技术发展最主要的动力。高速连铸技术已可成功地以4.3m/min的拉速生产130mm×130mm的连铸坯，即连铸机单流的产量已可达33t/h。以连铸机本身而言，无论从质量还是产量角度，都不需要更大断面的铸坯，小型棒材轧机更应充分利用连铸的成果，以减少机架数量和轧制过程的变形功。

With the improvement of alloy steel continuous casting technology, such as high-quality carbon steel, alloy structural steel, spring steel, austenitic stainless steel, bearing steel, etc. can be directly continuous cast now. The trend of transition from alloy steel continuous casting slab to medium section will be accelerated. More alloy steel grades and varieties are adopting (160mm×160mm)~(240mm×240mm) continuous casting, and the number of bloom above 300mm×300mm is gradually decreasing, so as to reduce the investment of continuous casting machine and small bar rolling mill, and promote the improvement of production level of small bar rolling mill.

随着合金钢连铸技术水平的提高，像优质碳素钢、合金结构钢、弹簧钢、奥氏体不锈钢、轴承钢等现在都可以直接进行连铸。合金钢连铸坯向中断面过渡的趋势将加快，更多的合金钢钢种和品种正在采用（160mm×160mm）~（240mm×240mm）的连铸坯，300mm×300mm以上的大方坯的数量在逐渐减少，以减少连铸机和小型棒材轧机的投资，推动小型棒材轧机生产水平的提高。

2.1.3 Hot Delivery and Hot Charging of Billet
2.1.3 坯料的热送热装

2.1.3.1 Overview
2.1.3.1 概述

The traditional process is to cool down the blooms or billets from the continuous caster. If surface defects are found through inspection, the billets must be cleaned and then loaded into the heating furnace of the rolling workshop. As a result, the metal loss is large, which consumes more manpower and energy.

传统的工艺是将初轧坯或来自连铸机的坯料冷却下来，经检查若发现有表面缺陷则必须对钢坯进行清理，然后再将钢坯装入轧钢车间加热炉。其结果是金属损失大，耗用人力和能源较多。

It has been the wish of metallurgical workers for many years to directly roll the continuous casting slab into material. As early as 1960s, many researches have been carried out abroad in this field, trying to directly roll the continuous casting slab into material, but it has not been successful because of the mismatch between the capabilities of the continuous casting machine and the rolling mill. After the world energy crisis in the 1970s, the small-scale EAF steel plant, adopting the short process steel production process, arranged EAF steelmaking, off furnace refining, continuous casting and rolling mill together in a compact way. The continuous casting machine is connected with the rolling mill equipment, and the continuous casting billet sent from the continuous casting machine is directly sent to the heating furnace of the rolling workshop for supplementary heating without cooling, and then sent to the rolling mill for rolling. This method is called direct hot charging process of continuous casting slab.

将连铸坯直接轧制成材是冶金工作者多年的愿望，早在 20 世纪 60 年代国外就进行了许多这方面的研究工作，试图把连铸坯直接轧制成材，但因连铸机与轧机的能力不匹配等问题而没有成功。在 70 年代世界能源危机以后发展起来的电炉小钢厂，采用短流程的钢铁生产工艺，将电炉炼钢、炉外精炼、连铸、轧机紧凑地布置在一起，连铸机与轧机用设备相联接，从连铸机送出的连铸坯不经冷却，直接送入轧钢车间的加热炉中补充加热，然后即送入轧机轧制。这种方法就称为连铸坯直接热装工艺。

In 1979, Konia plant, located in Austa, Italy, produced 200mm × 235mm alloy steel (stainless steel, valve steel, etc.) continuous casting billets on a 3-strand continuous casting machine. There was a heat preservation cover at the billet exit stand. After the heat preservation stand, the billets were directly heated in a walking beam furnace, and then rolled into (90mm × 90mm) ~ (150mm × 150mm) by a 650mm two roll reversible mill. This was the early hot delivery and hot charging Technology. After the preparation and technical preparation in 1980s, a set of "black box" plant with 100% hot delivery and hot charging of continuous casting slab was put into operation in vent plant of Italy in 1989. After two years of operation, it has been proved that it has a series of advantages such as saving investment and energy. Since 1992, this new energy-saving technology has been rapidly promoted all over the world. The advantages of direct hot charging process are as follows:

1979 年位于意大利奥斯塔的柯尼亚厂在 3 流连铸机上生产 200mm×235mm 的合金钢（不锈钢、阀门钢等）连铸坯，在钢坯出坯台架处设有保温罩，钢坯经保温台架后直接进入步进式加热炉加热，然后经 650mm 二辊可逆式轧机轧制成（90mm×90mm）~（150mm×150mm），这就是早期的热送热装工艺。经 20 世纪 80 年代的酝酿和技术准备，1989 年在意大利 Vent 厂投产了一套连铸坯 100%热送热装的"黑匣子"工厂，经两年的运行之后证明它具有节省投资、节约能源等一系列优点。1992 年以后这项节约能源的新技术在世界范围内得到迅速推广。采用直接热装工艺的优点是：

(1) Reduce the fuel consumption of heating furnace and increase the output of heating furnace. It can reduce fuel consumption by 40%~67%, save energy by 0.8~0.4GJ/t and increase furnace output by 20%~30% if the hot charging temperature is 900~600℃.

(1) 减少加热炉燃料消耗，提高加热炉产量。可降低燃料消耗 40%~67%，若热装温度为 900~600℃，则可节能 0.8~0.4GJ/t，加热炉产量提高 20%~30%。

(2) It can reduce the heating time and metal consumption by 0.3% compared with the cold charging.

(2) 减少加热时间，减少金属消耗，一般可比冷装减少 0.3% 的金属损耗。

(3) Reduce the amount of steel billets in stock, plant area and lifting equipment, reduce personnel, and reduce construction investment and production costs.

(3) 减少库存钢坯量、厂房面积和起重设备，减少人员，降低建设投资和生产成本。

(4) Shorten the production cycle, from order acceptance to delivery to users can be shortened to a few hours.

(4) 缩短生产周期，从接受订单到向用户交货可以缩短到几个小时。

Therefore, this technology has been widely used in small bar mill, wire mill, carbon steel plant and special steel plant. The advantages of special steel plant are more obvious because a large number of billet cleaning work and slow cooling facilities can be saved.

因此，这项技术在小型棒材轧机和线材轧机，在碳素钢厂和特殊钢厂均得到了广泛应用。而对特殊钢厂由于可省去大量钢坯清理工作和缓冷设施，优点更为明显。

In addition to the above-mentioned direct hot charging process, some plants can only use the method of hot delivery of billets from the continuous casting workshop with the holding car because the distance between the rolling cars and the continuous casting workshop is far. In this production mode, because it is not directly connected, the hot delivery temperature is not guaranteed, and it often fluctuates. Generally, a heat preservation pit needs to be set up, so the effect of hot charging is obviously not as good as that of direct hot charging.

除上述直接热装工艺外，有的厂由于轧钢车间距连铸车间较远，只能采用从连铸车间用保温车热送钢坯的方法。这种生产方式，由于不是直接联接，热送温度无保证，经常波动，一般还需设保温坑，热装效果显然不如直接热装。

The close connection and hot charging between rolling mill and continuous caster have revolutionized the concept of production management in rolling workshop. In the traditional cold charging process, the steel rolling workshop organizes production according to the specification, that is, each diameter product has a group of billets of different steel grades, which are put into the heating furnace and then rolled, so that the rolling groove can be fully utilized. In the hot charging process, the production plan is arranged according to the steel grades. Each casting unit of a steel grade is at least one heat of steel, and in most cases there is more than one heat. This batch of billets is not only rolling one specification, but rolling multiple specifications. Therefore, the rolling mill must change the roll several times. In particular, the order quantity of special steel plant is small, and roll change is more frequent. At the present production level, it is possible to roll 2~3 specifications for each furnace of steel. To produce more than 3 specifications, the production organization is quite difficult due to too frequent hole changing slots.

轧机与连铸机的紧密衔接和热装，使轧钢车间的生产管理概念产生了革命性的变化。在以往的传统冷装工艺中，轧钢车间是按规格来组织生产的，即每一直径的产品都有不同钢种的坯料编成一组，装入加热炉而后进行轧制，这样可以充分利用轧槽。而在热装工艺中生产计划是根据生产的钢种进行安排的，一个钢种的每一个浇铸单元最低为一炉钢，而大多数情况下均多于一炉，这批钢坯就不是仅轧制一个规格而是要轧制多个规格，因此轧机必须多次换辊。尤其是特殊钢厂订货批量较小，更需频繁换辊。以现在的生产水平每一炉钢轧制 2~3 个规格是可以达到的，要生产 3 个以上的规格，由于换孔槽过于频繁，生产组织有相当的难度。

2.1.3.2 Conditions for Hot Charging
2.1.3.2 采用热装的条件

The conditions for hot charging are:

采用热装的条件是:

(1) The steelmaking workshop should be equipped with necessary equipment and technology to ensure the stable and balanced production of flawless continuous casting slab and production process. These equipments and technologies include refining outside the furnace, slag free tapping, argon blowing and stirring, fine-tuning composition of wire feeding, submerged nozzle, gas seal tundish, protective casting, mould liquid level control, electromagnetic stirring, air mist cooling, multi-point straightening, etc. special steel production should also have vacuum degassing, soft reduction and other technologies. All unqualified billets shall be removed in the steelmaking workshop.

(1) 炼钢车间应具备必要的设备和技术，以保证生产出无缺陷连铸坯和生产过程的稳定均衡。这些设备和技术包括炉外精炼、无渣出钢、吹氩搅拌、喂丝微调成分、浸入式水口、气封中间包、保护浇铸、结晶器液面控制、电磁搅拌、气雾冷却、多点矫直等，特殊钢生产还应具有真空脱气、软压下等技术。不合格坯均在炼钢车间剔出处理。

In the steel killed by aluminum, the aluminum nitride precipitates along the grain boundary in the hot charging temperature range, and the surface cracks will be produced in the rolling process. In order to solve this problem, an effective method is to set up a water-cooled quenching device at the exit of the continuous casting machine, which can quickly cool the surface of the continuous casting slab to about 550℃, forming a certain depth of surface hardening layer, so as to avoid the precipitation of aluminum nitride on the surface.

用铝镇静的钢，连铸坯中氮化铝在热装温度范围沿晶界析出，在轧制过程中就会产生表面裂纹。为解决此问题，一个有效的方法是在连铸机的出口处设置水冷淬火装置，将连铸坯表面迅速冷却到550℃左右，形成一定深度的表面淬硬层，从而避免氮化铝在表面析出。

(2) The steelmaking and continuous casting workshop and rolling workshop shall organize

production according to the unified production plan and arrange the planned maintenance as much as possible.

（2）炼钢连铸车间与轧钢车间应按统一的生产计划组织生产，并尽可能统一安排计划检修。

（3）The hourly output of caster and rolling mill should be matched properly. If the hourly output of the rolling mill is less than the maximum hourly output of the continuous caster （regardless of the preparation time of the continuous caster）, many hot billets will not enter the rolling mill and must be separated from the rolling line to become cold billets; if the design capacity of the rolling mill is greater than the maximum hourly output of the continuous caster, the rolling mill capacity will not play and cause waste, so in principle, the hourly output of the rolling mill shall be balanced with the maximum hourly output of the continuous caster, and the rolling mill shall During the design, the hourly output of products of various specifications shall be as close as possible.

（3）连铸机与轧机小时产量应匹配得当。若轧机小时产量小于连铸机最大小时产量（不考虑连铸机准备时间），则将有许多热坯不能进入轧机而必须脱离轧线变成冷坯；若轧机设计能力大于连铸机最大小时产量，则轧机能力将不能发挥而造成浪费，故原则上轧机小时产量应与连铸机最大小时产量平衡，轧机设计时应力求各规格产品的小时产量尽量接近。

（4）In order to give full play to the hot charging effect, it is hoped that the cold billet will not be offline even when the mill stops rolling for a short time （roll change and groove change）, so a buffer zone should be set between the charging roller table and the heating furnace to temporarily store the billet. In order to absorb the accumulated hot billet after rolling again, the maximum hourly output of rolling mill （including heating furnace） should be 20%~25% higher than that of continuous casting machine.

（4）为充分发挥热装效果，希望即使在轧机短时停轧（换辊、换轧槽）时也不产生冷坯离线，故在装炉辊道与加热炉之间应设缓冲区，以暂时储存钢坯。为了在重新开轧后能吸收掉积存的热坯，轧机（包括加热炉）最大小时产量应高于连铸机最大小时产量20%~25%。

（5）The furnace shall be able to flexibly adjust the combustion system to adapt to the frequent fluctuation of hourly production of rolling mill and the frequent conversion between hot slab and cold slab.

（5）加热炉应能灵活调节燃烧系统，以适应经常波动的轧机小时产量以及热坯与冷坯之间的经常转换。

（6）The rolling mill shall have a reasonable pass design and adopt a common pass system to reduce the number of roll changes. In order to shorten the short-time stop time caused by roll change and groove change, the rapid roll change device, roll guide pre-adjustment and guide guide quick positioning technology should be adopted.

（6）轧机应有合理的孔型设计，采用共用孔型系统以减少换辊次数。为尽量缩短因换辊、换槽等引起的短时停车时间，应采用快速换辊装置、轧辊导卫预调、导卫快速定位技术等。

(7) A perfect computer system should be set up to control and coordinate between the steelmaking caster and the rolling workshop.

(7) 应设置完善的计算机系统,在炼钢连铸机与轧钢车间之间进行控制和协调。

2.1.3.3 Hot Charging Process
2.1.3.3 热装工艺

(1) Capacity matching of continuous caster and rolling mill.

(1) 连铸机与轧机能力的匹配。

It is easy to realize hot charging when a set of continuous casting machine is connected with a set of rolling mill. If one set of caster is matched with two sets of rolling mills or two sets of caster with smaller capacity are matched with one set of rolling mills, although there may be some seemingly feasible schemes in design, it is difficult to realize them. The biggest difficulty is steel grade management, and the next is that it is difficult to organize production accurately to balance the capacity of caster and rolling mill, so the advantage of hot charging will be greatly reduced. Because of the long distance between the equipment, there will be problems in the transportation of hot billets.

较容易实现热装的情况是一套连铸机与一套轧机相连。若是一套连铸机与两套轧机相配或两套能力较小的连铸机与一套轧机相配,虽然在设计上可有某些看似可行的方案,但实现起来就比较困难。最大的困难是钢号管理,其次是很难精确组织生产使连铸机与轧机能力达到均衡,因而热装的优越性将大打折扣。因设备之间距离远,热坯的运输也会有问题。

In the case of one–to–one (one electric furnace, one refining equipment outside the furnace, one continuous caster corresponding to one set of small or wire mill), the situation of small bar mill and wire mill is different. The hourly output of small bar mill is relatively low when rolling small-sized products, which can be solved by cutting and properly increasing the final rolling speed. In addition, the roll change time of this mill can be shortened to 7~8min by using the whole frame quick change device, which will not affect the continuous hot assembly. When the wire rod mill is rolling small-sized products (e.g. $\phi5.5mm$), even if the highest rolling speed is adopted, its hourly output is still far lower than that of larger specifications. Moreover, the roll changing rings of the wire rod's twistless finishing mill are still replaced manually one by one for a long time. If the time is more than 30min, the buffer device is full, and some hot billets must be cooled off-line (these billets must be separately numbered). Therefore, it is impossible for wire mill to achieve 100% hot charging.

在一对一的情况下(一座电炉、一座炉外精炼设备、一套连铸机对应一套小型或线材轧机),小型棒材轧机与线材轧机的情况又有所不同。小型棒材轧机轧小规格产品时轧机小时产量较低,可通过切分及适当提高终轧速度来解决,另外这种轧机换辊时间通过采用整机架快速更换装置可缩短到7~8min,不会影响热装的继续进行。而线材轧机在轧制小规格产品(例如$\phi5.5mm$)时,即使采用最高的轧制速度其小时产量仍远低于较大规格的小时产量。再者,线材的无扭精轧机换辊环仍采用人工逐个更换,时间较长,若超过30min,缓冲装置已充满,部分热坯必须离线冷却下来(这些坯料必须单独编炉号)。因此，线材轧机不可能实现100%的热装。

Obviously, the less the number of rolling mills on the rolling line, the easier it is to realize hot charging, because the accidents and roll changing times are greatly reduced. The first to realize 100% hot charging is a set of rolling mill which uses 240mm×240mm and 280mm×280mm continuous casting slab to produce 90~200mm square round bar steel. The rolling line has only 8 stands.

显然，轧线上轧机数量越少实现热装越容易，因为事故和换辊次数都大大减少。较早实现100%热装的是一套用240mm×240mm、280mm×280mm连铸坯生产90~200mm方圆棒钢的轧机，该轧线仅有8个机架。

(2) Setting of buffer device.

(2) 缓冲装置的设置。

The function of the buffer device is to adjust the hourly output of the continuous caster and the rolling mill, and to temporarily store the hot billet when the rolling mill stops due to accident or roll change, so as to keep the hot charging uninterrupted. The capacity of the general buffer device is designed to store the continuous casting billets (preferably a furnace of billets) produced in 30min, because the accidents and roll changes can be handled in 30min.

缓冲装置的作用是当连铸机与轧机小时产量不同时起调节作用，并在轧机因事故或换辊而停车时暂时储存热钢坯，以使热装不致中断。一般缓冲装置的能力设计成可贮存30min产出的连铸坯（最好是一炉钢坯），因为在30min内一般事故和换辊均可处理完。

There are many forms of buffer device. One is to set buffer zone outside the furnace——insulation room, which is constructed with insulation materials, but without burners. There are two types: one is that there is only one hot billet conveying roller table, and the hot billet enters first and then exits, so that the temperature of the billet entering the heating furnace is different; the other is that there are two roller tables, i.e. the conveying roller table and the output roller table, the hot billet can enter first and exit first, and the temperature of the hot billet entering the heating furnace can keep basically the same.

缓冲装置的形式有多种，一种是设炉外缓冲区——保温室，采用绝热材料构筑，但不设烧嘴。其中又分两种，一种是仅有一条热坯输送辊道，热坯先进后出，这样进加热炉的坯料温度就有差别；另一种是分设输入和输出两条辊道，热坯可先进先出，进加热炉的热坯温度可保持基本一致。

Another kind of buffer device is to use the buffer zone in the furnace, that is, the entrance zone of the heating furnace is 3~4m as the buffer zone, and a set of charging machine shall be set at the same time. If the hot billet is fed into the furnace, the charging machine shall be used to send the damaged steel to the heating zone over the distance of 3~4m. If the cold billet is fed into the furnace, it shall stay in the zone and heat slowly. In recent years, there is only a way to set up a bench without adding a heat preservation chamber, because after a period of practice, it is found that the temperature drop of continuous casting slab with a certain size is not large in the buffer zone.

还有一种缓冲装置是采用炉内缓冲区，即将加热炉入口区3~4m作为缓冲区，同时需设一套装料机，若是热坯进炉后即用此装料机将钢坯越过3~4m距离送至加热区，若是冷坯则在此区停留并缓慢加热。近年来，也有仅设台架不加保温室的做法，这是因为经过一段时期的实践，发现一定尺寸以上的连铸坯在缓冲区内温降不大。

(3) Hot charging process.

(3) 热装工艺。

Several hot charging processes are introduced as follows:

几种热装工艺介绍如下：

1) Normal hot charging. In normal rolling state, the hot continuous casting slab is taken out from the cold bed of the continuous casting machine one by one and placed on a single slab transport roller table, which is then sent to the vicinity of the heating furnace, where the length is measured and the unqualified slab is removed. After the qualified slab is lifted, it is dropped into the charging roller table, weighed and then loaded into the heating furnace for heating, and then sent to the rolling mill for rolling.

1) 正常热装。正常轧制状态下，热连铸坯自连铸机冷床处用取料机逐根取出放到单根坯运输辊道上，再由此辊道送到加热炉附近，在此进行测长并剔除不合格坯，合格坯提升后落入装料辊道，称重后装入加热炉加热，并随后送入轧机轧制。

2) Indirect hot charging. If the rolling mill stops rolling for a short time (roll change, groove change, general accident), the hot continuous casting slab will be sent to the buffer insulation room temporarily from another group of multi row conveying rollers, and the slab will be moved to the outlet of the insulation room by the transfer machine in the insulation room. When the rolling mill is put back into operation, the temporary hot continuous casting billets are sent out one by one from the output roller table, and then put into the furnace after length measurement, lifting and weighing. At the same time, the hot continuous casting slab from the cold bed of the continuous casting machine is also sent to the heating furnace along the same roller table. At this time, the heating furnace and rolling mill will produce with high hourly capacity until the blank accumulated in the buffer insulation room is completely empty. At this time, the rolling mill returns to normal rolling state. Because the hot billet is not directly from the continuous casting machine but through the insulation chamber into the heating furnace, it is called indirect hot charging.

2) 间接热装。若轧机短时停轧（换辊、换轧槽、一般事故），热连铸坯则从另一组多排料的输送辊道送入缓冲保温室暂存，并由保温室中的移送机将坯料移向保温室出口附近。当轧机重新运转后，暂存的热连铸坯从其输出辊道逐根送出，经测长、提升、称重后入炉。同时从连铸机冷床来的热连铸坯也沿同一辊道送入加热炉。此时加热炉和轧机将以较高的小时能力生产，直至缓冲保温室内积存的坯料完全出空。此时轧机又恢复到正常轧制状态。由于热坯不是直接从连铸机而是经由保温室进入加热炉，故称之为间接热装。

3) Cold charging. From the cold billets left by the continuous caster and the cleaned billets, they are hoisted in rows from the billet storage site to the cold loading platform by crane. When the billets arrive at the end of the platform, they are transported to the roller table one by one, and then measured, lifted and weighed into the furnace.

3) 冷装。从连铸机甩下的冷坯及清理后的连铸坯，用吊车从坯料堆存场地成排吊到冷装台架上，钢坯到达台架端部时，逐根被拨入运输辊道，然后测长、提升、称重入炉。

4) Mix hot charging. When some specifications require that the hourly output of rolling mill is larger than that of continuous casting machine, the mixed hot charging method can be adopted. The process is as follows: the hot billets sent from the continuous caster are transported to the buffering and heat preservation room along a plurality of billet transport roller beds for temporary storage, and the cold billets are loaded into the heating furnace at the same time, and rolled according to the required hourly output. When the buffer holding chamber is full of billets, the cooling billets will be stopped, and then the buffer holding chamber and the continuous casting machine will supply heat to the rolling mill heating furnace. At this time, the hourly production of the rolling mill can be higher than that of the continuous casting machine. When the hot billets in the buffer insulation room are empty, the hot billets are transferred to the buffer insulation room for temporary storage, and the cold billets are used for charging at the same time.

4）混合热装。当某些规格需要轧机小时产量大于连铸机的小时产量时，可采取混合热装方式操作。其过程如下：从连铸机送来的热坯沿多根坯运输辊道进入缓冲保温室暂存，同时将冷坯装入加热炉，并按要求的小时产量进行轧制。当缓冲保温室已存满钢坯时停止上冷坯，改由缓冲保温室和连铸机共同向轧机加热炉供热坯，此时的轧机小时产量就可高于连铸机的产量。当缓冲保温室内的热坯出空时，热坯又转向缓冲保温室暂存，同时改用冷坯装炉。

5) Delay hot charging. As mentioned before, in order to prevent surface cracks after rolling when hot charging is used for some steel grades, it is necessary to rapidly cool the continuous casting slab from above 900℃ to about 550℃ before charging. This hot charging process is called delayed hot charging.

5）延迟热装。如前所述，某些钢种为防止采用热装时在轧制后出现表面裂纹，需将连铸坯从900℃以上迅速冷却到550℃左右再装炉，这种热装工艺称为延迟热装。

Task 2.2 Heating of the Billet
任务 2.2 坯料的加热

Mission objectives
任务目标

(1) Master the heating process specification of billet, the common heating defects and their formation mechanism in the heating process;

（1）掌握坯料的加热工艺规程，掌握加热过程中常见加热缺陷及其形成机理；

(2) Be able to operate the heating furnace to complete the related operation of charging and tapping.

（2）能够操作加热炉完成装钢、出钢等加热炉相关操作。

2.2.1 Heating Equipment
2.2.1 加热设备

2.2.1.1 General
2.2.1.1 概述

Continuous heating furnace is the most common furnace in steel rolling workshop. Billets are loaded at the end of the furnace and sent out at the other end after heating. In pusher type continuous heating furnace, the billet is continuously moved forward along the furnace bottom slide by the thrust of pusher in the furnace; in mechanized furnace bottom continuous heating furnace, the billet is continuously moved forward in the furnace by the transmission machinery at the furnace bottom. The furnace gas produced by combustion generally flows towards the heated billet towards the furnace end, i.e. counter flow. When the billet is moved to the discharging end, it is heated to the required temperature, discharged through the tapping port, and then sent to the rolling mill along the roller table.

连续加热炉是轧钢车间应用最普遍的炉子。钢坯由炉尾装入，加热后由另一端送出。推钢式连续加热炉，钢坯在炉内是靠推钢机的推力沿炉底滑道不断向前移运；机械化炉底连续加热炉，钢坯则靠炉底的传动机械不停地在炉内向前运动。燃烧产生的炉气一般是对着被加热的钢坯向炉尾流动，即逆流式流动。钢坯移到出料端时，被加热到所需要的温度，经过出钢口出炉，再沿辊道送往轧钢机。

The work of the continuous heating furnace is continuous. The billet is continuously heated and continuously pushed out after heating. Under the condition of stable operation of the furnace, the temperature of each point in the furnace can be regarded as not changing with time, which belongs to the stable temperature field. The heat transfer in the furnace can be regarded as the stable heat transfer approximately, and the heat transfer guide in the billet belongs to the unstable heat transfer.

连续加热炉的工作是连续性的，钢坯不断地加热，加热后不断地推出。在炉子稳定工作的条件下，炉内各点的温度可以视为不随时间而变，属于稳定态温度场，炉膛内传热可近似地当作稳定态传热，钢坯内部热传导则属于不稳定态导热。

There are many furnaces with the thermal characteristics of continuous heating furnace. From the aspects of structure and thermal system, the continuous heating furnace can be classified according to the following characteristics:

具有连续加热炉热工特点的炉子很多，从结构、热工制度等方面看，连续加热炉可按下列特征进行分类：

(1) According to the temperature system, it can be divided into two-stage type, three-stage type and intensified heating type.

（1）按温度制度可分为两段式、三段式和强化加热式。

(2) According to the types of heated metal, it can be divided into: heating square billet, heating slab, heating round tube billet, heating special billet.

（2）按被加热金属的种类可分为加热方坯的、加热板坯的、加热圆管坯的、加热异型坯的。

(3) According to the type of fuel used, it can be divided into solid fuel, heavy oil, gas fuel and mixed fuel.

(3) 按所用燃料种类分为使用固体燃料的、使用重油的、使用气体燃料的、使用混合燃料的。

(4) According to the preheating mode of air and gas, it can be divided into: heat exchange type, heat storage type and non-preheating type.

(4) 按空气和煤气的预热方式可分为换热式的、蓄热式的、不预热的。

(5) According to the discharging mode, it can be divided into end discharging and side discharging.

(5) 按出料方式可分为端出料的和侧出料的。

(6) According to the way of steel moving in the furnace, it can be divided into push type continuous heating furnace, step type furnace, roller bottom type furnace, rotary bottom type furnace, chain type furnace, etc. Walking beam furnace is the fastest growing furnace type in various mechanized bottom furnaces, and it is the main furnace type to replace pusher furnace. Since the 1970s, almost all the large-scale rolling mills built in the world, such as the hot rolling mill, have adopted the walking beam furnace. In China, the walking beam furnace has been widely used in plate rolling, high-speed wire rolling and continuous small-scale rolling mills since the 1980s. The walking beam type heating furnace is used in small continuous rolling mill according to the process requirements, and its comparison with other furnace types and main features are as follows.

(6) 按钢料在炉内运动的方式可分为推钢式连续加热炉、步进式炉、辊底式炉、转底式炉、链式炉等。步进式加热炉是各种机械化炉底炉中使用发展最快的炉型，是取代推钢式加热炉的主要炉型。20世纪70年代以来，世界各国兴建的热轧等大型轧机，几乎都采用了步进式炉，在我国，步进式炉在80年代以来广泛用于轧板、高线、连续小型轧钢厂。小型连轧厂根据工艺要求采用步进梁式加热炉，其与其他炉型的比较及主要特点如下所述。

2.2.1.2 Comparison between Walking Beam Furnace and Pusher Furnace
2.2.1.2 步进梁式加热炉与推钢式炉的比较

(1) In the pusher furnace, the operation of the billet depends on the thrust of the pusher to slide on the slide rail. Therefore, scratches often occur on the lower surface of the billet, which has a negative impact on the surface quality of the billet. However, in the walking beam furnace, the operation of the billet is completed by the lifting of the walking beam and the lowering of the walking beam, so there is no scratch.

(1) 在推钢式炉内，钢坯的运行是靠推钢机的推力在滑轨上滑行的，因此，钢坯下表面往往产生划痕，对钢坯表面质量带来不利影响，但在步进梁式炉中，钢坯的运行是靠步进梁托起—前进—放下来完成的，所以不产生划痕。

(2) In pusher type furnace, the temperature of billet contacting water-cooled slide rail is

relatively low, and "black mark" is serious, which has a great influence on the dimensional deviation of rolled piece. Although walking beam furnace has water beam, the billet does not contact water beam continuously, but contact water beam intermittently and alternately. The phenomenon of "black mark" is light, and the temperature difference is small, which greatly improves the influence of dimensional deviation of products.

（2）在推钢式炉内，钢坯接触水冷滑轨部分温度较低，"黑印"严重，对轧件的尺寸偏差影响很大，而步进梁式炉虽有水梁，但钢坯并不连续接触水梁，而是间断、交替地接触水梁，"黑印"现象较轻，温差较小，对产品尺寸偏差的影响大有改善。

（3）In the pusher type furnace, the billets are close together, which is easy to produce "sticking steel" phenomenon under high temperature, and can only be heated on one or both sides. The heating speed is slow, and the temperature is not uniform. However, there is a large gap between each billets in the walking beam furnace, and the phenomenon of "sticking steel" is avoided in the walking beam furnace, and the four sides heating is realized, the heating speed is fast, and the temperature is uniform.

（3）在推钢式炉内，钢坯是紧紧靠在一起的，高温下易产生"粘钢"现象，并且只能单面或双面受热，加热速度慢，温度不够均匀，但在步进梁式炉中每根钢坯间都留有较大的间隙，步进避免了"粘钢"现象，而且实现了四面加热，加热速度快，温度均匀。

（4）The pusher furnace is prone to arch steel accident when pushing steel. The effective size of the furnace is limited by the section size of billet and the capacity of pusher, but the walking beam furnace will not have arch steel phenomenon. The design of the furnace does not need to consider the arch steel problem to limit the length of the furnace.

（4）推钢式炉在推钢时易发生拱钢事故，炉子有效尺寸受钢坯断面尺寸和推钢机能力的限制，但步进梁式炉不会发生拱钢现象，炉子设计也不用考虑拱钢问题而限制炉子长度。

（5）The pusher type furnace can not be empty, so the flexibility of adjustment and maintenance of empty furnace for different steel grades and different heating processes is very poor, but the walking beam type furnace is convenient to empty, the walking operation is flexible, and the adaptability of heating various steel grades is strong.

（5）推钢式炉不能空炉，因此，对不同钢种不同加热工艺的调整、检修空炉等灵活性很差，但步进梁式炉空炉方便，步进操作灵活，加热各种钢种的适应性强。

（6）The walking beam furnace has the advantages of fast heating speed, uniform temperature and flexible operation. Therefore, the burning loss of the billet is reduced. The scale of the pusher furnace accounts for 1% ~ 1.5% of the total weight of the billet, and the walking beam furnace only accounts for 0.5% ~ 0.8%, reducing the labor intensity of slag cleaning.

（6）步进梁式炉加热速度快，温度均匀，操作灵活，因此，减少了钢坯的烧损，推钢式炉的氧化铁皮占钢坯总重的1% ~ 1.5%，步进梁式炉的仅占0.5% ~ 0.8%，减轻了清渣劳动强度。

（7）The walking beam furnace is flexible in operation, which can adjust the steel loading according to the mill output, facilitate the change of steel grades, adapt to hot charging and hot feeding: accurately send the billet to the rolling center line, match with the full continuous rolling of the full continuous small bar mill; facilitate the realization of the computer control and billet tracking functions of the full-automatic steel in and out.

（7）步进梁式炉操作灵活，可根据轧机产量调节装钢量，便于更换钢种，适应热装、热进；能够准确地将钢坯送到轧制中心线，与全连续小型棒材轧机全连续轧制相匹配；便于实现全自动进出钢的计算机控制和钢坯跟踪等功能。

（8）The total water-cooling surface area of the fixed beam, walking beam and straight support tube of the walking beam furnace is about twice of that of the bottom tube of the pusher type furnace, so its theoretical heat consumption is 10%~15% larger than that of the pusher type furnace; however, because the walking beam does not vibrate when pushing steel, the inner fiber felt is used for the thermal insulation binding of the walking beam, and the outer layer is poured with castable, with a service life of several years; However, in the pusher furnace, the water pipe vibrates when pushing the steel, the insulation binding of water pipe is easy to fall off. The insulation binding of high-temperature section falls off 15% or more within one month. In order to maintain and use conveniently, rough insulation methods such as prefabricated insulation bricks of water beam are adopted. Therefore, in the actual operation process, the energy consumption of push steel furnace is much higher than that of walking beam furnace.

（8）步进梁式炉的固定梁、步进梁和直撑管总的水冷表面积，约比推钢式炉底管的冷却表面积大一倍，故其理论热耗比推钢式炉的大 10%~15%；但是由于步进梁没有推钢时的振动，步进梁绝热包扎采用了内层纤维毡，外层用浇铸料浇铸的办法，寿命长达数年；而推钢式炉由于推钢时水管振动，水管绝热包扎极易脱落，高温段绝热包扎使用一个月就脱落 15%，甚至更多，为维护方便使用单位又采用了如水梁预制绝热砖等粗糙的绝热方式，所以在实际运行过程中，推钢式炉比步进梁式炉的能耗反而高很多。

（9）The water consumption of walking beam furnace is about 60% higher than that of pusher type; in terms of investment, the total investment of heating furnace system is 25%~30% higher than that of pusher type furnace; in terms of maintenance, although the maintenance of bottom water beam binding is very small, the maintenance of electromechanical control, hydraulic drive of furnace bottom machinery and other equipment is greatly increased.

（9）步进梁式炉水耗比推钢式的大 60%左右；在投资方面，加热炉系统总投资步进梁式炉比推钢式炉的高 25%~30%；在维护方面，虽然炉底水梁包扎等的维护量很小，但机电控制、炉底机械的液压驱动等设备的维护量大大增加。

2.2.1.3 Comparison of Walking Beam Furnace with Walking Bottom Furnace and Beam Bottom Combined Furnace
2.2.1.3 步进梁式炉与步进底式炉、梁底组合式炉的比较

（1）The walking beam furnace is a furnace for heating up and down. During the whole heating process, the billet is basically in a symmetrical heating state (except for the contact point with the walking beam). The billet temperature is uniform and there is no internal and external surface, so it is suitable for heating the billet of various sections. In the whole heating process, the billet is basically in an asymmetric heating state, with uneven heating up and down; when heating the large section billet, the temperature difference between the upper and lower sections of the billet caused by asymmetric heating causes the billet to deform and bend up, affecting normal operation, so the walking bottom furnace is only suitable for heating under 120mm×120mm for small section billets, it is better to heat the billets below 100mm×100mm.

（1）步进梁式炉是上下加热的炉子，钢坯在整个加热过程中，基本处于对称加热状态（除与步进梁接触点外），钢坯温度均匀，无阴阳面，适于加热各种断面的坯料。而步进底式炉只有上加热，没有下加热，钢坯在整个加热过程中，基本处于非对称加热状态，上下加热不均匀；当加热大断面的钢坯时，由于非对称加热产生的钢坯断面上下温差，导致钢坯变形向上弯出，影响正常运行，因此步进底式炉只适于加热 120mm×120mm 以下的小断面坯料，最好是加热 100mm×100mm 以下的方坯。

（2）Walking beam furnace is a furnace that is heated up and down, with fast heating speed and large furnace output. When heating large section billet, the output per unit bottom steel passing area of walking beam furnace is generally about 350kg/(m²·h), and that per unit bottom steel passing area of beam bottom combined furnace is usually 400~450kg/(m²·h), while that per unit bottom steel passing area of energy-saving walking beam furnace is usually 600~650kg/(m²·h), that is, under the same effective furnace length. Although the output of the furnace can be increased by 45%~80%.

（2）步进梁式炉是上下加热的炉子，加热速度快，炉子产量大。在加热较大断面钢坯时，一般步进底式炉单位炉底过钢面积的产量常取 350kg/(m²·h) 左右，梁底组合式炉单位炉底过钢面积的产量常取 400~450kg/(m²·h)，而节能型步进梁式炉的单位炉底过钢面积的产量常取 600~650kg/(m²·h)，即在同样有效炉长的情况下，步进梁式炉的产量可提高 45%~80%。

（3）When the furnace temperature is too high, it is easy to produce slag at the bottom of the furnace; when the oxide scale and refractory materials at the bottom of the furnace fall off and enter the gap between the stepping bottom and the fixed bottom, it will produce the phenomenon of stuck, which will affect the production. The walking beam furnace has the advantages of short time in furnace, thin oxide scale, and easy to remove slag at the bottom of furnace.

（3）步进底式炉钢坯在炉时间长，氧化铁皮厚，增加了清渣次数；当炉温过高时，易产生炉底结渣；氧化铁皮和炉底耐火材料脱落进入步进底和固定底缝隙后，会产生卡死现象，影响生产。而步进梁式炉钢坯在炉时间短，氧化铁皮薄，易于清除炉底积渣。

（4）There is no water-cooling beam and heat loss in the furnace, so the energy consumption is low and the water consumption is small. The billet will not collapse on the bottom of the furnace, which is its biggest advantage. The walking beam and the fixed beam in the walking beam furnace are both water-cooled beams, which lead to a large loss of heat absorption and water consumption.

（4）步进底式炉炉内没有水冷梁，无水冷吸热损失，故能耗低、耗水量少，钢坯在步进底上不会"塌腰"，是其最大优点。而步进梁式炉炉内步进梁和固定梁均为水冷梁，水冷吸热损失大，导致能耗稍高，水耗大。

(5) The beam bottom combined walking beam furnace combines the advantages of walking beam furnace and walking beam furnace. Generally, walking beam is used in the low temperature section to eliminate the temperature difference of billet section, prevent the billet from bending and deformation, and the water cooling and heat absorption loss in the low temperature section is less; while walking beam is used in the high strength section to reduce the number of walking beam, which can not only reduce the water and heat loss, but also prevent the small section billet from being in the high temperature section. There is "waist collapse". However, because of the use of stepping bottom in high temperature section, there are many disadvantages of stepping bottom furnace, especially when heating large section billet. Its disadvantages are:

(5) 梁底组合式步进炉综合了步进底式炉和步进梁式炉两者的优点，一般低温段采用步进梁，以消除钢坯断面温差，防止钢坯弯曲变形，且低温段水冷吸热损失少；而高强段采用步进底，减少了步进梁的数量，既可以减少水和热损失，又可防止小断面方坯在高温段产生"塌腰"。但由于高温段采用步进底，就带来了步进底式炉的许多缺点，特别是在加热大断面钢坯时尤为显著。其缺点为：

1) The furnace bottom strength is low (such as the comparison in Item (2) above), and the length and investment of the walking beam furnace with the same output increase compared with the walking beam furnace;

1) 炉底强度低（如上面第（2）项中的比较），同样产量的步进底式炉与步进梁式炉比较增加了炉长和投资；

2) It is difficult to adapt to the requirements of modern wire and bar rolling mill for the uniformity of billet temperature (within 30℃) due to the large temperature difference of billet section;

2) 钢坯断面温差大，难以适应现代线棒材轧机对钢坯温度均匀性的要求（同条温差30℃以内）；

3) The thickness of oxide scale and decarbonization are increased;

3) 增大了钢坯氧化铁皮厚度，增加了钢坯脱碳；

4) It is troublesome to clean up the scale in high temperature section. If not handled in time, the scale will rise and the billet will deviate laterally in high temperature section.

4) 高温段氧化铁皮清理麻烦，处理不及时会造成氧化铁皮涨高，使钢坯在高温段横向跑偏。

Because of the above reasons, the beam bottom combined walking beam furnace is only suitable for heating the billets with the section size of (100mm×100mm)~(130mm×130mm), up to 150mm×150mm; generally, when the section size of the billets is larger than 130mm×130mm, the walking beam furnace should be used.

由于上述原因，梁底组合式步进炉只适合于加热断面尺寸为（100mm×100mm）~（130mm×130mm）的方坯，最大到150mm×150mm；一般在方坯断面尺寸大于130mm×130mm时，宜采用步进梁式炉。

2.2.1.4 Functions and Characteristics of Stepping Mechanism
2.2.1.4 步进机构的功能及特点

(1) Composition of Step Mechanism.
(1) 步进机构的组成。

The stepping mechanism of walking beam heating furnace is composed of driving system, stepping frame and control system. The stepping system can be generally divided into two types: electric type and hydraulic type. The driving of the moving part is generally realized by the hydraulic system. Some of the lifting parts adopt the electric cam type, some adopt the hydraulic curved rod type or the hydraulic inclined rail type. The electric cam type and the hydraulic curved rod type mechanism adopt the single-layer stepping frame. The hydraulic inclined rail type adopts double-layer stepping frame, which slides on the roller through the track. The constant power pump with flow and pressure compensation and the full hydraulic inclined rail type stepping mechanism with large capacity proportional valve are used for stable and reliable operation. When the lifting hydraulic cylinder drives the lifting frame to do lifting movement on the inclined rail, the stepping bottom and the stepping beam will move forward and backward accordingly. The action and movement speed of hydraulic cylinder are controlled by the opening, closing and opening degree of hydraulic valve, while the action of valve is controlled by PLC program. The step action signal is sent out by the proximity switch and programmable travel switch installed on the corresponding parts.

步进梁式加热炉的步进机构由驱动系统、步进框架和控制系统组成。步进系统一般可分为电动式和液压式两种，行进部分的驱动一般靠液压系统实现，升降部分有的采用电动凸轮式，也有的采用液压曲杆式或液压斜轨式。电动凸轮式和液压曲杆式机构采用单层步进框架。液压斜轨式采用双层步进框架，步进框架通过轨道在辊轮上滑动。采用流量、压力补偿的恒功率泵和大容量比例阀的全液压斜轨式步进机构，运行稳定可靠，升降液压缸带动升降框架在斜轨上做升降运动时，步进底和步进梁便随之前进和后退。液压缸的动作和其运动速度是由液压阀门的开、闭和开启度的大小来控制的，而阀门的动作则由PLC程序控制。步进动作的信号由安装在相应部位上的接近开关和可编程行程开关发出。

(2) Function and Characteristics of B Step Mechanism.
(2) 步进机构的功能及特点。

The step bottom and the step beam move up, forward, down and backward periodically with the movement of the step frame. Generally, the starting point is set at the front lower limit or the rear lower limit of the step beam, such as the starting point of the heating furnace is at the front lower limit of the step beam. The stepping mechanism can perform the following functions:

步进底和步进梁随着步进框架的运动做上升、前进、下降、后退的周期运动，一般将起始点设在步进梁的前下限或后下限，如加热炉的起始点在步进梁的前下限。步进机构可以完成如下功能：

1) Positive cycle. The walking beam moves backward, up, forward and down from the original position to complete a cycle, which can transport the billet one step forward; the billet before

the discharging cantilever roller table is fed to the cantilever roller table, and the billet after the feeding roller table is further advanced, which ensures the continuity of the feeding and discharging.

1) 正循环。步进梁由原始位置后退、上升、前进、下降完成一个周期,可以输送钢坯前进一个步距;出料悬臂辊道前一根钢坯被进到悬臂辊道上,而进料辊道后一根钢坯被向前进一步,保证了进料、出料的连续性。

2) The reverse cycle and the step beam complete a cycle from the original position up, back, down and forward, and the billet can be transported back one step. This function can realize accident steel pouring.

2) 逆循环、步进梁由原始位置上升、后退、下降、前进完成一个周期,可以输送钢坯后退一个步距。该功能可以实现事故倒钢。

3) Step. The walking beam only moves up and down, not forward and backward. When the rolling line stops rolling for a short time (less than 30min), in order to reduce the black mark of billet, this function is used to make the contact time of billet and walking beam and fixed beam equal, and the cycle time of each step is controlled at about 5min.

3) 踏步。步进梁只做上升和下降运动,不做前进和后退运动D当轧线短期停轧时(少于30min),为减少钢坯黑印,采用该功能,使钢坯与步进梁和固定梁的接触时间相等,每次踏步周期时间控制在5min左右。

4) Middle position hold. When the rolling mill stops rolling for more than 30min and the furnace does not discharge steel for a long time, in order to prevent the billet from bending down, it is required that the upper surface of the walking beam and the upper surface of the fixed beam stop at the same level, that is, the walking beam and the fixed beam support the billet at the same time.

4) 中间位置保持。当轧机停轧时间长于30min炉子长时间不出钢时,为防止钢坯下弯,要求步进梁上表面与固定梁上表面停在一个标高处,即步进梁与固定梁同时支撑钢坯。

5) Step and wait. Steel rolling requires that the tapping period of furnace is always longer than the minimum stepping period of stepping furnace equipment. In this case, the extra time is allocated to the turning point of stepping beam track as "waiting".

5) 步进等待。轧钢要求炉子的出钢周期总是比步进炉设备的最小步进周期要长,这时把多出的时间作为"等待"分配到步进梁轨迹拐点上。

2.2.1.5 Combustion System and Furnace Structure Characteristics of Heating Furnace
2.2.1.5 加热炉燃烧系统及炉子结构特点

(1) Use full flat flame burner. The top of the upper heating furnace adopts full flat flame burner, with uniform temperature field and high radiation intensity, which is easy to maintain positive pressure on the top of the furnace and prevent cold air from being inhaled. The heating speed is fast and the temperature is even, which is beneficial to reduce oxidation and decarbonization and prevent the temperature drop of the billets to be tapped at the tapping side.

（1）采用全平焰烧嘴。上加热炉顶采用全平焰烧嘴，温度场均匀，辐射强度大，易于维持炉顶正压，防止冷风吸入。加热钢坯速度快，温度均匀，有利于减少氧化和脱碳，防止出钢侧出钢坯温度下降。

（2）The lower heating adopts the end burner. The advantage is that the temperature on the furnace width is easy to adjust, and it is easy to ensure the temperature uniformity of the billet length direction.

（2）下加热采用端烧嘴。其优点是炉宽上温度便于调整，易于保证钢坯长度方向的温度均匀性。

（3）Reasonable partition structure in the furnace. The top partition wall and bottom partition wall are set at the heating part of the upper heating flat flame burner and the heating part of the lower heating end burner to form a choke, reduce the radiation heat transfer between the heating section and the soaking section, so as to ensure the single temperature control of the heating section and the soaking section; the s furnace bottom retaining wall is set between the lower heating section and the preheating section to distinguish the heating section and enhance the radiation.

（3）合理的炉内隔墙结构。在上加热平焰烧嘴供热部位和下加热端烧嘴供热部位设置炉顶隔墙和炉底隔墙，形成扼流，减少加热段与均热段之间的辐射传热，以保证加热段和均热段的单段温度控制；在下加热与预热段之间设 S 炉底挡墙，区分加热区段，增强辐射。

（4）The furnace adopts fully cast composite lining, the water beam in the furnace is wrapped with double-layer insulation, and the insulation of the whole furnace is good.

（4）炉子采用全浇铸复合内衬，炉内水梁包扎采用双层绝热，全炉绝热较好。

2.2.2 Fuel Selection
2.2.2 燃料选择

At present, the most widely used solid fuel for heating furnace in metallurgical enterprises is coal, liquid fuel is heavy oil, and gas fuel is mixed gas. The following focuses on gas fuel.

目前冶金企业加热炉最为广泛采用的固体燃料有煤，液体燃料有重油，气体燃料则使用混合煤气，下面重点介绍气体燃料。

2.2.2.1 Natural Gas
2.2.2.1 天然气

Natural gas is a kind of combustible gas extracted directly from underground, which is a gas fuel with high industrial economic value. Its main component is methane, the content (mass fraction) is generally 80%~90%, there are a small amount of heavy hydrocarbons and combustible gases such as H_2, CO, etc., the noncombustible component is very small, so the heat output is very high, mostly in $33500 \sim 46000 kJ/m^3$.

天然气是直接由地下开采出来的可燃气体，是一种工业经济价值很高的气体燃料。它的主要成分是甲烷，含量（质量分数）一般在 80%~90%，还有少量重碳氢化合物及 H_2、CO 等可燃气体，不可燃成分很少，所以发热量很高，大多在 $33500 \sim 46000 kJ/m^3$。

Natural gas is a colorless, slightly rotten gas with a density of about $0.73 \sim 0.80 \text{kg/m}^3$, lighter than air. The ignition temperature of natural gas ranges from 640℃ to 850℃, when it is mixed with air to a certain proportion (volume ratio is 4% ~ 15%), it will immediately ignite or explode in case of open fire. The amount of air required for natural gas combustion is very large. The combustion flame is bright and has strong radiation capacity, because a large number of solid particles are separated from methane and other hydrocarbons during combustion.

天然气是一种无色、稍带腐烂臭味的气体，密度约 $0.73 \sim 0.80 \text{kg/m}^3$，比空气轻。天然气着火温度范围在 640~850℃，与空气混合到一定比例（体积比为 4%~15%），遇到明火会立即着火或爆炸。天然气燃烧所需的空气量很大，燃烧火焰光亮，辐射能力强，因为燃烧时甲烷及其他碳氢化合物分解析出大量固体颗粒。

Natural gas has few inert gases and high calorific value, and can be transported for a long distance. It is an excellent heating furnace fuel. It is widely used in foreign countries and less used in domestic metallurgical enterprises for various reasons.

天然气含惰性气体很少，发热量高，并可以做长距离运输，是优良的加热炉燃料，国外使用较多，国内冶金企业因各种原因使用较少。

2.2.2.2 Blast Furnace Gas and Coke Oven Gas
2.2.2.2 高炉煤气和焦炉煤气

(1) Blast furnace gas is a by-product of iron making production. Generally, blast furnace gas used in heating furnace is cleaned gas. Because the gas from blast furnace has a high dust content, the dust is easy to deposit in the pipeline during transportation, and the burner is easy to be blocked during combustion. After cleaning, the dust content of gas (standard state) can be reduced to below 20mg/m^3.

（1）高炉煤气是炼铁生产的副产品，通常加热炉使用的高炉煤气都是经过清洗后的煤气，因为从高炉出来的煤气含尘量很高，在输送过程中灰尘容易沉积在管道中，燃烧时容易堵塞燃烧器等。清洗后煤气（标态）含尘量可降到 20mg/m^3 以下。

According to the macroscopic estimation, $3800 \sim 4000 \text{m}^3$ blast furnace gas can be produced per 1t coke consumed in the blast furnace, which shows the large quantity. Therefore, the comprehensive utilization of blast furnace gas is of great significance for energy conservation. The biggest characteristic of blast furnace gas is that it contains more N_2 and CO_2, so its calorific value is relatively low, usually only $3350 \sim 4200 \text{kJ/m}^3$, so the combustion temperature is low, so it is difficult to apply it on the heating furnace alone, which is often used in combination with other high calorific value gas, or the combustion supporting air and blast furnace gas are preheated to a higher temperature at the same time before combustion. The main combustible components of blast furnace gas are CO and H_2, in addition to a small amount of CH_4 and hydrocarbons. The amount of each component is related to the smelting method of blast furnace, the variety of pig iron, the situation of raw materials and other factors. With the continuous improvement of smelting technology, coke ratio continues to decline, the quality of blast furnace gas continues to decline.

据宏观估计，高炉每消耗 1t 焦炭可产生 $3800 \sim 4000 \text{m}^3$ 高炉煤气，可见数量之大，因此将高炉煤气加以综合利用对于节约能源有重要意义。高炉煤气的最大特点是含 N_2、CO_2 多,所以它发热量较低,通常只有 3350~4200kJ/m³,因此燃烧温度低,单独在加热炉上应用比较困难,往往是与其他高发热量的煤气混合使用,或者将助燃空气及高炉煤气同时预热到较高的温度后再燃烧。高炉煤气的主要可燃成分是 CO、H_2,此外尚有少量的 CH_4 及碳氢化合物等。各组成成分的多少与高炉冶炼方法、生铁的品种、原料情况等因素有关。随着冶炼技术的不断提高,焦比不断下降,高炉煤气的质量不断下降。

The dry composition (volume fraction) of blast furnace gas is roughly as follows:

高炉煤气的干成分(体积分数)大致如下:

CO^{dry}	H_2^{dry}	CH_4^{dry}	$CH_2^{dry}+SO_2^{dry}$	O_2^{dry}	N_2^{dry}
25%~30%	1.5%~3%	0.2%~0.6%	8%~15%	0.2%~0.3%	55%~58%
$CO^干$	$H_2^干$	$CH_4^干$	$CH_2^干+SO_2$	$O_2^干$	$N_2^干$
25%~30%	1.5%~3%	0.2%~0.6%	8%~15%	0.2%~0.3%	55%~58%

It can be seen that blast furnace gas contains more CO, so poisoning should be prevented during use.

由此可见,高炉煤气含 CO 多,使用时要防止中毒。

(2) Coke oven gas is a byproduct of coking production. The coke oven gas of 400~450m³ can be obtained for each ton of coke. Because the coal is carbonized in the process of coking under the condition of air isolation, the content of noncombustible in the by-product coke oven gas is very small. The main combustible components of coke oven gas are H_2, CH_2, CO and hydrocarbon. When the sulfur content in raw materials is high, the combustible components are H_2S. The nonflammable components include N_2, O_2 and CO_2, but the content is relatively low, so its heating capacity is very high, which belongs to high calorific value and high quality fuel.

(2) 焦炉煤气是炼焦生产的副产品。焦炉每炼 1t 焦炭能得到 400~450m³ 焦炉煤气。由于炼焦过程是在隔绝空气的情况下将煤进行干馏的,所以它的副产品焦炉煤气中非可燃物含量很少。焦炉煤气的主要可燃成分有 H_2、CH_2、CO 和碳氢化合物,当原料中硫含量高时,其可燃成分还有 H_2S。不可燃成分有 N_2、O_2、CO_2,但是含量较低,所以它的发热量很高,属于高热值优质燃料。

The dry composition (volume fraction) of coke oven gas is roughly as follows:

焦炉煤气的干成分(体积分数)大致如下:

H_2^{dry}	CH_4^{dry}	C_nH_m	CO^{dry}	O_2^{dry}	N_2^{dry}	SO_2
50%~60%	20%~30%	1.5%~2.5%	5%~9%	0.5%~0.8%	1%~8%	0.4%~0.5%
$H_2^干$	$CH_4^干$	C_nH_m	$CO^干$	$O_2^干$	$N_2^干$	SO_2
50%~60%	20%~30%	1.5%~2.5%	5%~9%	0.5%~0.8%	1%~8%	0.4%~0.5%

The calorific value of coke oven gas is 15490~18840kJ/m³, and the theoretical combustion temperature can reach 2100~2200℃. The main combustible component of coke oven gas is the ratio of H_2 and CH_4, so the density of coke oven gas is relatively small, and the flame has the phenomenon of floating up when burning, that is to say, the rigidity of flame is small. In a sense, buoyancy is not conducive to heating in the furnace.

焦炉煤气的发热量为 15490~18840kJ/m³，理论燃烧温度可达 2100~2200℃。焦炉煤气的主要可燃成分是 H₂ 和 CH₄，所以焦炉煤气的密度比较小，燃烧时火焰具有上浮现象，也就是说火焰的刚性小。从某种意义上来说，上浮是不利于加热炉内加热的。

(3) Blast furnace coke oven gas mixture. In the iron and steel complex, a large number of blast furnace gas and coke oven gas can be obtained at the same time. The ratio of coke oven gas to blast furnace gas output is about 1∶10. It is unreasonable to use coke oven gas alone from the perspective of the total energy distribution of the enterprise. Therefore, in the iron and steel joint enterprise, different proportions of blast furnace gas and coke oven gas can be used to prepare a variety of calorific mixed coal gas with a calorific value of 5900~9200kJ/m³, which can be used as fuel for various metallurgical furnaces in the enterprise.

(3) 高炉—焦炉混合煤气。在钢铁联合企业里，可以同时得到大量的高炉煤气和焦炉煤气。焦炉煤气与高炉煤气产量的比值大约为 1∶10，单独使用焦炉煤气从企业总的能量分配来看是不合理的，所以在钢铁联合企业里可以利用不同比例的高炉煤气和焦炉煤气配备成各种发热量的混合煤气，其发热量为 5900~9200kJ/m³，供企业内各种冶金炉作为燃料。

The calorific value of blast furnace gas and coke oven gas is Q_{high} and Q_{coke} respectively. The mixture of blast furnace gas and coke oven gas with calorific value $Q_{混}$ should be prepared. The ratio can be calculated by the following formula. If the percentage of coke oven gas in mixed gas is, the percentage of blast furnace gas is $(1-x)∶X$.

高炉煤气和焦炉煤气的发热量分别为 $Q_{高}$ 和 $Q_{焦}$，要配成发热量为 $Q_{混}$ 的混合煤气，其配比可用下式计算。设焦炉煤气在混合煤气中所占的百分比为 x，则高炉煤气所占的百分比为 $(1-x)∶X$。

$$Q_{混} = xQ_{焦} + (1-x)Q_{高}$$

According to the above formula:

整理上式得：

$$x = (Q_{混} - Q_{高})/(Q_{焦} - Q_{高})$$

2.2.3　Heating Process of Billet
2.2.3　坯料的加热工艺

2.2.3.1　General
2.2.3.1　概述

The heating quality of steel directly affects the quality, output, energy consumption and rolling mill life. The correct heating process can improve the plasticity of steel, reduce the deformation resistance during hot working, provide the billet with good heating quality for the rolling mill on time, and ensure the smooth production of the rolling mill. On the contrary, if the heating temperature is too high, the steel will overheat and overburn, which will result in scrap;

if the surface of steel is seriously oxidized and decarburized, it will also affect the quality of steel, even scrap. The heating process of steel includes: heating temperature and uniformity of steel, heating speed and time, furnace temperature system, furnace atmosphere, etc.

钢的加热质量直接影响到钢材的质量、产量、能源消耗以及轧机寿命。正确的加热工艺可以提高钢的塑性,降低热加工时的变形抗力,按时为轧机提供加热质量优良的钢坯,保证轧机生产顺利进行。反之,如果加热温度过高,发生钢的过热、过烧,就会造成废品;如果钢的表面发生严重的氧化和脱碳,也会影响钢的质量,甚至报废。钢的加热工艺包括:钢的加热温度和加热均匀性、加热速度和加热时间、炉温制度、炉内气氛等。

2.2.3.2 Heating Temperature of Steel
2.2.3.2 钢的加热温度

The heating temperature of steel refers to the surface temperature when the steel is heated in the furnace and discharged from the furnace. It is mainly determined according to the organization transition temperature in the iron carbon phase diagram. The specific determination of the heating temperature depends on the steel type, billet section specification and rolling process equipment conditions. From the angle of rolling, when the temperature is high, the plasticity of billet is good and the deformation resistance is small; when the temperature is low, the plasticity of billet is poor and the deformation resistance is large. However, with the increase of the heating temperature, the mechanical properties of the steel change, and the oxidation burning rate of the steel increases sharply with the increase of the heating temperature. If the oxide scale is not easy to fall off, it will cause the surface defects of the rolled piece during rolling; if the heating temperature is high, it will inevitably reduce the service life of the heating furnace, and also significantly increase the fuel consumption; in addition, if the heating temperature is too high, there will be overheating and over burning of the billet, resulting in waste products. Therefore, the heating temperature should be selected reasonably from the comprehensive consideration of process, steel type, specification, quality, yield, energy saving and consumption reduction. According to the heating practice of low carbon steel, high carbon steel and low alloy steel, the heating temperature of $1050 \sim 1180℃$ is suitable.

钢的加热温度是指钢料在炉内加热完毕出炉时的表面温度,其主要根据铁-碳相图中的组织转变温度来确定,具体确定加热温度还要看钢种、钢坯断面规格和轧钢工艺设备条件。从轧钢角度看,温度高时钢坯的塑性好,变形抗力小;温度低时钢坯的塑性差,变形抗力大。但随着加热温度的提高,钢材力学性能发生改变,而且钢的氧化烧损率也随着加热温度的升高而急剧增加,若氧化铁皮不易脱落,在轧制时会造成轧件的表面缺陷;加热温度高,必然降低加热炉的寿命,也明显增加燃料消耗;另外,加热温度过高,还会出现钢坯的过热和过烧,造成废品。因此,应从工艺、钢种、规格、质量、成材率和节能降耗等诸因素综合考虑,合理选择加热温度。从低碳钢、高碳钢及低合金钢的加热实践看,1050~1180℃的加热温度是比较适宜的。

2.2.3.3 Heating Speed and Time of Billet
2.2.3.3 钢坯的加热速度和加热时间

The heating speed of billet usually refers to the rising speed of billet surface temperature in unit time, unit: ℃/h. In the actual production, the heating speed of billet is expressed by the time (unit: min/cm) required for the unit thickness of billet to be heated to the specified temperature or the thickness (unit: cm/min) of billet heated in unit time. The heating time of billet usually refers to the total time required for the billet to reach the furnace temperature from normal temperature.

钢坯的加热速度通常是指单位时间内钢坯表面温度的上升速度，单位为℃/h。在实际生产中，钢坯的加热速度用单位厚度的钢坯加热到规定温度所需时间（单位为 min/cm）或单位时间内加热的钢坯厚度（单位为 cm/min）来表示。钢坯的加热时间通常指钢坯从常温加热达到出炉温度所需的总时间。

The heating speed and heating time are affected by the heat load and heat transfer conditions of the furnace, the billet specification and the temperature conductivity of the steel grades. When the heating speed is high, it can give full play to the heating capacity of the furnace. The furnace time is short, the burning loss rate is small, and the burning consumption is low. Therefore, the heating speed should be increased as much as possible to pursue more advanced production indicators. However, excessive temperature difference between the surface and the interior should be avoided, otherwise the billet will be bent and internal crack caused by thermal stress. Generally, the heating speed of carbon structural steel and low alloy steel is not limited, and the heating time is short; however, for large section billet and high carbon and high alloy steel, the heating speed must be controlled to avoid internal defects caused by large temperature difference between inside and outside; when heating the billet, the heating speed is not limited because there is no residual stress and it has entered the plastic state. The walking furnace can heat the three or four sides of the steel uniformly, and the heating conditions are greatly improved. For conventional high-speed wire and small-scale heating furnace, the empirical data of heating speed of low-carbon steel is: pusher furnace is about 6min/cm, while walking beam furnace is about 4.5min/cm.

加热速度和加热时间受炉子热负荷的大小和传热条件、钢坯规格和钢种导温系数大小的影响。加热速度大时，能充分发挥炉子的加热能力，在炉时间短，烧损率小，燃耗低。因此，在可能的条件下应尽量提高加热速度来追求较先进的生产指标。不过应避免表面和内部产生过大的温差，否则钢坯将会产生弯曲和由热应力引起的内裂。碳素结构钢和低合金钢一般可不限制加热速度，加热时间都较短；但对大断面钢坯和高碳、高合金钢，必须控制好加热速度，以免内外温差大造成钢坯内部缺陷；热装加热热坯时，由于不存在残余应力，而且已进入塑性状态，所以加热速度也可不受限制。步进式炉可使钢述三面或四面均匀受热，加热条件大大改善。对于常规的高线和小型加热炉，其低碳钢加热速度的经验数据为：推钢式炉为6min/cm左右，而步进梁式炉则为4.5min/cm。

The heating time of billet is the time of billet in furnace, which is the sum of preheating time, heating time and soaking time. The heating time obtained from theoretical calculation can not match the actual situation at present. The empirical formula and actual data are still the main basis for determining the heating time in production. For example, if a heating furnace is used to heat 150mm×150mm×10000mm square billet, the heating time is about 15×(4.5~5)= 67.5~75min, which is basically consistent with the actual situation.

钢坯的加热时间是钢坯的在炉时间,是预热时间、加热时间、均热时间的总和,由理论计算得出的加热时间目前还不能与实际相吻合,经验公式及实际资料仍是生产中确定加热时间的主要依据。如某加热炉加热 150mm×150mm×10000mm 方坯,套用经验公式,其加热时间约为 15×(4.5~5)= 67.5~75min,与实际情况基本相符。

2.2.3.4 Uniformity of Steel Heating
2.2.3.4 钢加热的均匀性

The best way to heat steel is to heat it to the same temperature inside and outside, but in fact it is difficult to do so. Therefore, according to the allowable range of processing, there is a certain degree of non-uniformity in the temperature inside and outside the billet after heating. Generally, the allowable temperature difference of section is:

钢加热最理想的情况是能把它加热到里外温度都相等,但实际上很难做到,所以根据加工的许可范围,允许加热终了的钢坯内外温度存在一定程度的不均匀性。一般规定断面允许温差为:

$$\Delta t_{终}/s = 100 \sim 300$$

Where Δt_{end}——Temperature difference of section during final heating of steel, ℃;
s——Penetration depth of steel during heating, m.

式中 $\Delta t_{终}$——钢最终加热时的断面温差, ℃;
s——钢加热时的透热深度, m。

The allowable temperature difference between inside and outside varies with the plasticity of steel. For low carbon steel, which has better plasticity, the value of $\Delta t_{end}/s$ can be larger; for high carbon steel and alloy steel, the value of $\Delta t_{end}/s$ should be smaller. In addition, its size is also related to the type of pressure processing, for example, the temperature difference of cross-section required for heating before piercing of tube blank is very small. The temperature difference of steel section specified above is achieved by controlling heating and soaking time in production, because the center temperature of billet cannot be measured online.

允许的内外温差随钢的可塑性不同而有所不同,对于低碳钢这一可塑性比较好的钢种来说,$\Delta t_{终}/s$ 的数值可大一些;对于高碳钢及合金钢,$\Delta t_{终}/s$ 的数值应该小一些。另外,它的大小还和压力加工的种类有关,例如管坯穿孔前加热要求断面温差很小。以上规定的钢断面温差在生产上是通过控制加热及均热时间来达到的,因为钢坯中心温度在线无法测量。

In addition to the temperature difference between the surface and the center, there is also a temperature difference between the upper and lower surfaces of the billet, which is directly related to the furnace shape. The uniformity of the upper and lower surfaces of the walking beam furnace is better than that of other furnace types. Through reasonable heating, high-performance heat-resistant slider and reasonable water beam distribution, the black mark on the lower surface can be basically eliminated, so that the upper and lower surface temperatures are basically the same.

除了表面和中心温差外，钢坯上下表面也具有温差（阴阳面），其大小与炉型有直接关系。步进梁式炉上下表面的均匀性好于其他炉型，通过合理的上下加热、高性能的耐热滑块、合理的水梁分配可以基本消除下表面黑印，使上下表面温度基本相同。

2.2.3.5 Heating System of Steel
2.2.3.5 钢的加热制度

For different kinds of steel, the heating process includes: heating temperature of steel, allowable temperature difference of cross section, heating speed, furnace temperature system and heating system. The latter two are collectively referred to as heating system. The heating system of steel can be divided into one-stage heating system, two-stage heating system, three-stage heating system and multi-stage heating system.

对于不同钢种，加热工艺包括：钢的加热温度、断面允许温差、加热速度以及炉温制度和供热制度，后两项统称为加热制度。钢的加热制度按炉内温度随时间的变化，可以分为一段式加热制度、二段式加热制度、三段式加热制度和多段式加热制度。

The one-stage heating system is to heat the steel in the furnace with the furnace temperature basically unchanged, which is characterized by large temperature difference between the furnace temperature and the steel surface, fast heating speed, short heating time, simple furnace structure and operation. The disadvantage is that the exhaust gas temperature is high and the thermal utilization rate is poor. Because there is no preheating period and soaking period, it is only suitable for heating steel or hot charging steel with small section size, good thermal conductivity and plasticity.

一段式加热制度是把钢料放在炉温基本不变的炉内加热，特点是炉温和钢料表面的温差大、加热速度快、加热时间短、炉子结构和操作简单。缺点是废气温度高、热利用率差，因没有预热期和均热期，只适合加热断面尺寸小、导热性好、塑性好的钢料或热装钢料。

The two-stage heating system is to heat the steel in two different temperature regions successively, which is composed of heating period and soaking period or preheating period and heating period. The two-stage heating system consisting of heating period and soaking period is to directly load the ingot into a high-temperature furnace for heating, characterized by fast heating speed, small cross-section temperature difference, high temperature of exhaust gas from the furnace, low heat utilization rate, which is only suitable for heating the steel with good thermal conductivity and low temperature stress. The two-stage heating system consists of preheating period and heating period. The temperature of exhaust gas from the furnace is low, and the heating speed of metal is slow. Because the temperature difference between the center and the surface is small, some steel with poor thermal conductivity is heated in the preheating period (with small strength stress). After the temperature rises into the plastic state of the steel, it is heated quickly in the high temperature area. Because there is no soaking period, the temperature on the section cannot be guaranteed, so it can not be used to heat billets with large section.

二段式加热制度是使钢料先后在两个不同的温度区域内加热,由加热期和均热期组成或由预热期和加热期组成。由加热期和均热期组成的二段式加热制度是把钢锭直接装入高温炉膛进行加热,特点是加热速度快、断面温差小、出炉废气温度高、热利用率低,只适合加热导热性好、快速加热温度应力小的钢料。由预热期和加热期组成的二段式加热制度,出炉废气温度低,金属的加热速度较慢,因为中心与表面的温差小,一些导热性差的钢先在预热段加热(强度应力小),待温度升高进入钢的塑性状态后再到高温区域进行快速加热,因没有均热期最终不能保证断面上温度的均匀性,所以不能用于加热断面大的钢坯。

The three-stage heating system is to heat the steel in three regions (or periods) with different temperature conditions, namely preheating period, heating period and soaking period. It combines the advantages of the above two heating systems. The billet is first preheated in the low temperature area. At this time, the heating speed is relatively slow and the temperature stress is small, which will not cause danger. When the central temperature of the metal exceeds 500℃, it enters the plastic range, and then it can be heated rapidly until the surface temperature rises rapidly to the temperature required by the furnace. At the end of the heating period, there is a large temperature difference on the metal cross section, which needs to enter the per capita heat period for soaking. At this time, the surface temperature of the steel will not rise any more, but the center temperature will rise gradually, reducing the temperature difference on the section. The three-stage heating system not only considers the risk of temperature stress in the early stage of heating, but also considers the rapid heating in the middle stage and the uniformity of the final temperature, taking into account both the output and the quality. When this heating system is used in the continuous heating furnace, because of the preheating section, the temperature of exhaust gas from the furnace is low, the utilization of heat energy is good, and the unit fuel consumption is low. The heating section can strengthen heating, fast heating, reduce oxidation and decarbonization, and ensure high productivity of furnace. This heating system is relatively perfect and reasonable, which is suitable for heating various sizes of carbon and alloy billets.

三段式加热制度是把钢料放在三个温度条件不同的区域(或时期)内加热,依次是预热期、加热期、均热期,它综合了以上两种加热制度的优点。钢坯首先在低温区域进行预热,这时加热速度比较慢,温度应力小,不会造成危险。等到金属中心温度超过 500℃ 以后,进入塑性范围,这时就可以快速加热,直到表面温度迅速升高到出炉所要求的温度。加热期结束时,金属断面上还有较大的温度差,需要进入均热期进行均热。此时钢的表面温度基本不再升高,而使中心温度逐渐上升,缩小断面上的温度差。三段式加热制度既考虑了加热初期温度应力的危险,又考虑了中期快速加热和最后温度的均匀性,兼顾了产量和质量两方面。在连续加热炉上采用这种加热制度时,由于有预热段,出炉废气温度较低,热能的利用较好,单位燃料消耗低。加热段可以强化供热,快速加热,减少了氧化与脱碳,并保证炉子有较高的生产率。这种加热制度是比较完善与合理的,适用于加热各种尺寸的碳素钢坯及合金钢坯。

The multi-stage heating system is used in the heat treatment process of some steel materials, including several heating, soaking (heat preservation) and cooling sections; it can also refer to the multi-point heating and multi section heating situation adopted in the modern large continuous heating furnace due to the large heating capacity. For the continuous heating furnace, although the multi-stage heating includes the first heating section and the second heating section in addition to the preheating section and the soaking section, it still belongs to the three-stage heating system from the point of view of heating system.

多段式加热制度用于某些钢料的热处理工艺中，包括几个加热、均热（保温）、冷却段；也可指现代大型连续加热炉中，由于加热能力大而采用的多点供热多区段加热的情况。对于连续式加热炉来说，多段式加热虽然除预热段和均热段外还包括第一加热段、第二加热段等，但从加热制度的观点上说仍属于三段式加热制度。

2.2.3.6 Heating Defects of Steel
2.2.3.6 钢的加热缺陷

In the heating process of steel, the temperature and atmosphere of the furnace must be properly adjusted. If the operation is improper, various heating defects will appear, such as oxidation, decarbonization, overheating, overburning, etc. These defects affect the heating quality of steel and even cause waste products, so we should try our best to avoid them in the heating process.

钢在加热过程中，炉子的温度和气氛必须调整得当，如果操作不当，会出现各种加热缺陷，如氧化、脱碳、过热、过烧等。这些缺陷影响钢的加热质量，甚至造成废品，所以加热过程中应尽力避免。

(1) Oxidation of the steel.

(1) 钢的氧化。

1) Formation of the oxide scale.

1）氧化铁皮的生成。

Steel can also be oxidized and rusted at room temperature, but the oxidation process is very slow. When the temperature continues to rise, the oxidation speed will be accelerated. When the temperature is above 1000℃, the oxidation will begin to be intense. When the temperature is over 1300℃, the oxidation is more intense. If the burning loss at 900t is taken as 1, it will be 2 at 1000℃, 3.5 at 1100℃, 7 at 1300℃. The oxidation process is the result of chemical reaction between the oxidizing gases (O_2, CO_2, H_2O, SO_2) in the furnace gas and the iron on the surface layer of steel. According to the different degree of oxidation, several kinds of iron oxides—FeO, Fe_3O_4 and Fe_2O_3 are formed. The formation process of iron oxide scale is also the diffusion process of oxygen and iron. Oxygen diffuses from the surface to the inside of iron, while iron diffuses to the outside. When the concentration of outer oxygen is large and the concentration of iron is small, the high valent oxide of iron is formed; when the concentration of inner iron is large and the concentration of oxygen is small, the low valent oxide of iron is formed. Therefore, the structure of iron oxide scale is actually layered. FeO is the closest to the iron layer, and Fe_3O_4 and Fe_2O_3 are successively outward. The proportion of each layer is about 40% FeO, 50% Fe_3O_4 and 10% Fe_2O_3. The melting point of such scale is about 1300~1350℃.

钢在常温下也会氧化生锈，但氧化进行得很慢。温度继续升高后氧化的速度加快，到了1000℃以上，氧化开始激烈进行。当温度超过1300℃以后，氧化进行得更加剧烈。如果以900t时烧损值作为1，则1000℃时为2，1100℃时为3.5，1300℃时为7。氧化过程是炉气内的氧化性气体（O_2、CO_2、H_2O、SO_2）和钢的表面层的铁进行化学反应的结果。根据氧化程度的不同，生成几种不同的铁的氧化物——FeO、Fe_3O_4、Fe_2O_3。氧化铁皮的形成过程也是氧和铁两种元素的扩散过程，氧由表面向铁的内部扩散，而铁则向外部扩散。外层氧浓度大，铁的浓度小，生成铁的高价氧化物；内层铁的浓度大而氧的浓度小，生成铁的低价氧化物。所以氧化铁皮的结构实际上是分层的，最靠近铁层的是FeO，依次向外是Fe_3O_4和Fe_2O_3。各层大致的比例是FeO占40%，Fe_3O_4占50%，Fe_2O_3占10%。这样的氧化铁皮其熔点约在1300~1350℃。

2) Factors affecting oxidation.

2) 影响氧化的因素。

①The effect of heating temperature. When the steel is heated, the higher the furnace temperature is, and the amount of iron oxide scale generated when the heating time is constant is more, because with the temperature rising, the diffusion speed of various components in the steel increases. It is pointed out that the relationship between the amount of scale formation and temperature and time is as follows:

①加热温度的影响。钢在加热时，炉温越高，而加热时间不变的情况下所生成的氧化铁皮量越多，因为随着温度的升高，钢中各成分的扩散速度加快。研究指出，氧化铁皮生成量与温度和时间的关系为：

$$W = a\sqrt{\tau}e^{\frac{b}{T}}$$

Where　a, b——Constant;

　　　　τ——Heating time;

　　　　T——The surface temperature of the steel.

式中　a, b——常数;

　　　τ——加热时间；

　　　T——钢的表面温度。

②The effect of heating time. It can be seen from the above formula that the longer the steel is heated, the more oxide scales are formed, and the more oxide scales are formed at high temperature.

②加热时间的影响。由上式可知，钢加热时间越长生成氧化铁皮越多，高温下生成氧化铁皮更多。

③Influence of furnace gas composition. The composition of furnace gas generally includes CO_2, CO, H_2O, H_2, O_2, N_2, etc. according to the different fuels, there are also 2, 4 and other gases, among which H_2O and O_2 have large oxidation capacity, and their concentration directly affects the formation of iron oxide scale.

③炉气成分的影响。炉气成分一般包括CO_2、CO、H_2O、H_2、O_2、N_2等，根据燃料的不同还存在SO_2、CH_4等气体，其中H_2O、O_2氧化能力较大，其浓度大小直接影响到氧

化铁皮的生成多少。

④Influence of chemical composition of steel. When the carbon content in the steel is large, the burning loss rate of the steel will decrease; when the steel contains elements such as Cr, Ni, Si, Mn, Al, etc., because these elements can form a very dense oxide film after oxidation, which hinders the outward diffusion of metal atoms or ions, the oxidation speed will be greatly reduced.

④钢的化学成分的影响。钢中碳含量大时,钢的烧损率有所下降;钢中含有 Cr、Ni、Si、Mn、Al 等元素时,由于这些元素氧化后能生成很致密的氧化膜,这样就阻碍了金属原子或离子向外扩散,结果使氧化速度大为降低。

3) Measures to reduce steel oxidation.

3) 减少钢氧化的措施。

The factors affecting oxidation are as mentioned above, in which the composition of steel is a fixed factor, so other factors are the main factors to reduce the amount of oxidation loss. Specific measures are as follows:

影响氧化的因素如上所述,其中钢的成分是固定的因素,因此要减少氧化烧损量主要从其他因素着手。具体措施有如下几种:

①According to the heating process, the furnace temperature and heating time shall be strictly controlled to reduce the residence time of the steel in the high-temperature area, and the high-temperature steel shall not be produced. The heat preservation steel to be rolled must be cooled to be rolled.

①根据加热工艺严格控制炉温,严格控制加热时间,减少钢在高温区域的停留时间,不出高温钢,该保温待轧的必须降温待轧。

②Control the atmosphere in the furnace. Under the premise of ensuring complete combustion, reduce the air consumption coefficient. Strictly control the furnace pressure, ensure the tightness of the furnace body, reduce the cold air suction, and especially reduce the cold air suction in the high temperature area of the furnace. In addition, the water content in the fuel should be minimized.

②控制炉内气氛。在保证完全燃烧的前提下,降低空气消耗系数。严格控制炉压力,保证炉体的严密性,减少冷空气吸入,特别是减少炉子高温区吸入冷空气。此外,还应尽量减少燃料中的水分等。

③Take special measures, such as direct heating with little or no oxygen. Its basic principle is to use a small air consumption coefficient in the high-temperature section, and supply necessary air in the low-temperature section, so that the combustion of incomplete combustion components is complete.

③采取特殊措施,如采用少或无氧直接加热。其基本原理是高温段采用小的空气消耗系数,而在低温段则供入必要的空气,使不完全燃烧的成分燃烧完全。

(2) Decarbonization of the steel.

(2) 钢的脱碳。

1) Reason for decarbonization.

1) 脱碳的原因。

Decarbonization refers to the phenomenon that the carbon content of a layer on the steel surface decreases during the heating process of the steel in the high temperature furnace. Carbon exists in the form of Fe_3C in steel, which is the component that directly determines the mechanical properties of steel. After decarbonization of steel surface, the mechanical properties will change, especially for high carbon steel, such as tool steel, ball bearing steel, spring steel, etc. Decarbonization is not expected, so decarbonization is considered as a defect of steel, and will be scrapped if it is serious. Decarbonization and oxidation of steel occur simultaneously and promote each other. If the depth of decarbonization layer is greater than that of oxide layer, the damage will be great. The decarbonization process of steel is the result of the reaction of H_2O, CO_2, H_2, and O_2 in furnace gas and Fe_3C in steel. Among these gas components, the decarbonization capacity of H_2O is the strongest, followed by CO_2, O_2, H_2. The oxidation and decarbonization of steel occur together at high temperature. The formation of iron oxide scale is helpful to restrain decarbonization and slow the diffusion. When dense iron oxide scale is formed on the surface of steel, it can hinder the development of decarbonization.

钢料在高温炉内加热过程中,钢表面一层碳含量降低的现象称为脱碳。碳在钢中以Fe_3C的形式存在,它是直接决定钢的力学性能的成分。钢表面脱碳后将引起力学性能发生变化,特别是高碳钢,如工具钢、滚珠轴承钢、弹簧钢等都不希望发生脱碳现象,因此脱碳被认为是钢的缺陷,严重时将予以报废。钢的脱碳与它的氧化是同时发生的,并且相互促进。若钢的脱碳层深度大于氧化层深度,危害就大了。钢的脱碳过程是炉气中的H_2O、CO_2、H_2、O_2和钢中的Fe_3C反应的结果,在这些气体成分中H_2O脱碳能力最强,其次为CO_2、O_2、H_2。高温下钢的氧化和脱碳是相伴发生的,氧化铁皮的生成有助于抑制脱碳,使扩散趋于缓慢,当钢的表面生成致密的氧化铁皮时,可以阻碍脱碳的发展。

2) Factors affecting decarbonization.

2) 影响脱碳的因素。

①The effect of heating temperature. For most steel grades, with the increase of temperature, it can be seen that the decarbonization layer almost increases in a straight line; for some steel grades, because the oxidation speed is greater than the decarbonization speed after a certain high temperature, the decarbonization layer will not increase but decrease after a certain high temperature.

①加热温度的影响。对多数钢种来说,随着温度的增加,可见脱碳层几乎呈直线增加;有的钢种因一定高温后氧化速度大于脱碳速度,脱碳层会在一定高温后不再增加而是减少。

②The effect of heating time. At low temperature, decarbonization is not significant even if the steel stays in the furnace for a long time. The longer the steel stays at high temperature, the thicker the decarbonization layer. Some easily detachable carbon steels are not allowed to be kept warm for rolling under high temperature for a long time. In case of too long rolling stop time due to failure, the billet in the furnace shall be removed from the outside of the furnace.

②加热时间的影响。在低温条件下即使钢在炉内时间较长,脱碳也不显著,在高温下

停留的时间越长，则脱碳层越厚。一些易脱碳钢不允许长时间在高温下保温待轧，遇到故障停轧时间过长时应把炉内钢坯退出炉外。

③Influence of furnace gas composition. It can be seen from the decarbonization process of steel that if there are H_2O, CO_2, O_2 and H_2 in the furnace gas, the steel must be decarburized, and the furnace gas is decarburized atmosphere. The concentration of these gases in the furnace gas is one of the main factors affecting the decarbonization speed. The content of these gases depends on the fuel type, combustion method, air consumption coefficient, furnace pressure, etc. It has been proved that the smallest visible decarbonization layer is obtained in an oxidizing atmosphere rather than a reducing atmosphere.

③炉气成分的影响。从钢的脱碳过程可以看出，若炉气中存在着 H_2O、CO_2、O_2 和 H_2，则钢必然脱碳，炉气都是脱碳气氛的，炉气中这几种气体的浓度大小是影响脱碳速度快慢的主要因素之一。而这些气体的含量决定于燃料种类、燃烧方法、空气消耗系数、炉膛压力等。实践证明，最小的可见脱碳层是在氧化性气氛中而不是在还原性气氛中得到的。

④Influence of composition of steel. The higher the carbon content of steel, the easier decarbonization of steel. Alloy elements have different effects on decarbonization, aluminum, diamond and tungsten can promote decarbonization, while chromium, manganese and boron can reduce decarbonization of steel. The easily decarburized steels include carbon tool steel, mould steel, high speed steel, etc.

④钢的成分的影响。钢的碳含量越高，钢的脱碳越容易。合金元素对脱碳的影响不一，铝、钴、钨这些元素能促使脱碳；铬、锰、硼则减少钢的脱碳。易脱碳的钢种有碳素工具钢、模具钢、高速钢等。

3）Measures to reduce decarbonization of steel.

3）减少钢脱碳的措施。

The above measures to reduce the oxidation of steel are basically applicable to reduce decarbonization. For example, fast heating can shorten the stay time of steel in the high temperature area; heating temperature can be selected correctly to avoid the range of decarbonization peak value which is easy to decarburize; atmosphere in the furnace can be adjusted and controlled properly to keep the oxidation atmosphere in the furnace, so that the oxidation speed is higher than the decarbonization speed; adopting reasonable furnace structure, the walking furnace is the best choice for easy decarbonization, because it can control the temperature of steel in the furnace. Once the rolling mill stops rolling due to failure, all the billets in the furnace can be withdrawn in time. Decarbonization is less important than oxidation for general steel grades, but attention should be paid to heating easy decarbonization and some heat treatment processes.

前述减少钢的氧化的措施基本适用于减少脱碳。例如进行快速加热，缩短钢在高温区域停留的时间；正确选择加热温度，避开易脱碳的脱碳峰值范围；适当调节和控制炉内气氛，对易脱碳钢使炉内保持氧化气氛，使氧化速度大于脱碳速度；采取合理的炉型结构，易脱碳钢最好采用步进式炉，因为它可以控制钢在高温区的停留时间，一旦轧机因故障停轧，可以把炉内全部钢坯及时退出。脱碳问题对一般钢种来说，比起氧化的问题是次要的,只是加热易脱碳钢和某些热处理工艺需要注意。

(3) Overheating and overburning of the steel.

(3) 钢的过热和过烧。

When the heating temperature of steel exceeds the critical heating temperature, the grains of steel begin to grow, that is to say, the steel overheats. Grain coarsening is the main feature of overheating. The grain grows too large, the mechanical properties of steel decline, and cracks are easy to occur during processing.

钢的加热温度超过临界加热温度时,钢的晶粒就开始长大,即出现钢的过热。晶粒粗化是过热的主要特征,晶粒过分长大,钢的力学性能下降,加工时容易产生裂纹。

Heating temperature and heating time have a decisive influence on grain growth. The higher the heating temperature and the longer the heating time, the more obvious the phenomenon of grain growth. In the heating process, the heating temperature and the residence time of the steel in the high temperature area should be mastered. In addition, most of the alloy elements can reduce the trend of grain growth, only carbon, phosphorus and manganese can promote the grain growth.

加热温度与加热时间对晶粒的长大有决定性的影响,加热温度越高、加热时间越长,晶粒长大的现象越显著;在加热过程中,应掌握好加热温度及钢在高温区域的停留时间。另外合金元素大多数是可以减少晶粒长大趋势的,只有碳、磷、锰会促进晶粒的长大。

When the steel is heated to a temperature higher than the superheat temperature, not only the grain of the steel grows, but the film around the grain begins to melt. Oxygen enters into the gap between the grains, which causes the metal to oxidize, and promotes its melting. As a result, the binding force between the grains is greatly reduced, and the plasticity is deteriorated. In this way, the steel will crack in the process of pressure processing, which is called overburning.

当钢加热到比过热更高的温度时,不仅钢的晶粒长大,晶粒周围的薄膜开始熔化,氧进入晶粒之间的间隙,使金属发生氧化,又促进了它的熔化,导致晶粒间彼此结合力大为降低,塑性变坏,这样钢在进行压力加工过程中就会裂开,这种现象就是过烧。

As with overheating, overburning is often caused by too long staying time in the high temperature area, such as rolling line failure, roll change, etc., In this case, measures should be taken in time. In addition, overburning not only depends on the heating temperature, but also on the atmosphere in the furnace. The stronger the oxidation capacity of furnace gas is, the more prone to overburning. In reducing atmosphere, overburning may also occur, but the temperature at the beginning of overburning is $60 \sim 70$℃ higher than that in oxidizing atmosphere. The higher the carbon content in the steel, the lower the temperature of overburning danger.

与过热相同,发生过烧往往也是由于在高温区域停留时间过长的缘故,如轧线发生故障、换辊等,遇到这种情况要及时采取措施。另外,过烧不仅取决于加热温度,也和炉内气氛有关。炉气的氧化能力越强,越容易发生过烧现象,在还原性气氛中,也可能发生过烧,但开始过烧的温度比氧化性气氛要高 $60 \sim 70$℃。钢中碳含量越高,产生过烧危险的温度越低。

The overheated steel can be reheated for pressure processing. The overheated steel can not be reheated in the furnace, but only smelted as scrap.

已经过热的钢可以重新加热进行压力加工，过烧的钢不能重新回炉再加热，只有作为废钢重新冶炼。

2.2.3.7 Production Capacity of Heating Furnace
2.2.3.7 加热炉的生产能力

(1) Expression method of production capacity of heating furnace.
(1) 加热炉生产能力的表示方法。

Furnace productivity is an indicator indicating the production capacity of the furnace, i.e. heating metal plate per unit time (unit: t/h or kg/h). For example, the production capacity of heating furnace in small continuous rolling mill is generally 90t/h.

炉子生产率是表示炉子生产能力大小的指标，即单位时间加热金属板（单位为 t/h 或 kg/h）。如小型连轧厂加热炉生产能力一般为 90t/h。

Furnace bottom strength is the amount of metal heated per unit furnace bottom area in unit time (unit: $kg/(m^2 \cdot h)$), which can be used to compare the production capacity of different furnaces. It can be expressed in two ways: one is the strength of the steel pressure furnace bottom, the other is the effective furnace bottom strength. The difference between them is that the bottom area of the former refers to that part of the area pressed by steel, and the bottom area of the latter refers to the whole effective bottom area. If the furnace productivity is G, the steel pressure area or effective furnace bottom area is A, the furnace bottom strength is:

炉底强度是单位时间内单位炉底面积所加热的金属量（单位为 $kg/(m^2 \cdot h)$），可用来比较不同炉子的生产能力。它有两种表示方法：一种是钢压炉底强度，另一种是有效炉底强度。两者之间的区别是前者的炉底面积是指钢压住的那一部分面积，后者的炉底面积是整个有效炉底面积。假设炉子生产率为 G，钢压面积或有效炉底面积为 A，则炉底强度为：

$$P = G/A$$

(2) Main factors affecting furnace capacity.
(2) 影响炉子能力的主要因素。

The main factors affecting furnace capacity are:
影响炉子能力的主要因素有：

1) Process factors. Process factors such as operation cycle, heating type, furnace temperature of steel material, tapping temperature, heating uniformity, process heat preservation, etc. determine the adoption of different furnace types and different heating processes, and determine the different production capacity of the furnace. For example, the bottom strength of energy-saving walking beam furnace for continuous small bar mill can reach $600 \sim 650 kg/(m^2 \cdot h)$, while the bottom strength of walking beam furnace for thin plate continuous rolling is only $370 \sim 560 kg/(m^2 \cdot h)$.

1）工艺因素。作业周期、加热品种、钢料入炉温度、出钢温度、加热均匀性、工艺保温等工艺因素，决定了不同炉型和不同加热工艺的采用，决定了炉子的不同生产能力。如连续小型棒材轧机用节能型步进梁式炉的炉底强度可达 $600 \sim 650 kg/(m^2 \cdot h)$，而薄板连轧用的步进式炉炉底强度只有370~560kg/(m²·h)。

2) Thermal factors. When the process factors are fixed, the heating load, temperature system, furnace pressure system, heating system, furnace heat exchange, waste heat utilization and other thermal factors of the furnace play a key role in the production capacity of the furnace. When the rolling line needs, the furnace capacity can be improved by increasing the heating load holding temperature system, or the furnace temperature can be appropriately increased to increase the furnace output and heating load. When increasing the furnace temperature, it is necessary to consider whether the furnace heat exchange is normal, whether the furnace pressure can maintain normal, and whether the heat exchanger can adapt.

2) 热工因素。工艺因素一定时,炉子的供热负荷、温度制度、炉压制度、供热制度、炉膛热交换、炉子余热利用等热工因素对炉子生产能力的大小起着关键性的作用。在轧线需要时,可通过提高供热负荷保持温度制度提高炉子能力,也可以适当提高炉温提高炉子产量,提高供热负荷。提高炉温时又必须考虑炉子热交换是否正常,炉压是否能够维持正常,换热器等是否能够适应。

3) Other factors. In the case of certain above factors, the temperature of furnace in and out, the mechanization of furnace and the level of automation equipment directly affect the capacity of furnace.

3) 其他因素。在上述因素一定的情况下,进出炉温度、炉子的机械化和自动化装备水平等直接影响炉子能力。

2.2.4 Thermal System
2.2.4 热工制度

2.2.4.1 Daily Work Requirements
2.2.4.1 日常工作要求

(1) The operation shall be adjusted according to the heating system according to the change of production rhythm and variety specifications.

(1) 操作中应根据生产节奏和品种规格的变化,按加热制度进行调节。

(2) During the work, the requirements of frequent inspection, adjustment and contact shall be strictly implemented, and the air-fuel ratio shall be adjusted in time to make the fuel fully burn in each section, so as to ensure the reduction of fuel consumption and the reduction of billet oxidation and burning loss.

(2) 工作中严格执行勤检查、勤调整、勤联系的要求,及时调节空燃比,使燃料在各段达到完全燃烧,确保降低燃料消耗,减少钢坯氧化烧损。

(3) All kinds of heating parameters transmitted to the IPC shall be accurate, and the parameter control of each section shall be stable, so as to make the IPC in good operation state.

(3) 向工控机输入的各种加热参数应准确,各段参数控制应稳定,使工控机处于良好的运行状态。

(4) Check according to the requirements of spot check, deal with the problems found in a

timely manner, and report the problems that cannot be dealt with by the post level by level, and make records. The existing problems shall be inspected, the causes shall be analyzed and preventive and corrective measures shall be formulated. After the problems are solved, they shall be verified and recorded to form a closed loop.

(4) 按照点检的要求进行检查，对发现的问题能处理的应及时处理，本岗位处理不了的应逐级汇报，并做好记录。对存在的问题应加强检查，分析原因并制定预防纠正措施，问题解决以后进行验证，记录形成闭环。

(5) The original records shall be filled in (or printed) accurately, completely and clearly as required.

(5) 班中应按要求准确、完整、清楚地填写（或打印）有关原始记录。

(6) Strictly implement the heating system and the system to be rolled. When heating up, first increase, add down, and then increase, add up and add down; when cooling down, first decrease, add up and add down, then decrease, add up and add down, and finally reduce the heat.

(6) 严格执行加热制度和待轧制度，升温时先升均热、一加、下加，待轧顺时再升二加、侧加；降温时先降侧加、二加，然后再降下加、一加，最后降均热。

(7) When the coal quality is poor, the air-fuel ratio is below 1.3, or the rolling rhythm is too fast, and the temperature of the steel out of the furnace can not meet the production requirements, timely feedback the scheduling, suggest the production workshop to control the rolling rhythm, and make a record of the waiting temperature.

(7) 当煤质差、空燃比在1.3以下或轧制节奏太快、出炉钢温不能满足生产要求时，及时反馈调度，建议生产车间控制轧制节奏，并做好待温记录。

(8) In the process of heating, we should pay close attention to the temperature of each section of furnace and steel temperature, and compare them. When the furnace temperature exceeds the heating requirements, corrective measures shall be taken immediately, and "△" shall be marked on the top of the record to indicate the reason.

(8) 加热过程中应密切关注各段炉温和钢温的情况，并加以比较。当炉温超过加热要求时，应立即采取纠正措施，并在记录上方打上"△"，注明原因。

(9) Adjust the flue ram frequently. It is strictly prohibited to have fire at the furnace head and tail or absorb air. It is better to control the furnace pressure at 10~30Pa.

(9) 勤调节烟道闸板，严禁炉头炉尾冒火或吸风，炉膛压力控制在10~30Pa为宜。

(10) No matter in normal production or accident shutdown, the air butterfly valve in front of the burner shall not be closed. In normal production, the air disc valve in front of the burner shall be fully opened, and the air disc valve in front of the burner not used shall be kept at 1/5 opening.

(10) 无论正常生产还是事故停产时，烧嘴前的空气碟阀均不得关死。正常生产时，所使用烧嘴前的空气碟阀应全开，不使用烧嘴前的空气碟阀应保留1/5开度。

2.2.4.2　Gas Delivery Procedure of Heating Furnace
2.2.4.2　加热炉送煤气程序

(1) For the newly built or modified furnace, the gas pipeline system, valve and flange shall be tested for leakage to ensure that there is no leakage.

(1) 对新建的或改造的炉子,应对煤气管道系统、阀门、法兰进行试漏,确保严密无漏气。

(2) Check one by one to confirm that all gas burner valves must be closed.

(2) 逐一检查确认所有的煤气烧嘴阀门必须处于关闭状态。

(3) Check that the gas release valve of each section must be in full open state.

(3) 检查各段煤气放散阀必须处于全开状态。

(4) Each section of gas, air actuator valve position must be kept a certain degree of opening.

(4) 各段煤气、空气执行器阀位必须保留一定的开度。

(5) Fully open the smoke gate and start the fan.

(5) 全打开烟闸,启动风机。

(6) Before gas delivery, the gas pipeline shall be cleaned with nitrogen, and the gas can be delivered only after the air in the pipeline is discharged, and the gas shall be delivered to the furnace head.

(6) 送煤气前应先用氮气清扫煤气管道,将管道内的空气排干净后方可送煤气,并要把煤气送到炉头。

2.2.4.3　Heating Furnace Ignition Procedure
2.2.4.3　加热炉点火程序

(1) Prepare the torch before ignition, check that the gas and air pressure must be in normal state and the water cooling system is normal.

(1) 点火前准备好火把,检查煤气和空气压力必须处于正常状态,水冷系统正常。

(2) Before ignition, take samples from the test cylinder at the test valve at the end of the gas pipeline for gas explosion test, and the ignition can be started only after the test is qualified.

(2) 点火前在煤气管道末端试验阀处用试验筒取样做煤气爆发试验,试验合格后方可点火。

(3) During the ignition operation, there must be a special person to command, one person to hold the torch, one person to open the valve, one person to contact.

(3) 点火作业时,必须有专人指挥,一人执火把,一人开阀门,一人联系。

(4) Open 1/5 of the air valve of the burner to supply air to the furnace.

(4) 烧嘴空气阀门开1/5往炉内送风。

(5) Send open fire to the furnace, about 100mm away from the specified burner brick.

(5) 往炉内送明火，距离指定烧嘴砖约 100mm。

(6) Slowly open the gas cock in front of the burner until it is ignited.

(6) 缓慢打开烧嘴前煤气旋塞阀直至点燃。

(7) If the burner cannot be ignited or is ignited and extinguished, the ignition shall be stopped, the gas valve of the burner shall be closed immediately, the cause shall be found out, the gas valve shall be emptied for 15min after treatment, and then the ignition shall be carried out according to the above steps.

(7) 如果烧嘴点不着或点着又灭，则停止点火，立即关闭该烧嘴煤气阀门，查明原因，处理完毕后排空 15min，再按上述步骤点火。

(8) After the burner is ignited, properly adjust the air-fuel ratio to make the burner burn normally.

(8) 烧嘴点燃后，适当调整空燃比，使烧嘴燃烧情况达到正常。

(9) Close the gas vent valve of each section.

(9) 关闭各段煤气放散阀。

(10) The burners must be ignited one by one, and the adjacent burners must be supervised by a specially assigned person before ignition. After all burners are ignited, adjust one by one, and switch to the industrial computer control after the combustion is normal.

(10) 烧嘴必须逐个点燃，有临近的烧嘴必须有专人监护方可引燃。全部烧嘴点燃后逐个调节，待燃烧正常后切换至工控机控制。

(11) Adjust the position of flue gate to keep the furnace micro positive pressure.

(11) 调整烟道闸板位置，保持炉膛微正压。

2.2.4.4 Shutdown the Procedure of Heating Furnace
2.2.4.4 加热炉闭火程序

(1) Close all gas burner valves. After confirming the correct closing one by one, open all flue rams.

(1) 关闭所有煤气烧嘴阀门。逐一确认关闭无误后，全部打开烟道闸板。

(2) The valve position of each section of gas and air actuator must be kept at a certain opening, and the manual operators are all in the manual state.

(2) 各段煤气、空气执行器阀位必须保持一定的开度，手操器全处于手动状态。

(3) After receiving the notice of closing the main gas valve, first connect the nitrogen gas pipe with the nitrogen gas pipe to supply nitrogen, then open the gas relief valve of each section to purge the gas in the pipe, and then close the nitrogen valve and block the blind plate at the gas pipe valve after confirming that the purging is clean.

(3) 接到关闭煤气总阀门通知后，先用氮气管接通煤气管送氮气，然后打开各段煤气放散阀对管道内的煤气进行吹扫，确认吹扫干净后，关闭氮气阀门，在煤气管阀门处堵盲板。

(4) Open the hot air vent valve.

(4) 打开热风放散阀。

(5) The fan can be stopped only when the furnace temperature drops below 500℃.

(5) 炉温降到500℃以下方可停风机。

2.2.4.5　Process Monitoring
2.2.4.5　过程监控

(1) Check the structure of the furnace, the tightness of gas and air valves and the flexibility of burner valves on time, and take measures to deal with any problems found.

(1) 按时检查炉体结构，以及煤气、空气阀门的严密性和烧嘴阀门的灵活性，发现问题及时采取措施进行处理。

(2) Check the operation of bottom water pipe and water beam on time to ensure the outlet water temperature is less than 55℃.

(2) 按时检查炉底水管及水梁的运行情况，确保出水温度小于55℃。

(3) Observe the gas and air pressure at any time. If the gas pressure is not stable, contact the dispatcher in time to understand the cause and development trend. For example, when the pressure of the gas main before adjustment is lower than 4000Pa, the emergency plan of the air inlet is on alert. When the pressure of the air main is lower than 5000Pa, check whether the fan operates normally, whether the air inlet is blocked, whether the valves on the pipeline are closed, and whether the hot air is released whether the valve is closed.

(3) 随时观察煤气、空气压力情况，如果煤气压力不稳定应及时与调度联系，了解原因及发展趋势，如煤气总管调前压力低于4000Pa时即进入事故预案戒备状态，空气总管压力低于5000Pa时应检查风机运行是否正常，进风口有无堵塞、管道上各阀门有无关闭，热风放散阀是否关闭。

(4) Regularly check the combustion condition of the burner and the temperature before and after the heat exchanger, and take measures in time in case of any abnormality.

(4) 定期检查烧嘴燃烧情况和换热器前后温度，发现异常及时采取应对措施。

2.2.4.6　Emergency Measures
2.2.4.6　应急措施

(1) When the pressure of main gas pipe before adjustment is lower than 4000Pa, the adjustment of each section shall be switched to manual state. When the gas pressure is reduced, the air pressure shall be reduced in the same proportion. The air control valve shall keep at least 1/5 of the valve position.

(1) 当煤气总管调前压力低于4000Pa时，各段调节转换至手动状态，在煤气压力降低的同时要同比例降低空气压力，空气调节阀应保留至少1/5的阀位。

(2) The pressure of gas main pipe shall be no less than 3200Pa before adjustment and no less than 2800Pa after adjustment. If the pressure drops to the lowest limit and there is a trend of continuous decline, the number of burners shall be determined according to the falling range val-

ue; if the pressure of coal before adjustment is less than 2500Pa, the alarm shall be started immediately and all burners shall be closed.

(2) 煤气总管压力应保证调前不小于 3200Pa、调后不小于 2800Pa, 如果压力下降到最低限并有继续下降的趋势时, 视下降的幅度值确定关闭烧嘴的个数; 调前煤压低于 2500Pa 时, 立即启动报警, 关闭全部烧嘴。

(3) When the coal pressure is not greater than 3200Pa before the adjustment, it is strictly forbidden to use the method of regulating the gas branch pipe actuator to limit the gas flow rate. First, reduce the air volume, then adjust the gas volume, and strictly prohibit the proportion imbalance, so as to ensure the normal combustion of the burner.

(3) 当调前煤压不大于 3200Pa 时, 严禁使用调节煤气支管执行器限制煤气流量的方法, 应先减少空气量, 后调煤气量, 严禁比例失调, 以保证烧嘴的正常燃烧。

(4) In case of steel sticking accident, the furnace can not be cooled, the furnace temperature in the soaking section can be properly increased, and the speed of steel tapping can be accelerated; the billets that have been bonded and cannot be opened shall be lifted away in time to avoid affecting normal steel tapping. After the accident is handled, the fire can be adjusted normally.

(4) 发生粘钢事故时, 炉子不能降温, 可适当提高均热段炉温, 加快出钢速度; 已经黏结而处理不开的坯料, 应及时吊走以免影响正常出钢。待事故处理好后, 方可正常调火。

2.2.4.7　Treatment of Gas Fire Accident
2.2.4.7　煤气着火事故的处理

(1) If the diameter of the gas pipeline is less than 150mm, the gas butterfly valve can be directly closed for flameout.
(1) 煤气管道直径在 150mm 以下, 可直接关闭煤气碟阀熄火。

(2) If the diameter of the gas pipeline is more than 150mm, turn down the gas butterfly valve gradually to reduce the gas pressure at the ignition point, but not less than 50~100Pa; after the fire is reduced, turn off the gas with nitrogen and do not turn off the gas dish suddenly (Note: when the fire accident lasts too long and the gas equipment is red, do not use water for cooling).

(2) 煤气管道直径在 150mm 以上, 应逐渐关小煤气碟阀, 降低着火处的煤气压力, 但不得低于 50~100Pa; 火势减小后, 再通入氮气熄火, 严禁突然关死煤气碟 (注意: 当着火事故时间太长、煤气设备烧红时, 不得用水冷却)。

(3) When the gas leaks, implement the prevention measures of the dangerous source point.
(3) 煤气泄漏时执行危险源点的预防措施。

(4) In case of power failure, water failure, gas failure and explosion-proof plate collapse, the emergency plan of heating furnace shall be implemented immediately.
(4) 停电、停水、停煤气、防爆板崩时应立即执行加热炉突发事故预案。

Task 2.3　Rolling
任务2.3　轧材轧制

Mission objectives
任务目标

(1) Familiar with the common pass system and its characteristics, and familiar with the structure and application of the guide and guard device commonly used in the rolling process;

(1) 熟悉螺纹钢常用孔型系统及其特点；熟悉螺纹钢轧制过程中常用的导卫装置的构造和用途；

(2) Familiar with different types of rolling mill structure, understand the roll changing mode and device, master the inspection and online adjustment of rolling mill;

(2) 熟悉不同类型的轧机结构，了解换辊方式及换辊装置，掌握轧机的检查与在线调整；

(3) Be able to select the correct pass system according to different products, be able to install, dismantle and adjust the guide rail on the rolling line;

(3) 能够根据不同的产品挑选正确的孔型系统，能够对轧线上的导卫进行安装、拆卸并调整；

(4) It can dismantle and replace the roll correctly, operate the rolling mill for rolling production, and operate the on-line pre-adjustment of the rolling mill.

(4) 能够正确拆卸和更换轧辊，操作轧机进行轧制生产，能够操作轧机在线预调整。

2.3.1　Rolling Process System
2.3.1　轧制工艺制度

Rolling process system mainly includes temperature system, rolling system and cooling system.

轧钢工艺制度主要包括温度制度、轧制制度和冷却制度。

2.3.1.1　Temperature System
2.3.1.1　温度制度

The purpose of heating the raw material before rolling is to improve the plasticity of the billet, reduce the deformation resistance and improve the internal structure and properties of the metal, so as to facilitate rolling. That is to say, the steel should be heated to the temperature range of the austenite single-phase solid solution structure, with higher temperature and enough time to homogenize the structure and dissolve the carbide, so as to obtain high plasticity and deformation metal structure with low shape resistance and good processability.

在轧钢之前要将原料进行加热，其目的在于提高钢坯的塑性，降低变形抗力及改善金属内部组织和性能，以便于轧制加工，即一般要将钢加热到奥氏体单相固溶体组织的温度范围内，并使之有较高的温度和足够的时间以均化组织及溶解碳化物，从而得到塑性高、变形抗力低、加工性能好的金属组织。

Generally, in order to better reduce the deformation resistance and improve the plasticity, the processing temperature should be as high as possible. However, high temperature and incorrect heating system may cause strong oxidation, decarburization, overheating, overburning and other defects of steel, reduce the quality of steel, and lead to scrap. Therefore, the heating temperature of steel should be determined mainly according to the characteristics of various steels and the requirements of pressure processing technology to ensure the quality and output of steel.

一般情况下为了更好地降低变形抗力和提高塑性，加工温度应尽量高一些好。但是高温及不正确的加热制度可能造成钢的强烈氧化、脱碳、过热、过烧等缺陷，降低钢的质量，导致废品。因此，钢的加热温度主要应根据各种钢的特性和压力加工工艺要求，从保证钢材质量和产量出发进行确定。

The temperature system defines the temperature range during rolling, i.e. the start rolling temperature and the finish rolling temperature. The start rolling temperature is the rolling temperature of the first pass in the rolling process, and the finish rolling temperature is the rolling temperature of the last pass. In continuous small-scale production, the start rolling temperature is generally 1050~1150℃, and the finish rolling temperature is generally 950~1000℃.

温度制度规定了轧制时的温度范围，即开轧温度和终轧温度。开轧温度是轧制过程中第一道次的轧制温度，终轧温度是轧制最后一道次的轧出温度。在连续小型生产中，开轧温度一般在1050~1150℃，终轧温度一般在950~1000℃。

The heating time of billet not only affects the production capacity of heating equipment, but also affects the quality of steel. Even if the heating temperature is not high, heating defects will be caused due to too long time. The reasonable heating time depends on the steel grade, size, loading and unloading temperature, heating speed and the performance and structure of heating equipment.

坯料的加热时间长短不仅影响加热设备的生产能力，同时也影响钢材的质量。即使加热温度不过高，也会由于时间过长而造成加热缺陷。合理的加热时间取决于原料的钢种、尺寸、装卸温度、加热速度及加热设备的性能与结构。

2.3.1.2 Rolling System
2.3.1.2 轧制制度

Rolling system mainly includes deformation system and speed system.
轧制制度主要包括变形制度和速度制度。

（1）Transverse deformation during rolling——spread.
（1）轧制过程中的横变形——宽展。

In rolling, the height direction of the rolled piece is compressed by the roller. The compressed metal will move to the longitudinal and transverse direction according to the law of minimum resistance, and the deformation of the width of the rolled piece caused by the volume moving to the transverse direction is called spreading. It is an important step to estimate the spread of rolling correctly to ensure the quality of cross section. In bar production, if the grid spread is larger than the actual spread, the hole filling is not enough, resulting in a large ovality; if the calculated spread is smaller than the actual spread, the hole filling is too full, forming ears. Both of the above conditions result in rolling waste.

在轧制中轧件的高度方向受到轧辊的压缩作用，压缩下来的金属，将按最小阻力定律移向纵向及横向，由移向横向的体积所引起的轧件宽度的变形称为宽展。正确估计轧制中的宽展是保证断面质量的重要一环。在棒材生产中，如果计算宽展大于实际宽展，则孔型充填不满，造成很大的椭圆度；如果计算宽展小于实际宽展，则孔型充填过满，形成耳子。以上两种情况均造成轧制废品。

Under different rolling conditions, the spreading form of billets in rolling process is different. According to the degree of freedom of metal flow along the transverse direction, the spread can be divided into free spread, limited spread and forced spread.

在不同的轧制条件下，坯料在轧制过程中的宽展形式是不同的。根据金属沿横向流动的自由程度，宽展分为自由宽展、限制宽展和强迫宽展。

Free spread refers to the possibility of free movement of the metal mass points of the volume of metal to be pressed in the rolling process to the two sides perpendicular to the rolling direction. At this time, the metal flow is not hindered or restricted by any other factors (such as the side wall of the pass) except the contact friction. The result shows that the width of the rolled piece increases. It is the simplest rolling condition because of its uniform deformation.

自由宽展是指坯料在轧制过程中，被压下的金属体积其金属质点横向移动时，具有向垂直于轧制方向的两侧自由移动的可能性。此时，金属流动除受接触摩擦的影响外，不受其他任何（如孔型侧壁等）的阻碍和限制，结果表现出轧件宽度尺寸的增大。自由宽展时变形比较均匀，它是最简单的轧制情况。

Limited spread refers to that in the rolling process, when the metal particles move laterally, they are not only affected by the contact friction, but also limited by the side wall of the pass, thus destroying the free flow conditions. The resulting spread is called limited spread. In general, the amount of the limited spread is smaller than that of the free spread.

限制宽展是指坯料在轧制过程中，金属质点横向移动时，除受接触摩擦的影响之外，还受孔型侧壁的限制，因而破坏了自由流动条件，这时产生的宽展称为限制宽展。限制宽展中形成的宽展量一般小于自由宽展时所形成的宽展量。

Forced spread means that in the rolling process, when the metal particles move laterally, they are not hindered by any obstacles, but strongly promoted, resulting in an additional increase in the width of the rolled piece. At this time, the generated spread is forced spread. The forced spread is larger than the free spread because of the conditions favorable to the lateral movement of metal particles.

强迫宽展是指坯料在轧制过程中，金属质点横向移动时，不仅不受任何阻碍，反而受到强烈的推动作用，致使轧件宽度产生附加的增长，此时产生的宽展为强迫宽展。由于出现有利于金属质点横向移动的条件，所以强迫宽展大于自由宽展。

（2）Longitudinal deformation in rolling process——forward and backward sliding.

（2）轧制过程中的纵变形——前滑和后滑。

It has been proved that in rolling, the part of metal compressed in the height direction flows longitudinally, which makes the rolled piece extend, while the other part flows transversely, forming spread. The extension of the rolled piece is the result of the flow of the pressed metal towards the entrance and exit of the roll.

实践证明，轧制中在高度方向受到压缩的那部分金属，一部分向纵向流动，使轧件形成延伸，而另一部分金属向横向流动，形成宽展。轧件的延伸是由于被压下金属向轧辊入口和出口两个方向流动的结果。

In the rolling process, the phenomenon that the exit speed v_h of the rolled piece is greater than the linear speed of the roll at this point is called forward slip, while the phenomenon that the speed v_H of the rolled piece entering the roll is less than the horizontal component $v\cos\alpha$ of the linear speed v of the roll at this point is called backward slip.

在轧制过程中，轧件出口速度 v_h 大于轧辊在该处的线速度的现象称为前滑；而轧件进入轧辊的速度 v_H 小于轧辊在该点处线速度 v 的水平分量 $v\cos\alpha$ 的现象称为后滑。

Generally, the ratio of the difference between the exit speed v_h of the rolled piece and the linear speed v of the roller's peripheral speed at the corresponding point to the linear speed of the roller peripheral speed is called the forward slip value, namely：

通常将轧件出口速度 v_h 与对应点的轧辊圆周速度的线速度 v 之差与轧辊圆周速度的线速度之比，称为前滑值，即：

$$S_h = (v_h - v)/v \times 100\%$$

Where　S_h——Front slip value;

　　　v_h——The speed of the rolled piece at the exit of the roll, m/s;

　　　v——The linear speed of the roller peripheral speed, m/s.

式中　S_h——前滑值；

　　　v_h——在轧辊出口处轧件的速度，m/s；

　　　v——轧辊圆周速度的线速度，m/s。

The latter slip value refers to the ratio of the difference between the speed v_H of the workpiece at the entry section of the workpiece and the horizontal component $v\cos\alpha$ of the roller peripheral speed at this point to the horizontal component $v\cos\alpha$ of the roller peripheral speed, namely：

而后滑值是指轧件入口断面轧件的速度 v_H 与轧辊在该点处圆周速度的水平分量 $v\cos\alpha$ 之差同轧辊圆周速度水平分量 $v\cos\alpha$ 之比，即：

$$S_H = (v\cos\alpha - v_H)/(v\cos\alpha) \times 100\%$$

Where　S_H——Backward slip value;

　　　v_H——The speed of the rolled piece at the roll population, m/s;

　　　α——Bite angle, (°).

式中　S_H——后滑值；

v_H——在轧辊入口处轧件的速度，m/s；

α——咬入角，(°)。

Under the condition of equal flow per second, there are:

按秒流量相等的条件，则有：

$$F_H v_H = F_h v_h$$

or

或

$$v_H = F_h/F_H v_h = v_h/\mu$$

Where μ——Elongation coefficient.

式中 μ——延伸系数。

According to the definition formula of forward slip $v_h = v(1+S_h)$, substituting in can get:

根据前滑值定义公式 $v_h = v(1+S_h)$，代入可得：

$$v_H = v/\mu(1 + S_h)$$

The definition formula of backward slip value can be obtained as follows:

$$\mu = (1 + S_h)/[(1 - S_h)\cos\alpha]$$

It can be seen from the above formula that forward and backward sliding are the components of extension. When the elongation coefficient μ and the peripheral speed v of the roll are known, the actual speed of the rolling piece in the roll v_H and out of the roll v_h is determined by the front slip value S_h, or the back slip value S_h can be calculated when the front slip value is known. In addition, it can be seen that when the elongation coefficient μ and biting angle α are fixed, the front slip value increases and the back slip value decreases. There are many factors that affect the forward slip, mainly including:

由以上公式可知，前滑和后滑是延伸的组成部分。当延伸系数 μ 和轧辊圆周速度 v 已知时，轧件进出轧辊的实际速度 v_H 和 v_h 决定于前滑值 S_h，或知道前滑值便可求出后滑值 S_h。此外还可看出，当延伸系数 μ 和咬入角 α 一定时，前滑值增加，后滑值就必然减小。影响前滑的因素很多，主要表现在：

1) Reduction rate. The forward slip increases with the reduction rate. The reason is that because of the increase of multi-directional compression deformation, the longitudinal and transverse deformation increase, so the forward slip value increases.

1) 压下率。前滑随压下率的增加而增加。其原因是多向压缩变形增加，纵向和横向变形都增加，因而前滑值增加。

2) Thickness of rolled piece. After rolling, the thickness of the rolled piece increases with the decrease of the front slip.

2) 轧件厚度。轧后轧件厚度越小时前滑增加。

3) Roll diameter. The forward slip value increases with the increase of roll diameter. This is because under the same conditions of other conditions, when the roll diameter increases, the biting angle will decrease, while the friction angle remains constant, so the residual friction in the stable rolling stage will increase, which will lead to the increase of metal plastic flow speed, that is, the increase of forward slip.

3) 轧辊直径。前滑值随辊径增加而增加。这是因为在其他条件相同的条件下,当辊径增加时,咬入角就要减小,而摩擦角保持常数,所以稳定轧制阶段的剩余摩擦力就增加,由此将导致金属塑性流动速度的增加,也就是前滑的增加。

4) Coefficient of friction. Under the same reduction and other process parameters, the larger the friction coefficient is, the larger the forward slip value is.

4) 摩擦系数。在压下量及其他工艺参数相同的条件下,摩擦系数越大,其前滑值越大。

5) Tension. When the front tension increases, the resistance of metal flow forward decreases, thus increasing the front slip area and increasing the front slip. On the contrary, when there is back tension, the back slip zone increases.

5) 张力。前张力增加时,金属向前流动的阻力减小,从而增加前滑区,使前滑增加。反之,存在后张力时,则后滑区增加。

6) Pass shape. Because the linear speed of the roller is different at each point around the pass, but because each point on the cross section of the integrated rolling piece must be rolled at the same speed, this will inevitably cause different forward slip values at each point around the pass, so the pass shape used in rolling has an impact on the forward slip values.

6) 孔型形状。因为沿孔型周边各点轧辊的线速度不同,但由于金属的整体性轧件横断面上各点又必须以同一速度出辊,这就必然引起孔型周边各点的前滑值不一样,所以轧制时所使用的孔型形状对前滑值有影响。

There are many factors that affect the flow of metal in the deformation zone, but they are all based on the law of minimum resistance and the law of constant volume.

影响金属在变形区内沿纵向及横向流动的因素很多,但都是建立在最小阻力定律及体积不变定律的基础之上的。

(3) Movement Speed of Rolling Piece on Each Section in Deformation Area.

(3) 轧件在变形区内各断面上的运动速度。

When the metal is rolled from the height H before rolling to the height h after rolling, as the height decreases gradually after entering the deformation area, the movement speed of metal particles in the deformation area cannot be the same according to the constant volume condition. Relative motion may occur between metal particles and between metal surface particles and tool surface particles.

当金属由轧前高度 H 轧到轧后高度 h 时,由于进入变形区后高度逐渐减小,根据体积不变条件,变形区内金属质点运动速度不可能一样。金属各质点之间以及金属表面质点与工具表面质点之间就有可能产生相对运动。

It is assumed that the rolled piece has no spread, and the particle deformation along each height section is uniform, and the horizontal velocity of its movement is the same. In this case, according to the condition of constant volume, the speed of the rolled piece in the front sliding area is ahead of that of the roll, while the speed at the exit v_h is the maximum; in the back sliding area, the speed of the rolled piece lags behind the horizontal speed of the linear speed of the roll, and the speed at the entrance v_H is the minimum; on the neutral surface, the horizontal speed of the rolled piece and the roll is the same, and is expressed as the roll horizontal speed on the neutral surface. It can be concluded that:

设轧件无宽展，且沿每一高度断面上质点变形均匀，其运动的水平速度一样。此情况下，根据体积不变条件，轧件在前滑区相对于轧辊来说，超前于轧辊，而在出口处的速度 v_h 为最大；在后滑区，轧件速度落后于轧辊线速度的水平分速度，并在入口处的轧件速度 v_H 为最小；在中性面上，轧件与轧辊的水平分速度相等，并用表示在中性面上的轧辊水平分速度。由此可得出：

$$v_h > v > v_H$$

The horizontal velocity of the workpiece at any point in the deformation zone can be calculated under the condition of volume invariance, that is, the volume of metal passing through any section in the deformation zone in unit time should be a constant, and that is, the metal flow rate per second is equal. The metal flow through population section, exit section and any section in deformation area per second can be expressed as follows:

变形区任意一点轧件的水平速度可以用体积不变条件计算，也就是在单位时间内通过变形区内任一断面上的金属体积应为一个常数，即金属秒流量相等。每秒通过入口断面、出口断面及变形区内任一横断面的金属流量可用下式表示：

$$F_H v_H = F_x v_x = F_h v_h = 常数$$

Where　F_H, F_h, F_x——Refers to the area of population section, exit section and any section in deformation area respectively, mm^2;

　　v_H, v_h, v_x——The average movement speed of metal on any section of population section, exit section and deformation area respectively, m/s.

式中　F_H, F_h, F_x——分别为入口断面、出口断面及变形区内任一断面的面积，mm^2;

　　v_H, v_h, v_x——分别为入口断面、出口断面及变形区内任一断面上的金属平均运动速度，m/s。

2.3.2　Pass System
2.3.2　孔型系统

2.3.2.1　Pass Type and Its Composition
2.3.2.1　孔型及其构成

In order to roll products with different cross-sections with billets (or ingots) as raw materials, it is usually necessary to pass several passes on a group of rolling mills to gradually deform the metal, and finally get the products with the required shape and size. Therefore, the rolling groove must be machined on the roll as required, which is formed by the projection of two or more rolling grooves on the plane passing through the roll axis which is called pass.

以钢坯（或钢锭）为原料来轧制各种不同断面的产品，通常要在一组（架）轧机上经若干道次轧制，使金属逐渐变形，最后得到所需形状与尺寸的产品，为此必须在轧辊上按需要加工出轧槽，这种由两个或两个以上的轧槽在通过轧辊轴线的平面上投影所构成的形状称为孔型。

The pass mainly consists of the following parts:
孔型主要由以下几部分构成:

(1) Roll gap. The roll gap can prevent the direct contact of the rolls with each other, avoid the mutual wear and the increased energy consumption. In many cases, the size of pass can be changed by adjusting the roll gap value, which is of great value in improving the commonality of pass and saving the spare location of roll. However, if the gap value is too large, the groove will become shallow, which will not limit the metal flow and make the shape of the rolled piece incorrect.

(1) 辊缝。辊缝可以防止轧辊彼此的直接接触,避免互相磨损和由此增加的能量消耗。在许多情况下,调整辊缝值的大小可改变孔型的尺寸,这在提高孔型的共用性和节约轧辊备用方面是很有价值的。但辊缝值过大会使轧槽变浅,起不了限制金属流动的作用,使轧件形状不正确。

(2) Slope of hole side wall. The side wall of any pass should keep a certain slope, so that after the wear of pass, the shape and size can be restored by turning slightly at the original groove position. In addition, the side wall slope of the pass is also beneficial to the entry and exit of the slot.

(2) 孔型侧壁斜度。任何孔型的侧壁都需保持一定的斜度,以便在孔型磨损后,能在原有轧槽位置上稍经车削即可使形状和尺寸得到复原。此外,孔型侧壁斜度还有利于使轧件进、出槽孔。

(3) Fillet of pass. Unless there are special requirements, the inner and outer corners of the pass must be properly rounded. Fillets can be divided into inner fillets and outer fillets. The inner fillet can prevent the sharp cooling of the corner and reduce the stress concentration. The external fillet has the function of adjusting the fullness degree of the rolled piece in the pass, preventing the "ears" formed due to the increase of the spreading amount and the overfilling in the pass, so as to avoid the folding of the rolled piece during the continuous rolling, and the external fillet can also prevent the sharp edges and corners from scratching the rolled piece.

(3) 孔型的圆角。除有特殊要求者外,孔型的内、外棱角处通常都必须进行适当的圆化。圆角又可分为内圆角和外圆角。内圆角可防止轧件角部的急剧冷却,减轻应力集中。外圆角有调节轧件在孔型中充满程度的作用,防止由于宽展量的增加在孔型内过充满而形成"耳子",这样可避免轧件在继续轧制时形成折叠,同时外圆角也可起到避免尖锐的棱角划伤轧件的作用。

(4) Lock. When using closed pass and rolling some special-shaped steel, lock is used to control the section shape of the rolled piece. With the hole type of lock mouth, the lock mouth of the adjacent hole type is generally alternating up and down.

(4) 锁口。当采用闭口孔型及轧制某些异形型钢时,为控制轧件的断面形状而使用锁口。用锁口的孔型,其相邻道次孔型的锁口一般是上下交替出现的。

2.3.2.2　Common Extended Pass System and Its Characteristics
2.3.2.2　常用延伸孔型系统及其特点

The pass system to be adopted shall be determined according to the specific rolling conditions, including billet shape, size, rolling product, steel type, rolling mill form, motor capacity, auxiliary equipment, roll diameter, technical equipment level, etc.

采用何种孔型系统，要根据具体的轧制条件，包括坯料形状、尺寸、轧制产品、钢种、轧机形式、电机能力、辅助设备、轧辊直径、技术装备水平等来确定。

Common pass systems are:

常用孔型系统有：

(1) Box pass system. It is used in roughing of small bar or wire mill. Its characteristics are as follows:

(1) 箱形孔型系统。它运用于小型棒材或线材轧机的粗轧。其特点有：

1) It can reduce the number of passes, reduce the number of roll changing and groove changing, and improve the operation rate.

1) 用改变辊缝的方法轧制多种尺寸不同的轧件，共用性好；可减少孔型数量，减少换辊换槽次数，提高作业率。

2) Compared with other passes of the same area, the box pass has shallow groove, so the roll strength is higher, which can meet the larger deformation.

2) 与等面积的其他孔型相比，箱形孔型刻槽浅，故轧辊强度较高，可满足较大的变形量。

3) The deformation along the width direction of the rolled piece is even, so the wear of the pass is even, and the deformation energy consumption is small.

3) 沿轧件宽度方向的变形均匀，故孔型磨损均匀，且变形能耗小。

4) It is easy to peel off the oxide scale on the rolled piece and improve the surface quality of the rolled piece.

4) 易于脱落轧件上的氧化铁皮，改善轧件表面质量。

5) The section temperature of rolled piece is relatively uniform.

5) 轧件断面温度比较均匀。

6) Because of the shape characteristics of the box pass, it is difficult to produce the workpiece with accurate geometry.

6) 因为箱形孔型的形状特点，难以轧出几何形状精确的轧件。

7) Because the rolled piece is only compressed in two directions in the pass, its side surface is not easy to be straight.

7) 由于轧件在孔型中仅受两个方向的压缩作用，故其侧表面不易平直。

8) Because of the improper design of the side wall slope of the rolled piece in the box pass, it is easy to produce the phenomenon of steel pouring, increase the consumption of guide and guard, and produce the defect of wire scraping.

8) 箱形孔型中轧制的轧件，因侧壁斜度设计不合适，易产生倒钢现象，增加导卫消耗，并产生刮丝缺陷。

The elongation coefficient of box pass is generally 1.15~1.6, and the average elongation coefficient can be 1.15~1.4.

箱形孔型中的延伸系数一般为1.15~1.6，其平均延伸系数可取1.15~1.4。

(2) Oval round pass system. Its characteristics are as follows:

(2) 椭圆—圆孔型系统。其特点有:

1) The finished screw steel can be produced by the middle pass, so the roll changing operation can be reduced.

1) 可由中间道次孔型出成品螺纹钢,因此可减少换辊操作。

2) The rolled piece has no obvious edges and corners, and the temperature is even.

2) 轧件无明显棱角,温度均匀。

3) The deformation of the rolled piece in the pass is more uniform, the shape transition is smooth, and the local stress concentration can be reduced.

3) 轧件在孔型中的变形较均匀,形状过渡平滑,可减少局部应力集中。

4) Rolling in this pass is good for peeling off the surface oxide scale.

4) 在这种孔型中轧制有利于脱落表面氧化铁皮。

5) The small elongation coefficient leads to the increase of rolling passes.

5) 延伸系数较小,导致轧制道次增加。

6) When the ellipse enters into the round hole rolling, the rolling piece is unstable, easy to pour steel, and the requirements of guide and guard are strict.

6) 椭圆进入圆孔型轧制时轧件不稳定,易倒钢,对导卫要求严格。

7) The surface defects such as ears are easy to appear when the rolled piece is rolled in the round hole mold.

7) 轧件在圆孔型中轧制易出现耳子等表面缺陷。

The elongation coefficient of this pass system is generally 1.1~1.5.

这种孔型系统的延伸系数一般为 1.1~1.5。

(3) Rolling without pass. The pass free rolling is to reduce the cross-section to a certain extent through the square rectangular deformation process in the flat roll without groove, and then pass a certain number of finishing passes, and finally roll into square, round, flat and other cross-section rolled pieces. When rolling without pass, the height of roll gap is the width of rolled piece after free spread, and there is no effect of pass sidewall on rolled piece. The characteristics of pass free rolling are as follows:

(3) 无孔型轧制。无孔型轧制是在无刻槽的平辊中,通过方—矩形变形过程,断面减小到一定程度,再通过一定数量的精轧孔型,最终轧制成方、圆、扁等断面轧件。无孔型轧制时,辊缝高度即为自由宽展后的轧件宽度,没有孔型侧壁对轧件的作用。无孔型轧制的特点有:

1) Because there is no pass on the roll, when the product specification is changed, it is only realized by adjusting the roll gap, so the mill operation rate is increased.

1) 因轧辊上无孔型,改变产品规格时,仅通过调节辊缝即可实现,故提高了轧机作业率。

2) There is no groove on the roll, and the roll body, especially the outer hardness layer, can be fully utilized, which can increase the effective utilization length of the roll body to 75%~80%. The durability of each rolling part can be 1.5 times higher than that of pass rolling, which can increase the service life of the roll by 3~4 times.

2）轧辊上无刻槽，轧辊的辊身特别是外层硬度层能充分利用，可使辊身的有效利用长度提高到 75%~80%，每个轧制部位的耐久性可比有孔型轧制提高 1.5 倍，使轧辊使用寿命提高 3~4 倍。

3) When the rolled piece is rolled on the flat roll, there will be no defects such as ear, unfilled, wrong pass axis and so on.

3）轧件在平辊上轧制，不会出现耳子、欠充满、孔型轴错等有孔型轧制中的缺陷。

4) It can greatly reduce the metal consumption during the roll turning, reduce the working hours of the roll processing by 5 times, reduce the processing cost by 1.5~2 times, and the turning process is simple.

4）可大幅度降低轧辊车削时的金属消耗量，使轧辊加工的工时减少 5 倍，加工成本降低 1.5~2 倍，且车削加工简单。

5) Due to the reduction of the limiting effect of the side wall of the pass and the uniform deformation along the width direction, the deformation resistance can be reduced and the energy consumption can be saved.

5）因减少了孔型侧壁的限制作用，沿宽度方向变形均匀，因此降低变形抗力，可节约能耗。

6) It is beneficial to remove the scale on the surface of rolled piece.

6）有利于去除轧件表面的氧化铁皮。

7) When rolling between a pair of flat rolls, the side wall of the pass loses its clamping effect, which is prone to skew and square off.

7）轧件在一对平辊间轧制，失去了孔型侧壁对其夹持作用，易出现歪扭脱方现象。

2.3.2.3　Analysis of Pass Design of Typical Products
2.3.2.3　典型产品孔型设计的分析

Taking the production of 14mm Round Steel by 150mm×150mm square as an example, the pass pattern of this typical product is analyzed.

以 150mm×150mm 方式生产 ϕ14mm 圆钢产品为例，分析这种典型产品的孔型。

For 14mm Round steel products, the rolling speed guarantee value is 18m/s, the hourly production capacity is 73.7t, the rolling interval time of each billet is 5s, the pure rolling time is 79.6s, the total rolling duration is 139.5s, the total elongation coefficient is 146.2, and the average elongation coefficient is 1.32. This product needs 18 passes of rolling. Figure 2-1 is the schematic diagram of the pass system of 14mm product.

圆钢 ϕ14mm 产品，其轧制速度保证值为 18m/s，小时生产量为 73.7t，每支钢坯轧制间隔时间为 5s，纯轧时间为 79.6s，轧制总延续时间为 139.5s，总延伸系数为 146.2，平均延伸系数为 1.32，该产品需轧制 18 道次。图 2-1 为 ϕ14mm 产品孔型系统示意图。

(1) Box pass.

（1）箱形孔型。

The box pass is used in roughing mill 1 and 2. The groove of this pass is shallow on the roll, which reduces the stress on the roll, improves the strength of the roll and increases the reduction.

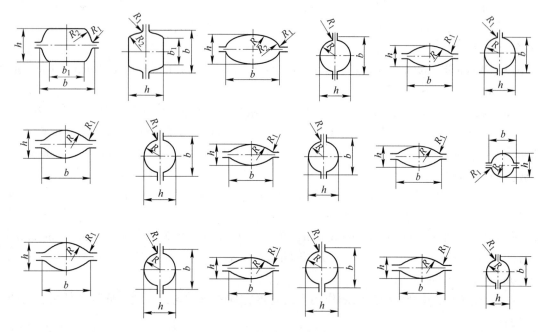

Figure 2-1　Pass system of 14mm product
图 2-1　14mm 产品孔型系统

粗轧 1、2 架采用箱形孔型，该种孔型在轧辊上的刻槽较浅，这样降低了轧辊所受应力，相对地提高了轧辊的强度，可增大压下量。

It is advantageous to use box hole several times before rough rolling for rolling large section workpiece, and the deformation in the width direction of the workpiece in the pass is relatively uniform, and the roll groove is shallow, which can meet the requirement of large reduction of Dong workpiece. However, in this pass, the metal can only be processed in two directions, and because of the side slope of the pass, the rectangular section rolled out is not regular enough. The side wall slope of the pass is $y = 15.0\% \sim 20.0\%$. The slope of the side wall of the pass has a righting effect on the rolled piece. If the design of the value is reasonable, it can not only improve the stability of the rolled piece in the pass, easily make the rolled piece out of the groove, but also improve the bite angle and increase the bite ability.

粗轧前几道次采用箱形孔对轧制大断面的轧件是有利的，而且在孔型中轧件宽度方向上的变形比较均匀，轧辊刻槽较浅，可满足大断面的轧件。但在这种孔型中轧制，金属只能受两个方向的加工，且由于该孔型存在有侧壁斜度，轧出的矩形断面不够规整。该孔型采用的孔型侧壁斜度为 $y = 15.0\% \sim 20.0\%$。孔型的侧壁斜度对轧件有扶正的作用，其值如果设计合理，不仅可提高轧件在孔型中的稳定性，易使轧件脱槽，而且还可提高咬入角，增加咬入能力。

The width value b_1 of the bottom of the box pass groove shall make the rolled piece first contact with the four points of the side wall of the pass at the beginning of biting, generate certain side pressure to clamp the rolled piece, and improve the stability and biting ability. However, if the value of b_1 is too large, there will be no side pressure, resulting in poor stability; if the value of b_1 is too small, the side pressure will be too large, which will make the pass wear too fast or out of the ear, thus affecting the quality of the rolled piece.

箱形孔型槽底宽度 b_1 值要使咬入开始时轧件首先与孔型侧壁四点接触，产生一定的侧压以夹持轧件，提高稳定性和咬入能力。但 b_1 值太大会产生无侧压作用，导致稳定性差；而 b_1 值过小，侧压过大，会使孔型磨损太快或出耳子，从而影响轧件质量。

In the design, the elongation coefficient μ of box pass is 1.25~1.5.

在设计中，箱形孔型的延伸系数选用 $\mu=1.25$~1.5。

（2）Flat oval pass.

（2）平椭圆孔型。

The third roughing mill adopts the flat elliptical pass with arc bottom, which is the transition pass from the box pass to the subsequent elliptical circular pass, and it is the abnormal elliptical pass. It can reduce the great change of the section shape caused by the rolling of the box hole into the round hole, and the excessive wear of the round hole. Moreover, the next round hole into the round hole has better stability and larger elongation coefficient than the oval one. Set the pass extension coefficient $\mu=1.42$~1.46. At the same time, the flat elliptical pass is conducive to further removing the oxide scale on the surface of the rolled piece and improving the surface quality of the rolled piece.

粗轧第3架采用弧底平椭圆孔型，这道孔型是由箱形孔型进入后续的椭圆—圆孔型的过渡孔型，是变态的椭圆孔。它减轻了由箱形孔进入圆孔型轧制而引起的轧件断面形状巨变，以及由此产生的圆孔型的过度磨损，而且进入下一道圆孔型比椭圆断面轧件进入圆孔型有较好的稳定性和较大的延伸系数。设道次延伸系数 $\mu=1.42$~1.46。同时，平椭圆孔型有利于进一步除去轧件表面的氧化铁皮，改善轧件表面质量。

（3）Oval and round hole type.

（3）椭圆、圆孔型。

From the 5th roughing mill to the 18th finishing mill, the elliptical round pass system is adopted. In this system, the cross-section shape of the rolled piece before and after rolling is relaxed, and the cross-section rolled out is smooth without edges. However, in this system, the round hole pattern has poor adaptability to the fluctuation of the size of the incoming material, is easy to come out of the ear and unfilled, has high requirements for adjustment, and has a small elongation coefficient, especially in the finishing pass. For $\phi14mm$ deformed steel products, the average elongation coefficient $\mu=1.21$. Because of the small extension of elliptical round pass system, it was not widely used in the past, but in the rolling of high quality or high alloy steel, the adoption of this pass system can improve the surface quality of products. Although the number of passes has increased, it can reduce the finishing work and improve the yield, which is reasonable economically. With the development of bar continuous rolling technology, the application of ellipse round hole system has gradually expanded, and flying shear is set in the rolling line to cut off the defects of the head of the rolling piece, which is more conducive to the realization of rolling automation.

从粗轧第5架至精轧第18架，采用椭圆—圆孔型系统。此系统中轧件在轧制前后的断面形状过渡缓和，所轧出的断面光滑无棱。但这种系统中圆孔型对来料尺寸波动适应能力差，易出耳子和欠充满，对调整要求较高，而且延伸系数也不大，特别是在精轧道次。对于 $\phi 14mm$ 螺纹钢产品平均延伸系数 $\mu_p = 1.21$。因椭圆—圆孔型系统的延伸小，以往应用不太广泛，但在轧制优质或高合金钢时，采用这种孔型系统能提高产品的表面质量，虽然轧制道次有所增加，但可减少精整工作量和提高成品率，从经济上来说是合理的。随着棒材连轧技术的发展，椭圆—圆孔系统的应用已逐渐扩展，而且在轧线上设置飞剪，切去轧件头部的缺陷，更有利于实现轧制的自动化。

In addition, for the rolled round steel and the deformed steel with the same size, the front hole of the finished product is only different from the front hole of the finished product, and the front hole of the finished product becomes a flat ellipse pass, rather than an ellipse pass. See the Table 2-1 for the comparison between the elongation coefficient and the elongation coefficient during rolling.

另外，对于轧制圆钢与轧制相同尺寸的螺纹钢仅在成品前孔不同，其成品前孔变为平椭圆孔型，而不是椭圆孔型。其延伸系数与螺纹钢轧制时延伸系数比较见表2-1。

Table 2-1　Comparison of elongation coefficient between round steel and midge grain steel
表2-1　圆钢与螺纹钢延伸系数比较

Billet size/mm×mm 坯料尺寸/mm×mm	Pass 道次	$\mu_{circular}$ $\mu_{圆}$	μ_{snail} $\mu_{螺}$	Billet size/mm×mm 坯料尺寸/mm×mm	Pass 道次	$\mu_{circular}$ $\mu_{圆}$	μ_{snail} $\mu_{螺}$
120×120	K_1	1.15~1.17		150×150	K_1	1.16~1.18	1.2~1.4
	K_2	1.20~1.22			K_2	1.10~1.30	1.1~1.2

From Table 2-1, it can be seen that for round steel products, the elongation coefficient of K_2 is slightly smaller than that of K_1; for threaded steel products, the elongation coefficient of K_1 hole is larger than that of K_2 hole, making the metal fill in the threaded hole K_1, forming normal ribs that meet the requirements.

从表2-1可得：对于圆钢产品，K_2 与 K_1 相比延伸系数变化不大，K_1 略小；对于螺纹钢产品，K_1 孔延伸系数较 K_2 孔延伸系数要大，使得金属在螺纹孔型 K_1 中充满，形成正常的符合要求的筋肋。

When the ellipse rolled piece is rolled into the round hole type, the side wall of the pass has little clamping force on the rolled piece. When the axis of the rolled piece is slightly skewed, the steel will be inverted, which has poor stability and high requirements for the guide and guard. Therefore, in the production line of small bar continuous rolling, the entrance guide is rolling type to provide enough clamping force to ensure that the rolled piece enters the next rolling in the right way. Moreover, the side wall of the pass has little effect on the width spread, and the width spread in the round pass is large. However, compared with other passes, the size of the width spread space left in the round pass is small, and the allowable deformation of the width spread is small. Therefore, on the one hand, the extension coefficient is limited, on the other hand, it is easy to get out of the ear.

椭圆轧件进入圆孔型轧制，孔型侧壁对轧件夹持力小，当轧件轴线稍有偏斜时即产生倒钢，稳定性差，对导卫要求较高。因此，小型棒材连轧生产线中，椭圆进圆孔型轧制时，入口导卫都采用滚动式，以提供足够的夹持力，保证轧件以正确的方式进入下一道次轧制。再者，孔型侧壁对宽展的限制作用小，圆孔型中的宽展大，但与其他孔型相比，在圆孔型中留有的宽展空间尺寸小，允许宽展的变形量也就小，因此，这一方面限制了延伸系数，另一方面容易出耳子。

The relationship between the parameters h and b of the elliptical pass and the parameter d of the back round pass is different because of the experience data used. The designed pass is no more than a thin, wide or thick, narrow ellipse. As long as the relationship between the reduction and the spread is mastered, it can be used flexibly, and the qualified products can be rolled through the adjustment during rolling.

椭圆孔型的参数 h、b 与其后圆孔型参数 d 的关系由于所用的经验数据不同，所设计的孔型不外乎是薄而宽或厚而窄的椭圆，只要掌握压下与宽展的关系，灵活运用，通过轧制时的调整，都能轧出合格的产品。

For 14mm products:

对于 ϕ14mm 的产品：

$$h = (0.65 \sim 0.85)d$$
$$b = (1.50 \sim 2.30)d$$

It has been proved that it is difficult to produce qualified deformed steel bars by drawing the pass with only one radius. This is because in this pass, small fluctuations of rolling conditions such as rolling temperature, wear of pass and size of incoming material will form ears or unfilled. At this time, in order to get qualified products, it is necessary to constantly adjust, thus making adjustment difficult. In order to eliminate the above shortcomings, the threaded steel hole should be designed as a round hole with a hole height less than the width of the hole, that is, with less expansion angle. However, the commonly used round hole pattern is the one with arc sidewall, and this kind of round hole pattern with straight sidewall increases the sensitivity of ear out due to the straight shape of both sidewalls.

实践证明，只用一个半径绘制出的螺纹钢孔型，是难以轧出合格螺纹钢的，这是因为在这种孔型中，轧制条件如轧制温度、孔型磨损以及来料尺寸等的微小波动，都会形成耳子或欠充满，此时，为得到合格成品，就必须不停地调整，从而使调整操作困难。为消除上述缺点，应将螺纹钢孔设计成孔型高度小于孔型宽度，即带有扩张角少的圆孔型。但现常用的圆孔型则是带有弧形侧壁的孔型，而这种带直线侧壁圆孔型，由于两侧壁为直线形状而增加了出耳子的敏感性。

Pass height:

孔型高度为：

$$h = \alpha d$$

In the formula α——linear expansion coefficient, for ordinary carbon steel $\alpha = 1.011 \sim 1.015$, the final rolling temperature is high, take the upper limit.

式中 α——线膨胀系数，对于普碳钢 $\alpha = 1.011 \sim 1.015$，终轧温度高，取上限。

The radius of pass arc is:

孔型圆弧半径为:

$$R = h/2$$

Notch width:

槽口宽度为:

$$b = 2R/\cos\phi - s\tan\phi$$

Where s——roll gap, mm;
       ϕ——Dilation angle, (°).

式中 s——辊缝, mm;
       ϕ——扩张角, (°)。

Expansion angle $\phi = 30°$, then:

扩张角 $\phi = 30°$, 则:

$$b = 2.31R - 0.577s$$

For the design of finished circular hole K_1, a single radius circular hole is used. Smaller values shall be selected for slot fillet and roll gap. Through the precise rolling of the extended pass and the front hole of the finished product, a smaller elongation coefficient (such as ϕ4mm product, the elongation coefficient of K_1 hole $\mu = 1.16$) is adopted in this pass, which is also conducive to the adjustment and rolling of the qualified finished product.

对于成品圆孔 K_1 的设计, 采用单一半径的圆孔。槽口圆角和辊缝选用较小的数值。通过延伸孔型和成品前孔精确的轧制, 在此道次采用较小的延伸系数 (如 ϕ4mm 产品, K_1 孔的延伸系数 $\mu = 1.16$), 这样也有利于调整而轧制出合格的成品。

The roll gap value has the functions of compensating the roll bounce, ensuring the height of the rolled piece after rolling, compensating the wear of the rolling groove, increasing the service life of the rolling coil, and improving the commonality of the pass. That is to say, the pass with different section sizes can be obtained by adjusting the roll gap; meanwhile, it is convenient for the adjustment of the rolling mill and reduces the depth of the roll groove cutting.

辊缝值具有补偿轧辊弹跳、保证轧后轧件高度、补偿轧槽磨损、增加轧辊使用寿命、提高孔型共用性的作用, 即通过调整辊缝可得到不同断面尺寸的孔型; 同时方便轧机的调整, 且减小轧辊切槽深度。

Under the condition of not affecting the section shape and rolling stability of the rolled piece, the larger the roll gap value s is better. But in several passes close to the finished pass, the roll gap cannot be too large, otherwise it will affect the correctness of the section shape and size of the rolled piece.

在不影响轧件断面形状和轧制稳定性的条件下, 辊缝值 s 越大越好, 但在接近成品孔型的几个孔型中, 辊缝不能太大, 否则会影响轧件断面形状和尺寸的正确性。

The relationship between the roll gap value s of the finished pass and the product specification is as follows:

成品孔型辊缝值 s 与产品规格的关系如下:

Product specification/mm 产品规格/mm	s/mm
φ10~17	1.0
φ18~30	1.5
φ32~40	2.0

The groove fillet can avoid forming sharp ears when the rolled piece is slightly filled in the pass. At the same time, when the rolled piece is not in the right time of entering the pass, it can prevent the scraping defects caused by scraping the side surface of the rolled piece by the roll ring.

槽口圆角可避免轧件在孔型中略有过充满时,形成尖锐的耳子,同时当轧件进入孔型不正时,它能防止辊环刮切轧件侧表面而产生的刮丝缺陷。

The relationship between groove fillet r and product specification of finished threaded steel pass is as follows:

螺纹钢成品孔型的槽口圆角 r 与产品规格的关系如下:

Product specification/mm 产品规格/mm	s/mm
φ10~11	1.5
φ12~25	2.0
φ26~30	2.5
φ32~40	3.0

2.3.2.4　Splitting Pass Rolling
2.3.2.4　切分孔型轧制

(1) Concept of slitting rolling.

(1) 切分轧制的概念。

Slitting rolling is to use the pass of a billet in the rolling process. Rolling into two or more parallel rolling pieces with the same shape, and then using the slitting equipment or roll ring to cut the parallel rolling piece into two or more single rolling pieces along the longitudinal direction.

切分轧制,就是在轧制过程中把一根钢坯利用孔型的作用。轧成具有两个或两个以上相同形状的并联轧件,再利用切分设备或轧辊的辊环将并联轧件沿纵向切分成两个或两个以上的单根轧件。

In the production of small bar, the diameter of bar less than 16 mm accounts for about 60% of the total output. The productivity of bar decreases with the decrease of diameter. Moreover, because the productivity of bars fluctuates with the product specifications, it is difficult to realize the continuous casting and rolling process. Because one of the important conditions of continuous casting and rolling is that the production capacity of steel-making, continuous casting and rolling must match each other, in order to make the productivity of bars of various diameters basically equal and realize continuous casting and rolling of bars, the productivity of small bars must be increased.

在小型棒材的生产中,直径小于 16mm 的钢筋占总产量的 60% 左右。而棒材的生产率随直径的减小而降低。再者,由于棒材的生产率随产品规格的不同而波动,使连铸连轧工艺的实现变得困难。因为连铸连轧的一个重要条件是炼钢、连铸和轧钢的生产能力必须相匹配,所以要使轧制各种直径棒材的生产率基本相等,以实现棒材的连铸连轧,就必须提高小规格棒材的生产率。

Recently, the newly-built small bar production line often uses two-line, three line or four line slitting rolling in the production of small products such as $\phi 10 \sim 16mm$. When the products of $\phi 10mm$ and $\phi 12mm$ are rolled by two-line slitting, their hourly output is more than 75t/h, which is close to that of other single line products. This not only facilitates the balance of rolling rhythm, but also improves the output without increasing rolling passes, and gives full play to the production capacity of rolling mill equipment.

近来新建的小型棒材生产线在生产小规格产品如 $\phi 10 \sim 16mm$ 时常采用两线、三线或四线切分轧制。$\phi 10mm$、$\phi 12mm$ 产品采用两线切分轧制时,其小时产量在 75t/h 以上,与其他单线生产的产品小时产量相接近。这样既便于轧制节奏的均衡,又在不增加轧制道次的前提下提高了产量,且充分发挥了轧机设备的生产能力。

The key technology of slitting rolling lies in the reliability of the slitting device, the rationality of pass design, the correctness of the shape of the rolled piece after slitting and the stability of the product quality.

切分轧制的工艺关键在于切分装置工作的可靠性、孔型设计的合理性、切分后轧件形状的正确性以及产品实物质量的稳定性。

(2) Characteristics of slitting rolling.

(2) 切分轧制的特点。

Slitting rolling has the following characteristics:

切分轧制具有以下特点:

1) It can greatly increase the output of rolling mill. For small size products, the length of the rolled piece and the rolling cycle are shortened by using multi line slitting rolling, so the productivity can be increased. Even if lower rolling speed is adopted, high rolling mill output can be obtained.

1) 可大幅度提高轧机产量。对小规格产品,用多线切分轧制缩短了轧件长度,缩短了轧制周期,从而可提高生产率。即使采用较低的轧制速度,也能得到高的轧机产量。

2) It can balance the production capacity of different products and create conditions for continuous casting and rolling. Because the capacity of steel-making and continuous casting is relatively stable, and the rolling capacity fluctuates greatly, adopting slitting rolling can ensure that the rolling capacity of various specifications of bars is basically equal, thus creating favorable conditions for continuous casting and rolling production.

2) 可使不同规格产品的生产能力均衡,为连铸连轧创造条件。因为炼钢连铸的能力相对稳定,而轧钢能力波动大,采用切分轧制可以保证多种规格棒材的轧制能力基本相等,从而为连铸连轧生产创造有利条件。

Slitting rolling can not only balance the rolling capacity of different products, but also give full play to the production capacity of rolling mill, cooling bed, heating furnace and other auxiliary equipment.

切分轧制不仅使不同规格产品的轧制生产能力均衡，而且可使轧机、冷床、加热炉及其他辅助设备的生产能力得到充分发挥。

3) Under the same rolling conditions, the return of large section can be adopted: or under the same section, the rolling passes and equipment investment can be reduced.

3) 在轧制条件相同的情况下，可以采用较大断面的材料；或在相同材料断面情况下，减少轧制道次，减少设备投资。

4) Save energy and reduce cost. About 80% of the total energy consumption of steel rolling is used for billet heating. Because slitting rolling provides the possibility for continuous casting and rolling, it can save a lot of energy. Moreover, due to the small number of passes, the exit temperature of billet can be reduced properly, which creates favorable conditions for low temperature rolling. During slitting rolling, fuel can be saved by 20%~30%, electric energy can be saved by 15%, and consumption index of water and other tons of steel can be reduced.

4) 节约能源，降低成本。轧钢总能耗的80%左右用于钢坯加热，由于切分轧制为连铸连轧提供了可能性，因此可节约大量能源。而且，因轧制道次少，钢坯的出炉温度可适当降低，为低温轧制创造了有利条件。切分轧制时燃料可节约20%~30%，电能可节约15%，水和其他吨钢消耗指标都有所降低。

However, there are still some problems in slitting rolling, so it is necessary to use this technology in strict accordance with the technological system. The main problems are as follows:

但切分轧制仍存在一定的问题，采用此项技术必须严格按工艺制度进行操作。存在的主要问题有：

1) The cut belt is easy to form burr, if not handled properly, it may form fold. Therefore, the slitting rolling is often used in the production of rebar.

1) 切分带容易形成飞边，如果处理不当有可能形成折叠。因此，棒材连轧生产中切分轧制多用于轧制螺纹钢筋产品。

2) The shrinkage cavity, inclusion and segregation of the billet are mostly located in the center, which are easy to be exposed to the surface and form surface defects.

2) 坯料的缩孔、夹杂和偏析多位于中心部位，经切分后易暴露至表面，形成表面缺陷。

3) When the splitting device is used to separate the parallel rolling piece, because of the shearing force of the splitting blade, the rolling piece is easy to twist after shearing, which affects the section shape and the splitting quality of the rolling piece. Therefore, it is necessary to adjust the position of inlet and outlet guide and the distance between slitting devices to ensure that the rolled piece is not cut off.

3) 当用切分装置分开并联轧件时，由于轧件受切分刀片的剪切力，剪切后轧件易扭转，影响轧件断面形状和切分质量。因此，应当调整好进、出口导卫位置和切分装置间距，保证轧件不被切偏。

The problems to be noticed in the design of slitting pass are as follows:

切分孔型设计中需注意的问题是：

1) Fully consider rolling mill bounce. Because the connecting belt of parallel rolling piece is required to be very thin, generally 0.5~4mm. If the spring value is too large, the requirement of the cutting size can not be guaranteed.

1) 充分考虑轧机弹跳。因为要求并联轧件的联接带很薄，一般为 0.5~4mm，如果弹跳值过大，则不能保证切分尺寸的要求。

2) The wedge angle of the cutting hole shall be greater than that of the precutting hole to ensure that the side wall of the wedge has enough reduction and horizontal component. The wedge angle is about 60°.

2) 切分孔型的楔角应大于预切孔的楔角，以保证楔子侧壁有足够的压下量和水平分力。楔角取值 60°左右。

3) The design of wedge angle and tip shall meet the requirements of wear resistance and impact resistance of wedge head to prevent damage.

3) 楔子角度和尖部的设计要满足楔子头部耐磨损、耐冲击的要求，防止破损。

4) The cutting wedge tip shall be lower than the roll surface (0.4mm lower) to ensure that the tip is not damaged.

4) 切分楔子尖部应低于辊面（低 0.4mm），保证尖部不被碰坏。

5) During continuous rolling slitting, it is necessary to accurately calculate the cross section of the rolled piece to ensure that the second flow of the rolled piece between each stand and two rolled pieces on the same stand after slitting is equal, or make the stacking coefficient reach the set value, so as to reduce the production accidents caused by the stacking between the rolled pieces.

5) 连轧切分时，要精确计算轧件断面，确保切分后轧件在各机架间和两根轧件在同一机架上的秒流量相等，或使堆拉系数达到所设定的数值，减少轧件间相互推拉而产生的生产事故。

In addition, it is necessary to consider the equipment conditions of double line or multi line rolling after slitting, as well as the billet quality condition, so as to prevent the internal defects of the metal after slitting from being exposed to the external surface of the finished product.

此外，还要考虑切分后有采用双线或多线轧制的设备条件；同时还要考虑钢坯质量状况，以防止切分后金属内部缺陷暴露于成品外表面。

2.3.3 Guide Device
2.3.3 导卫装置

2.3.3.1 Rolling Mill Guide Device
2.3.3.1 轧机导卫装置

(1) Overview.

(1) 概述。

Guide and guard device is an indispensable guide device in the production of section steel, which is installed at the population and outlet of roll pass. The function of the guide device is to guide the rolling piece into and out of the pass according to the required direction, so as to ensure that the rolling piece is rolled according to the established deformation conditions.

导卫装置是型钢生产中必不可少的诱导装置，安装在轧辊孔型的入口和出口处。导卫装置的作用是引导轧件按所需的方向进出孔型，确保轧件按既定的变形条件进行轧制。

Whether the design and use of the guide device are correct or not will directly affect the quality of the product and the ability of the unit. Using the correct guide and guard device can effectively avoid the occurrence of scraping, steel extrusion, roll winding and other accidents, improve the working conditions of the roll, and in some cases, it can also make the metal deform and turn over. The guide and guard device can be divided into inlet guide and outlet guide according to its use, and can be divided into sliding guide and rolling guide according to its type.

导卫装置的设计和使用正确与否，直接影响产品的质量和机组能力的发挥。使用正确的导卫装置可以有效地避免刮切、挤钢、缠辊等事故的发生，改善轧辊的工作条件，个别情况下还有使金属变形和翻转的作用。导卫装置按其用处可分为入口导卫和出口导卫；按其类型可分为滑动导卫和滚动导卫。

The main components of the guide device for bar rolling mill are: guide beam, guide plate, guide box, conduit, guide box, guide roller, rolling guide box, torsion roller and other devices that can make the rolled piece deform and twist outside the pass.

构成棒材轧机用导卫装置的主要部件有：导卫梁、导板、导板盒、导管、导管盒、导辊、滚动导卫盒、扭转辊等其他能使轧件在孔型之外产生变形和扭转的装置。

（2）Sliding guide.

（2）滑动导卫。

The guide and guard device that guides the pass in and out of the rolling piece by sliding friction can be called sliding guide and guard. The slide guide is simple in structure, convenient in maintenance and low in cost. In bar production, the sliding guide is usually used to guide the simple sections such as box and circle.

单纯以滑动摩擦的方式引导轧件进出孔型的导卫装置都可以称之为滑动导卫。滑动导卫结构简单、维护方便、造价低廉，使用中存在磨损快、精度低的缺点。在棒材生产中滑动导卫一般用于引导箱形和圆形等简单断面的轧件。

According to different design methods, the sliding guide of bar rolling mill can be divided into rough rolling guide and middle and finish rolling guide.

棒材轧机的滑动导卫按其设计方法的不同可分为：粗轧滑动导卫和中、精轧滑动导卫。

1) Rough rolling sliding guide.

1) 粗轧滑动导卫。

The rolling section guided by rough rolling sliding guide rail is large, and the guide rail bears large lateral force. The guide rail is designed with integral welded steel plate structure and fixed with high-strength bolts.

粗轧滑动导卫所引导的轧件断面较大，导卫所承受的侧向力大，导卫的设计采用整体焊接钢板结构，用高强度螺栓固定。

The substructure is designed as a double dovetail structure, 45° and 60° respectively. It is fixed on the guide rail with iron pressing and two sets of bolts. Its length is determined according to the shape of the guide rail. Other important dimensions are determined as follows：

底座设计为双燕尾结构，分别为45°和60°，用压铁和两组螺栓固定于导卫梁上，其长度根据导卫梁的形状来确定。其他重要尺寸确定如下：

H：For the size related to the height of guide rail, the side plate of guide rail shall be 30~50mm higher than the rolled piece.

H：与导卫高度有关的尺寸，入口导卫侧板应高于轧件30~50mm。

Z：The height dimension related to the bottom surface of the pass and the guide steel surface. Generally, the bottom surface of the guide steel surface of rough rolling shall be 10~15mm lower than the bottom surface of the pass to ensure that the rolled piece can enter and leave the guide steel smoothly.

Z：导卫过钢面与孔型底面有关的高度尺寸，一般粗轧导卫底面应低于孔型底面10~15mm，保证轧件能顺利地进出导卫。

ΔR：the gap value between the roller ring and guide plate, usually $\Delta R = 15~30$mm, so that the roller has enough adjustment range.

ΔR：辊环与导板的间隙值，通常取$\Delta R = 15~30$mm，使轧辊有足够的调整范围。

The length and width of the guide rail are determined according to the layout characteristics of the rolling mill and the distance between the stands.

导卫的长度和宽度根据轧机的布置特点、机架间的距离来确定。

The thickness of steel plate for guide and guard is generally 30~60mm according to the strength requirements.

导卫用钢板的厚度根据强度要求取值，一般为30~60mm。

2) Middle and finishing rolling sliding guide.

2) 中、精轧滑动导卫。

In small-scale plants, the inlet of the sliding guide is mainly composed of guide plate and guide box, and the outlet is composed of guide tube and guide box.

小型厂中、精轧滑动导卫入口主要由导板、导板盒构成，出口由导管、导管盒构成。

Guide plate box and guide tube box are used to fix and adjust the whole frame of guide plate and guide tube. They are locked by bolts on both sides and above. Its structure depends on the size of guide plate and conduit.

导板盒与导管盒是用来固定和调整导板和导管的整体框架，在其两侧和上方采用螺栓锁紧。其结构依靠导板和导管的尺寸来确定。

The pass design of sliding guide and guard in medium and finishing rolling is square pass, which is used to hold round rolling pieces. The main dimensions are determined as follows：

中、精轧滑动导卫孔型设计为方孔型，用于夹持圆形轧件，主要尺寸确定如下：

The value of guide plate pass width a：generally be 10~15mm larger than the diameter of round rolled piece.

导板孔型宽度 a 的取值：一般比圆形轧件直径大 10~15mm。

The value of pass height of guide plate b: generally b = diameter of rolled piece/2+(1~3) mm, and the upper limit of b is taken for rolled piece with large section.

导板孔型高度 b 的取值：一般为 b = 轧件直径/2+(1~3) mm，对断面较大的轧件 b 取上限。

The length of straight part of guide plate pass L_z: the size of L_z in the design of guide plate is very important, generally depends on the size and shape of rolled piece. If the induced rolled piece is round or with small cross section, L_z can be shortened appropriately. For oval rolled pieces, it depends on the situation. Under the condition that the rolled piece can be straightened, the shorter the better, if L_z is longer, the resistance of the rolled piece in the guide plate will be greater, and it will be difficult to enter the pass, but if L_z is too short, it will be difficult to straighten the rolled piece. In general, the guide plate L_z for medium and finishing rolling is 60~120mm, and the lower limit for round small section rolling is taken.

导板孔型直线部分长度 L_z 的取值：L_z 的尺寸在导板设计中极为重要，一般取决于轧件的大小和形状。如果诱导的轧件为圆形或断面较小，L_z 可适当取短一些。对于椭圆形轧件则应视情况而定。在能扶正轧件的情况下，越短越好，L_z 过长，会使轧件在导板中受到的阻力过大，进入孔型困难，但 L_z 过短，则难以扶正轧件。一般中、精轧用导板 L_z 的取值为 60~120mm，圆形小断面轧件取下限。

The length and height of the guide plate shall be determined according to the layout characteristics of the rack and the strength of the guide plate.

导板的长度和高度应根据机架的布置特点和导板的强度来确定。

The key point of the design and use of sliding guide is to choose the right material and have good commonality. The traditional guide plate material is generally nodular cast iron. At present, with the application of powder metallurgy, composite coating, laser quenching and other new technologies, the guide plate material tends to be diversified, but generally towards the direction of high nickel chromium alloy and high wear resistance.

滑动导卫设计和使用的重点是选择合适的材质和具有良好的共用性。传统的导板材质一般为球墨铸铁，目前随着粉末冶金、复合镀层、激光淬火等新技术的应用，导板的材质趋于多样化，但大体上向着高镍铬合金、高的耐磨性方向发展。

Good commonality can greatly reduce spare parts storage, save costs, and increase the number of repair and reprocessing.

良好的共用性，可以大大减少备件储备，节省费用，并使修复与重加工的次数增加。

(3) Rolling guide.

(3) 滚动导卫。

Rolling guide is a kind of guide device which mainly uses rolling friction and can make comprehensive use of sliding friction. According to different purposes, there are two kinds of rolling guide guards: twist guide guard and syncopation guide guard.

滚动导卫是一种以滚动摩擦为主并能将滑动摩擦加以综合利用的导卫装置。滚动导卫按不同的用途还有扭转导卫和切分导卫两种。

1) Structure of rolling guide.

1）滚动导卫的结构。

The structure of rolling guide is complex, but it has high precision and small wear, which has a good effect on improving product quality. In bar production, rolling guide is used to hold ellipse and other special-shaped rolling pieces. It is used for vertical stand in bar mill of a small factory.

滚动导卫结构较为复杂，但其精度高，磨损小，对提高产品质量有良好的效果。在棒材生产中，滚动导卫用来夹持椭圆轧件及其他异形轧件，在某小型厂的棒材轧机上用于立式机架的入口。

The rolling guide consists of the guide box, box, bracket, guide roller and guide plate, as shown in Figure 2-2.

滚动导卫主要由下面介绍的导卫盒、箱体、支架、导辊和导板等构成，如图 2-2 所示。

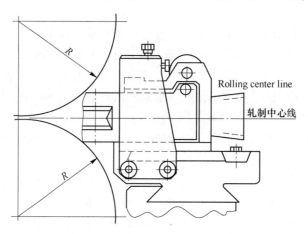

Figure 2-2　Structure of rolling guide
图 2-2　滚动导卫的结构

①Guide box. The guide box is a box used to install the rolling guide, which is fixed on the guide beam by pressing plate. The pressure block bolt above the guide box is used to fix the guide. The guide can be adjusted longitudinally in the box, so that the distance between the guide and the roll is appropriate.

①导卫盒。导卫盒是用于安装滚动导卫的箱体，通过压板固定在导卫梁上。导卫盒上方的压块螺栓用来固定导卫。导卫可在盒子中纵向调整，使导卫与轧辊间距合适。

②Box. The box body is used to install guide roller bracket and guide plate. It does not contact with red steel, but it must bear impact and vibration. The selection of box material shall fully consider the impact toughness and deformation resistance, so as to keep the adjustment size of guide plate and guide roller stable.

②箱体。箱体用来安装导辊支架与导板，自身不与红钢接触，但要承受冲击和振动。箱体材质的选择应充分考虑耐冲击韧性和抗变形能力，以使导板和导辊的调整尺寸保持稳定。

In the design, it should be ensured that the guide roller bracket and guide plate are installed firmly, the bracket has enough adjustment range, and has good water cooling and lubrication system. In addition, there should be a device for adjusting the balance of the guide roller. The box body is an assembly structure, and the determination of the size is related to the size of the rolled piece and the arrangement form of the rolling mill. There are various forms, but the general principle and requirements are to facilitate the use and maintenance.

在设计中,应保证导辊支架与导板安装牢固,支架有足够的调整范围,并要有良好的水冷和润滑系统。此外还应有导辊调整平衡的装置。箱体为装配结构,尺寸的确定与轧件的大小、轧机的布置形式有关,形式可多种多样,但总的原则和要求是要便于使用和维护。

③Bracket. The bracket is used to install the guide roller, which is positioned on the box body by the rotating shaft, and there is a disc spring balancing device below, which can adjust the gap and balance of the guide roller. The support is directly subjected to the impact of elastic deformation and thickness change of rolled piece, so there is a higher requirement on material selection. Generally, the support is forged from spring steel.

③支架。支架用于安装导辊,在箱体上靠转动轴定位,下方有碟簧平衡装置,可调节导辊的间隙与平衡。支架直接承受弹性变形和轧件厚度变化的冲击,因而在材质选择上有较高的要求,一般支架由弹簧钢锻造而成。

④Guide roller. The guide roller is an important part of the rolling guide. When it is combined with the rolling guide plate, the rolling piece can be clamped and guided better than the sliding guide. The guide roller usually has elliptical and diamond pass, which is in close contact with red steel to prevent torsion and deflection. The material requires high strength, rigidity, hardness, wear resistance, cold shock resistance and sufficient toughness. Therefore, the guide roller should be very careful in material selection, otherwise it will be difficult to meet the production requirements. The cold work die steel has been used in a factory. In addition, the guide roller should have good cooling.

④导辊。导辊为滚动导卫的重要部件,与滚动导板配合,能使轧件得到比滑动导卫更好的夹持和导入作用。导辊常带有椭圆、菱形孔型,与红钢紧密接触,防止其扭转和偏斜,材质要求有高的强度、刚度、硬度、耐磨性和抗激冷激热性能以及足够的韧性。因此导辊在材质的选择上要非常慎重,否则将难以满足生产的要求。某厂经多方比较,选用冷作模具钢使用效果较好。此外,导辊还要有良好的冷却。

The structure of guide roller is shown in Figure 2-3, and the specific dimensions are determined as follows:

导辊的结构如图 2-3 所示,具体尺寸确定如下:

H: The height of the guide roller depends on the height of the bracket opening.

H: 导辊的高度,依据支架开口的高度而定。

ϕ: The diameter of guide roller shall be determined according to the opening range of guide and guard bracket, and enough adjustment allowance and rework allowance shall be reserved.

ϕ：导辊的直径，根据导卫两支架所能张开的范围而定，并要留有足够的调整余量和重加工余量。

R：The pass radius of guide roller is the same as that of the guided rolling piece.

R：导辊的孔型半径，与所引导轧件的孔型半径相同。

b：Pass depth of guide roller. If the rolled piece to be guided is elliptical, the size of b shall be determined according to the following formula：
$$b = H/2 - (1 \sim 20)\text{mm}$$

b：导辊的孔型深度，若所要引导的轧件为椭圆形，则 b 的尺寸按下式确定：
$$b = H/2 - (1 \sim 20)\text{mm}$$

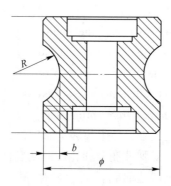

Figure 2-3　Structure of guide roller

图 2-3　导辊结构

⑤Guide plate. The guide plate for rolling guide and guard consists of two half blocks. The front Figure 2-3 structural surface of guide roller is close to the guide roller to guide the rolled piece into the pass type and protect the guide roller from severe impact. There is a block at the bottom of the guide box to prevent the guide plate from impacting the guide roller. The structure of the guide plate is shown in Figure 2-4. The size and material selection are the same as that of the guide plate for sliding guide.

⑤导板。滚动导卫用导板由两个半块组成，前图 2-3 导辊结构面紧挨着导辊，引导轧件进入孔型，保护导辊免受严重冲击。导卫箱体底部有挡块，防止导板冲击导辊。导板的结构如图 2-4 所示，尺寸确定、材质的选择与滑动导卫用导板相同。

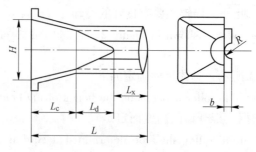

Figure 2-4　Structure of rolling guide plate

图 2-4　滚动导板结构

⑥Guide roller bearing and lubrication. The guide roller of rough rolling should bear great impact and lateral force, and single row tapered roller bearing should be selected. Light single row tapered roller bearing is selected for the guide rail of medium rolling and some finishing rolling. The single row ball bearing with high limit speed can be selected for finishing rolling. There are two lubrication methods: dry oil lubrication and oil air lubrication. It is easy to control the quantity and effect of dry oil lubrication, and it does not need to increase the investment of equipment, but it has the disadvantages of large waste and environmental pollution in use. Oil air lubrication is clean, easy to maintain, able to cool the bearing, and no need to stop when refueling. However, it is not easy to detect the presence or absence of oil and gas in production due to its oil supply through pipelines. Once the oil is cut off or the oil quantity is insufficient, it will directly lead to the burning loss of the guide guard, and the equipment needs a certain investment. At present, dry oil lubrication is used in a factory.

⑥导辊轴承和润滑。粗轧导辊要承受较大的冲击和侧向力，选用单列圆锥滚柱轴承。中轧和部分精轧导卫选用轻型单列圆锥滚柱轴承。精轧成品道次考虑导辊高速转动的需要，可选用极限转速高的单列滚珠轴承。润滑方式通常有干油润滑和油气润滑两种。干油润滑加油量和效果易于控制，且无须增加设备投资，但在使用中存在浪费大、污染环境的缺点。油气润滑方式干净，维护方便，能够冷却轴承，且加油时无须停车。但因其通过管路供油，油气量的有无生产中不易察觉。一旦断油或油量不足，将直接导致导卫的烧损，且设备需一定置的投资。某厂选用的润滑方式目前为干油润滑。

2) Use and maintenance of rolling guide.

2) 滚动导卫的使用和维护。

Rolling guide plays an important role in bar production, and the working environment is bad. It is affected by the impact of rolling piece, high temperature, water cooling, friction and many other factors, which directly determine the role of guide and service life. If there is any deviation, it will lead to accidents of stacking and steel jamming or scraping of rolled pieces, and quality problems such as folding and ears will occur. This requires great attention to the use and maintenance of rolling guide.

滚动导卫在棒材生产中地位重要，并且工作环境恶劣，在工作中受轧件很大的冲击和高温、水冷、摩擦等诸多因素的影响，这些都直接决定导卫作用的发挥和使用寿命。如有偏差就会导致堆、卡钢事故或刮伤轧件，出现折叠、耳子等质量问题。这就要求非常重视滚动导卫的使用和维护。

Attention shall be paid to the use and maintenance of Rolling Guide:

滚动导卫使用和维护时要注意：

①To ensure the machining quality and high dimensional accuracy of each part is the premise of the guide.

①确保各部件加工质量，高的部件尺寸精度是导卫发挥作用的前提。

②Ensure assembly quality and adjustment quality. Assembly should be careful to ensure smooth water and oil routes. The guide roller shall rotate flexibly, all parts shall be free of deformation and damage, and dirt such as oil stain and iron oxide scale on the surface shall be cleaned.

②确保装配质量和调整质量。装配时要认真仔细，保证水路、油路畅通。导辊要转动灵活，各部件要确认无变形、无损坏，表面油污、氧化铁皮等脏污要清理干净。

③The assembled guide guard shall check whether it conforms to the frame number, pass number and rolling specification.

③装配完的导卫要检查是否符合机架号、孔型号和轧制的规格。

④When installing the guide guard, make sure it is aligned with the rolling line.

④安装导卫时要确保与轧制线对中。

⑤Adhere to frequent inspection and adjustment in production, and solve problems in time if found. Damaged guide guards and excessively worn parts shall be replaced in time.

⑤坚持生产中勤检查、勤调整，发现问题及时解决。已损坏的导卫和过度磨损的部件要及时更换。

⑥The replaced guide guard shall be inspected and maintained in a timely manner to ensure that the guide guard put into use next time is free of defects and hidden dangers.

⑥换下来的导卫要及时进行全面检查和维护，保证下次投入使用的导卫无缺陷、无隐患。

In addition, the adjustment of guide roller gap is very important in the use of rolling guide. If the gap is too small, the rolling piece will be difficult to pass, resulting in steel piling, jamming accident, or excessive wear of guide roller. If the gap is too large, there may be torsion and inclination of the rolled piece, which will not play the proper clamping role and lead to quality accidents.

此外，导辊间隙的调整在滚动导卫的使用中至关重要。若间隙过小，轧件将难以通过，导致堆、卡钢事故，或导辊过度磨损。间隙过大，就有可能出现轧件扭转和倾斜，起不到应有的夹持作用，发生质量事故。

There are three ways to adjust the guide roller gap:

导辊间隙的调整方法通常有以下三种：

①Test rod adjustment method. The adjustment method of test bar is to use the same shape and size of the test bar to adjust the roll gap. The center line of the adjustment is the seam of the guide plate, and the adjustment accuracy depends on the experience and feeling of the operator. This method is convenient, not limited by the site, and has a fast adjustment speed. However, due to the interference of human factors, the accuracy is relatively low. The key of the adjustment method of the test bar is the quality of the test bar, that is, the size and shape of the test bar should be consistent with the size and shape of the rolled piece in the actual production.

①试棒调整法。试棒调整法适用于该道次轧件形状、尺寸相同的试棒去试调辊缝。调整时以导板的合缝为中心线，调整精度依靠操作者的经验和感觉。此方法比较方便，不受场地限制，调整速度快，但由于人为因素干扰多，精度相应较低。试棒调整法的关键是试棒的质量，即试棒尺寸和形状应与实际生产中轧件尺寸和外形相符合。

②Optical calibrator method. The principle of optical calibrator is to project the pass of guide roller on the screen with a light source.

②光学矫正仪法。光学矫正仪法的原理是利用光源将导辊的孔型投影于屏幕上。

With cursor and scale on the screen, adjust the guide according to the hole projection value to meet the requirements. This method requires optical correction equipment, increased investment, more precise instruments, high requirements for use and maintenance, and long adjustment time of guide and guard. However, it has high adjustment accuracy and is suitable for adjusting the guide of finishing mill.

屏幕上带有光标和刻度,按照孔型投影值调整导卫至符合要求即可。此法要求配有光学矫正设备,增加了投资,且仪器较精密,使用与维护要求高,导卫的调整时间长。但其调整精度高,适用于调整精轧机的导卫。

③Internal caliper measurement method. According to the shape and size of the guide roll pass and the shape and size of the rolled piece, the internal caliper measurement method calculates the roll gap value of the guide roll and completes the measurement and adjustment with the internal caliper. This method is suitable for the adjustment of large guide and special pass guide in rough rolling.

③内卡尺测量法。内卡尺测量法依据导辊孔型的形状和尺寸及轧件形状和尺寸,推算出导辊的辊缝值,用内卡尺完成测量调整。此法适用于粗轧大型导卫和特殊孔型导卫的调整。

(4) Twist guide.

(4) 扭转导卫。

The function of the twist guide is to turn over the rolled piece of this pass, so that the next pass can achieve the pressing direction of 45° or 90° with the pressing direction of this pass. The twist guide is generally located at the exit of the rolling mill. The selection of its type depends on the shape of the rolled piece to be twisted and the angle to be twisted. See Figure 2-5 for the structure of torsional guide and guard selected by a small factory, which is mainly composed of torsional roll, rotating body and conduit.

扭转导卫的作用是将本道次的轧件进行翻转,以便下一道次实现与本道次压下方向呈45°或90°方向地压下。扭转导卫一般位于轧机的出口,其类型的选择,依据所要扭转轧件的形状和所要扭转的角度来确定。某小型厂选用的扭转导卫结构如图2-5所示,主要由扭转辊、旋转体、导管组成。

The twist body of the guide body extends out of the guide body, which is conducive to the falling off of the iron oxide scale and the rapid disassembly and assembly of the roller. The roller shaft has eccentricity to adjust the gap between the two rollers. The twist angle is adjusted by the rotation of the rotating body.

此种导卫的扭转体伸出导卫体之外,利于氧化铁皮的脱落和辊子的快速拆装。辊轴带有偏心距,用以调整两辊的间隙。扭转角度依靠旋转体的转动来调整。

The main dimension of the twist guide is the twist angle. In the design, the first step is to determine the turning point of the rolled piece. The distance between the turning point and the center line of the roll depends on the distance between the guide beam and the roll. The angle is calculated as follows:

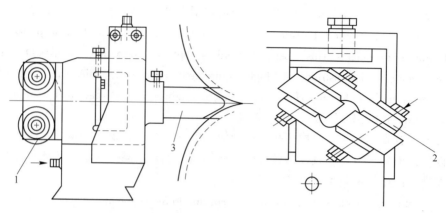

Figure 2-5 Structure of twist guide
1—Twist roll; 2—Rotating body; 3—Conduit
图 2-5 扭转导卫的结构
1—扭转辊；2—旋转体；3—导管

扭转导卫的主要尺寸是扭转角度。设计中首先要确定轧件开始扭转的扭转点，扭转点的距离根据导卫梁距轧辊中心线的距离而定。角度的计算方法如下：

$$\beta \approx L_a \alpha / (L_b - L_c)$$

Where β——The angle of the torsion roll relative to the rolling piece, (°);
  L_a——The distance from the twist roll to the center line of the rolling mill, mm;
  L_b——Distance between two racks, mm;
  L_c——The distance from the roller guide at the entrance of the next frame to the roller center line of the next frame, mm;
  α——The angle at which the rolled piece needs to be twisted into the next frame, (°).

式中 β——扭转辊相对于轧件扭转的角度，(°)；
  L_a——扭转辊到该轧机中心线的距离，mm；
  L_b——两机架之间的距离，mm；
  L_c——下一机架入口滚动导卫到下一机架轧辊中心线的距离，mm；
  α——轧件进入下一机架需要扭转的角度，(°)。

In use, when the size of the rolled piece changes or the twist roll is worn, the contact point between the rolled piece and the twist roll will change, and then the twist point and the twist angle will change, and the rolled piece cannot be turned over correctly. Therefore, in production, it is necessary to observe and adjust the twist guide frequently, and pay attention to water cooling and lubrication.

使用中，当轧件尺寸变化或扭转辊磨损时，轧件与扭转辊的接触点会发生变化，随之扭转点和扭转角度发生变化，轧件不能正确翻转。因此生产中针对扭转导卫要勤观察、勤调整，并应注意水冷和润滑情况。

(5) Division guide.

(5) 切分导卫。

The slitting guide is exactly the guide with slitting wheel, which can divide two or more parallel rolling pieces into a single rolling piece, generally located at the outlet of the slitting pass.

切分导卫确切地说是带有切分轮的导卫,能将两个或多个并联的轧件分成单根轧件,一般位于切分孔型的出口。

A factory uses a two-line slitting guide guard, whose structure is shown in Figure 2-6, mainly composed of a slitting wheel, a plug-in and a distributing box. The slitting wheel is a pair of driven wheels, the edge has an angle, the edge is sharp, and the remaining friction of the rolled piece cuts the rolled piece. The cantilever type is adopted for the installation of the splitting wheel, which is conducive to the rapid treatment of steel stacking accidents. The clearance of the cutting wheel is adjusted by the eccentricity of worm and shaft. In this way, the two cutting wheels can move at the same time, ensuring the stability of the rolling line and high adjustment accuracy. The guide design is symmetrical from top to bottom and can be used interchangeably, which can solve the problem of inconvenient installation and adjustment when using the roller side groove.

某厂使用的是二线切分导卫,结构如图 2-6 所示,主要由切分轮、插件、分料盒构成。切分轮是一对从动轮,刃部有斜角,边缘锋利,靠轧件的剩余摩擦力切分轧件。切分轮的安装采用悬臂式,有利于快速处理堆钢事故。切分轮间隙的调整采用蜗杆和轴的偏心距调节。这种方式可使两切分轮同时移动,保证轧制线的稳定,调整精度高。导卫设计上下对称,可调换使用,能解决轧辊边槽使用时安装、调整不方便的难题。

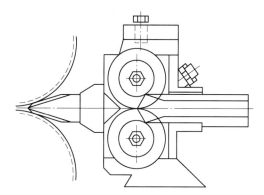

Figure 2-6   Segmentation Guide
图 2-6   切分导卫

The key to the use of the slitting guide is to adjust the clearance of the slitting wheel. In use, if the gap is too large, the rolled piece may not be cut; if the gap is too small, the cut rolled piece is easy to run to both sides, and the walking is not stable. Both will lead to steel piling accidents. In addition, the slitting guide should be strictly aligned with the rolling line in use. A slight deviation will lead to uneven slitting of the rolled piece and quality accident. Therefore, it is necessary to observe the cutting quality and the wearing condition of the cutting wheel at any time in production, adjust and deal with the problems found in time, and ensure the quality of water cooling and lubrication, so as to avoid the phenomenon of steel sticking of the cutting wheel, which will lead to steel stacking accident and the burning loss of the guide guard.

使用切分导卫的关键是切分轮间隙的调整。使用中,若间隙过大,则有可能切不开轧件;若间隙过小,则切开的轧件易向两边跑,行走不稳定。两者都会导致堆钢事故。此外,切分导卫在使用中要严格保证与轧制线对中,稍有偏差,将导致轧件切分不均匀,产生质量事故。所以生产中要随时观察切分质量和切分轮的磨损状况,发现问题及时调整、处理,并应保证水冷和润滑质量,以免发生切分轮粘钢现象,导致堆钢事故和导卫的烧损。

2.3.3.2　Guide and Guard Adjustment

2.3.3.2　导卫调整

(1) Rough rolling guide adjustment.

(1) 粗轧导卫调整。

1) When installing and adjusting the guide beam, ensure that the beam surface is level, with moderate height and firm fixation.

1) 安装调整导卫梁时,要保证梁面水平,高低适中,固定牢靠。

2) The center line of the inlet and outlet guide guard shall be aligned with the center line of the rolling groove and firmly fixed.

2) 进、出口导卫的中心线应与轧槽的中心线对正,固定牢靠。

3) The center line of the guide groove of the front and rear roller table of the rolling mill shall be consistent, the inlet and outlet of the guide groove shall be aligned with the rolling groove, and the height shall be moderate.

3) 轧机前、后辊道导槽中心线保持一致,导槽进、出口对正轧槽,高低适中。

4) The front end of the guard board must be consistent with the rolling groove, and the lower guard board shall be 5~10 mm lower than the rolling groove.

4) 卫板前端必须与轧槽吻合,下卫板要低于轧槽5~10mm。

5) The upper and lower rolls shall be adjusted horizontally with proper spacing.

5) 粗轧扭转辊应把上、下辊调整水平,间距适中。

6) The center line of the outlet pipe of the rolling mill shall be aligned with the pass, and the gap between the front end and the rolling groove shall not be greater than 5mm, which shall be fixed firmly.

6) 轧机出口管子中心线要与孔型对准,前端与轧槽间隙不大于5mm,固定牢固。

(2) Adjustment of guide and guard of medium rolling.

(2) 中轧导卫调整。

1) When installing and adjusting the guide guard beam, ensure that the beam is level, high and low, and fixed firmly.

1) 安装调整导卫梁时,要保证横梁水平,高低适中,固定牢靠。

2) The center line of the inlet and outlet guide guard shall be aligned with the center line of the rolling groove and firmly fixed. The center line of the guide groove of the front and rear roller table of the rolling mill shall be consistent with the rolling line, and the inlet and outlet of the guide groove shall be aligned with the rolling groove, and the height shall be moderate.

2）进、出口导卫的中心线应与轧槽中心线对正，固定牢靠。轧机前、后辊道导槽中心线与轧制线保持一致，导槽进、出口对正轧槽，高低适中。

3) The front end of the guard board must be consistent with the rolling groove, and the lower guard board shall be 5~10mm lower than the rolling groove.

3）卫板前端必须与轧槽吻合，下卫板要低于轧槽 5~10mm。

4) The torsion pipe shall be aligned and level with the pass, and the gap between the front end and the rolling groove shall not be greater than LMM. It shall be firmly fastened and the torsion angle shall be moderate.

4）扭转管应与孔型对正、水平，前端与轧槽间隙不大于 1mm，紧固牢靠，扭转角度适中。

5) The center line of the pipe at the outlet of the rolling mill shall be aligned with the pass, and the gap between the front end and the rolling groove shall not be more than 1mm, which shall be fixed firmly.

5）轧机出口管子中心线要与孔型对正，前端与轧槽间隙不大于 1mm，固定牢固。

6) It is forbidden to roll low-temperature steel. In case of black head, black mark or uneven temperature in the production process, the No. 4 platform shall be informed to stop in time.

6）禁止轧制低温钢，在生产过程中发现钢坯带黑头、黑印或温度不均匀时应及时通知 4 号台停车。

7) In case of any accident of steel jamming and roll wrapping during production, the cutter shall be used to cut open and release the guide guard, and command the reversing treatment.

7）生产中发现卡钢、缠辊事故时，用切割器割开松开导卫，指挥倒车处理。

8) According to the rolling specifications and passes, the qualified guide plates for import and export shall be selected. The guide plate shall be free of burr, hard point and uneven convex and concave defects.

8）导卫选择要根据轧制规格、道次挑出合格的进出口导卫板。导卫板不允许有飞边、硬点和凸凹不平等缺陷。

（3）Guide adjustment of finishing rolling.

（3）精轧导卫调整。

1) After the roller is adjusted to be qualified, the guide guard shall be installed and adjusted.

1）轧辊调整合格后进行导卫安装调整。

2) When installing and adjusting the guide beam, ensure that the beam surface is level, with moderate height and firm fixation.

2）安装调整导卫梁时，要保证梁面水平，高低适中，固定牢靠。

3) The center line of the inlet and outlet guide guard shall be aligned with the rolling tip and firmly fixed.

3）进、出口导卫的中心线应与轧梢对准，固定牢靠。

4) The front end of the guard pipe shall be consistent with the rolling groove, the gap shall not be greater than LMM, and the lower guard pipe shall be 5~10mm lower than the rolling

groove.

4）卫管前端应与轧槽吻合，间隙不大于1mm，下卫管要低于轧槽5~10mm。

5）The torsional pipe shall be aligned with the rolling groove, and the gap between the front end and the rolling groove shall not be more than 1mm, and it shall be firmly fixed.

5）扭转管应与轧槽对准，前端与轧槽间隙不大于1mm，固定牢固。

6）During the installation of the split guide rail, the special sample plate shall be used for inspection and adjustment to ensure that the population guide rail and the outlet guide rail are aligned with the rolling groove, to ensure that the outer wall clearance of the torsion pipe is ensured, and the sample plate shall be used for inspection to ensure that the two lines are parallel.

6）安装切分导卫时，用专用样板进行检查、调整，确保入口导卫和出口导卫对正轧槽，保证扭转管外壁间隙，并用样板检验，保证双线平行。

7）When installing the cutting wheel, the cutting cone and the cutting wheel are in the same straight line.

7）安装切分轮时，切分锥、切分轮在同一条直线上。

8）When a gasket is required in front of the inlet and outlet guide boxes, the thickness of the gasket shall be appropriate without burr or deformation.

8）进、出口导卫盒体前需加垫片时，垫片厚度适宜，且无飞边、无变形。

2.3.4 Common Rolling Defects and Prevention
2.3.4 常见的轧制缺陷及其预防

2.3.4.1 Pits
2.3.4.1 凹坑

Pit refers to the irregular, different size and depth of pits on the steel surface.

凹坑是指在钢材表面呈现无规则的、大小及深浅不一的凹点。

Causes:

形成原因：

(1) There are defects in roll pass during transportation and assembly;

(1) 轧辊孔型在运输、装配时存在缺陷；

(2) When rolling low temperature steel or piling steel, the pass will be worn or cut when cutting steel;

(2) 轧制低温钢或堆钢打滑时将孔型磨坏或割钢时割坏；

(3) Iron oxide scale, guide parts and other foreign matters are bitten into the groove, which is attached to the rolling groove (i.e. glued groove), resulting in the defect of the groove, or the outlet guide is installed too low, and the front end is caused by friction with the rolling groove.

(3) 氧化铁皮、导卫零件等异物被咬入孔型，附着在轧槽上（即粘槽）造成孔型缺损，或出口导卫安装过低，前端与轧槽摩擦所致。

Elimination measures:

消除措施：

(1) Check whether the rolling groove is damaged before putting on line;

(1) 上线前检查轧槽是否缺损；

(2) Non rolling low temperature steel;

(2) 不轧低温钢；

(3) Carry out spot check frequently, eliminate the remaining foreign matters in the guide and guard in time and replace them in time;

(3) 勤点检，发现导卫中残存异物及时消除，及时更换；

(4) Incorrect installation of outlet guide guard;

(4) 出口导卫安装不正确；

(5) Standardize the material type, and replace it in time when the rolling groove starts to line up or wears.

(5) 规范料型，轧槽起线或磨损时，及时更换。

2.3.4.2 Folding

2.3.4.2 折叠

Folding is a kind of folding body with various angles formed on the steel surface.

折叠是一种在钢材表面形成的各种角度的折体，长短不一。

Causes：

形成原因：

(1) If the material type is not suitable or the roll is not adjusted properly, the metal is too full in the pass to form ears or not filled in the pass (lack of meat) and rolled into folds in the next pass;

(1) 料型不合适或轧辊调整不当，金属在孔型中过充满形成耳子或没有填满孔型（缺肉）而在下一孔型中轧成折叠；

(2) The ears are produced due to the installation of the entrance guide and the deflection of the entrance guide, and are rolled into folds in the next pass;

(2) 因入口导卫安装、调整偏斜产生耳子，在下一孔型中轧成折叠；

(3) One side of the population guide plate is seriously worn and loses the supporting function. The rolled piece is not supported in the pass and the steel is poured, which results in the ears, forming the fold in the next pass.

(3) 入口导板一边磨损严重，失去扶持作用，轧件在孔型中扶不正倒钢而产生耳子，在下一孔型中形成折叠。

Elimination measures：

消除措施：

(1) Make proper roll adjustment and standardize material type;

(1) 进行适当的轧辊调整，规范料型；

(2) The guide guard is installed to align the rolling groove, and the guide roller is not deviated;

(2) 导卫安装对正轧槽，导辊不偏；

(3) Replace the guide plate regularly to prevent the use of unqualified guide plate;

(3) 定时更换导板，防止使用不合格导板；

(4) During adjustment, check the material type first, and then check the guide guard.

(4) 调整时，先检查料型，后检查导卫。

2.3.4.3 Ears
2.3.4.3 耳子

Ear is a defect that the metal is too full in the pass and overflows from the roll gap along the rolling direction. There are two kinds of ears, one side and two sides.

耳子是金属在孔型中过充满，沿轧制方向从辊缝中溢出而产生的缺陷，有单边耳子和双边耳子两种。

Causes:

形成原因：

(1) Overfill;

(1) 过充满；

(2) Improper roll gap adjustment;

(2) 辊缝调整不当；

(3) Inlet guide plate deflection;

(3) 入口导板偏斜；

(4) Unreasonable pass design;

(4) 孔型设计不合理；

(5) The temperature of rolled piece is low.

(5) 轧件温度低。

Elimination measures:

消除措施：

(1) The population guide plate is aligned and the hole pattern is fixed;

(1) 入口导板对正，孔型固定；

(2) Standard material type;

(2) 规范料型；

(3) Non rolling low temperature steel;

(3) 不轧低温钢；

(4) Pass design should be reasonable.

(4) 孔型设计要合理。

2.3.4.4 Scratches
2.3.4.4 刮伤

Scratch is a long and thin concave defect along the rolling direction. Its shape, depth and width are different for different reasons.

刮伤是沿轧制方向上纵向的细长凹下缺陷，其形状和深浅、宽窄因原因不同而有所不同。

Causes:

形成原因:

(1) The oxide scale or other foreign matters of the rolled piece are gathered in the guide device to contact with the rolled piece with high temperature and high speed movement;

(1) 轧件的氧化铁皮或其他异物聚集在导卫装置内与高温、高速运动的轧件相接触;

(2) The guide device is abnormally worn. The steel separator baffle for the above unloading or there are foreign matters and welding nodules in it that are not cleaned.

(2) 导向装置异常磨损,如上卸钢分钢挡板或其内有异物、焊瘤未清除干净。

Elimination measures:

消除措施:

(1) Check the guide and guard points in place, and replace the hanging thorns in time in case of any;

(1) 导卫点检到位,发现挂刺及时更换;

(2) Install guide guard correctly to prevent contact between guide guard and rolling piece;

(2) 正确安装导卫,防止导卫与轧件产生点线接触;

(3) Choose the material which is not easy to produce thermal bond to make the guide.

(3) 选择不易产生热黏结的材质制作导卫。

2.3.4.5 Scarring

2.3.4.5 结疤

Scarring is a defect caused by the rolling of the oxide scale remaining in the guide rail together with the rolled piece.

结疤是残留在导卫内的氧化铁皮与轧件一起进行轧制而产生的缺陷。

Causes:

形成原因:

(1) The rolled piece has a large tail and ears;

(1) 轧件尾巴大,带耳子;

(2) The guide roller is loose and the tail has big ears (i.e. "airplane");

(2) 导辊松,尾巴有大耳子(即"飞机");

(3) The guide rail is damaged, and the oxide scale left after pricking is rolled together with the rolled piece.

(3) 导卫坏,挂刺后留下的氧化铁皮与轧件一起轧制。

Elimination measures:

消除措施:

(1) Standardize the material type, use the material reasonably, prevent the tail from being big;

(1) 规范料型,合理用料,防止尾巴大;

(2) Carry out spot check frequently and replace the damaged guide in time.

（2）勤点检，导卫损坏及时更换。

2.3.4.6　Slope
2.3.4.6　斜面

Bevel is a common defect that two pairs of geometric dimensions on the cross section of steel are not equal due to the wrong roll.

斜面是指由于轧辊辊错造成的钢材横断面上的两对几何尺寸不相等的一种常见缺陷。

Causes:

形成原因：

(1) The roller screw is not fixed firmly;

（1）轧辊螺钉固定不牢；

(2) The axial screw is not fixed firmly;

（2）轴向螺钉固定不牢；

(3) Single side pressing of roller;

（3）轧辊单面压紧；

(4) The guide roller of the finished product is too loose or too tight;

（4）成品导辊过松或过紧；

(5) Back and forth movement of roller bearing;

（5）轧辊轴承来回窜动；

(6) The torsional guide roller at the outlet of the front frame of the finished product is damaged;

（6）成品前架出口扭转导辊损坏；

(7) The crossbeam of finished product population is high and uneven.

（7）成品入口横梁高低不平。

Elimination measures:

消除措施：

(1) Check regularly whether the roller screw and axial screw are firm;

（1）勤点检，检查轧辊螺钉、轴向螺钉是否牢固；

(2) When pressing the finished product, keep the same amount of reduction from north to South;

（2）成品压料时，应保持南北相等的压下量；

(3) If the guide roller is not tight, it should be adjusted in time;

（3）导辊松紧不当时，应当及时调整；

(4) Ensure the level of the beam.

（4）保证横梁的水平。

2.3.4.7　Pitted surface
2.3.4.7　麻面

Pockmarked surface refers to the defect caused by the uneven distribution of pockmarks on the steel surface.

麻面是指在钢材表面上出现的大小分布不均匀的麻点而造成的缺陷。

Causes：

形成原因：

(1) Water shortage of finished product tank；

(1) 成品槽缺水；

(2) The groove is seriously worn.

(2) 轧槽磨损严重。

Elimination measures：

消除措施：

(1) Ensure that the water pipe is aligned with the rolling groove；

(1) 确保水管对正轧槽；

(2) When finding pitted surface, change the groove in time.

(2) 发现有麻面时及时换槽。

2.3.4.8　Cracks

2.3.4.8　裂纹

Crack refers to the crack of different shapes on the steel surface.

裂纹是指钢材表面不同形状的破裂。

Causes：

形成原因：

(1) Overheating of raw materials；

(1) 原料过热；

(2) The surface quality of billet is poor；

(2) 钢坯表面质量差；

(3) Uneven deformation；

(3) 变形不均匀；

(4) Low rolling temperature or improper cooling.

(4) 轧件温度低或冷却不当。

Elimination measures：

消除措施：

(1) Check the blank strictly, and forbid charging when there is crack or subcutaneous bubble in the blank；

(1) 严格检查坯料，发现坯料存在裂纹或皮下气泡时，禁止装炉；

(2) Strictly implement the heating system, and prohibit overheating on the surface of billet；

(2) 严格执行加热制度，禁止出现坯料表面过热现象；

(3) It is strictly prohibited to roll steel with low temperature, and pay attention to uniform cooling water.

(3) 严禁轧制温度过低的钢，同时注意冷却水均匀。

Task 2.4 Control Cooling Process and Equipment
任务 2.4 控制冷却工艺及设备

Mission objectives
任务目标

(1) Understand the concept of controlled rolling and controlled cooling, the influence of controlled cooling on the properties of screw steel;

(1) 理解控制轧制和控制冷却的概念;理解控制冷却对螺纹钢性能的影响;

(2) It can set the process parameters of controlled rolling and cooling of bar, analyze the causes of common accidents in the process of controlled rolling and cooling of bar, and put forward measures to solve the problems.

(2) 能设定棒材控制轧制、控制冷却工艺参数;会分析棒材控轧控冷过程中出现的常见事故的原因,能提出解决问题的措施。

2.4.1 Theoretical Basis of Controlled Cooling Technology after Rolling
2.4.1 钢材轧后控制冷却技术的理论基础

As a strengthening method of steel, controlled rolling and controlled cooling are often used in steel rolling. This is an advanced rolling technology that simplifies the process and saves energy. It can fully tap the potential of steel by technological means, greatly improve the comprehensive properties of steel, and bring huge economic effects to metallurgical enterprises and society. Because of its comprehensive effect of deformation strengthening and transformation strengthening, it can not only improve the strength of steel, but also improve the toughness and plasticity of steel.

作为钢的强化手段在轧钢生产中常常采用控制轧制和控制冷却工艺。这是一项简化工艺、节约能源的先进轧钢技术。它能通过工艺手段充分挖掘钢材潜力,大幅度提高钢材综合性能,给冶金企业和社会带来巨大的经济效应。由于它具有形变强化和相变强化的综合作用,所以既能提高钢材强度又能改善钢材的韧性和塑性。

In the past few decades, the steel has been strengthened by adding alloy elements or reheating after hot rolling. These measures not only increase the cost but also extend the production cycle. In terms of performance, most of them improve the strength, reduce the toughness and deteriorate the welding performance.

过去几十年来通过添加合金元素或者热轧后进行再加热处理来完成钢的强化。这些措施既增加了成本又延长了生产周期,在性能上,多数情况是提高了强度的同时降低了韧性、焊接性能变坏。

In the past 20 years, the technology of controlled rolling and controlled cooling has been paid great attention by the international metallurgical industry. The metallurgists have comprehensively studied the relationship between various structures and properties of ferrite and pearlite steel, applied the laws of fine grain strengthening, precipitation strengthening and subcrystalline strengthening to the production of hot rolled steel, and controlled the grain size, subcrystalline strengthening size and properties of steel by adjusting the rolling process parameters. Because of the combination of hot rolling deformation and heat treatment, hot rolled steel with good strength and toughness is obtained, which greatly improves the performance of carbon steel. However, the control of rolling process generally requires low finishing temperature or large deformation, which will increase the mill load. Therefore, the control cooling process came into being. The purpose of controlled cooling after hot rolling is to improve the steel structure, properties, cooling time and production capacity. Controlled cooling after rolling can also prevent the steel from twisting and bending due to uneven cooling during the cooling process, and reduce the scale of oxide.

近20年来，控制轧制、控制冷却技术得到了国际冶金界的极大重视，冶金工作者全面研究了铁素体—珠光体钢各种组织与性能的关系，将细化晶粒强化、沉淀强化、亚晶强化等规律应用于热轧钢材生产，并通过调整轧制工艺参数来控制钢的晶粒度、亚晶强化的尺寸与数量。由于将热轧变形与热处理有机地结合起来，所以获得了强度、韧性都好的热轧钢材，使碳素钢的性能有了大幅度的提高。然而，控制轧制工艺一般要求较低的终轧温度或较大的变形量，因而会使轧机负荷增大，为此，控制冷却工艺应运而生。热轧钢材轧后控制冷却是为了改善钢材组织状态，提高钢材性能，缩短钢材的冷却时间，提高轧机的生产能力。轧后控制冷却还可以防止钢材在冷却过程中由于冷却不均而产生的钢材扭曲、弯曲，同时还可以减少氧化铁皮。

The strengthening and toughening properties of controlled cooling steel depend on the change of material factors such as phase transformation, precipitation strengthening, solution strengthening and ferrite recovery degree caused by rolling and water cooling conditions. Especially the rolling and water cooling conditions have a great influence on the transformation behavior.

控制冷却钢的强韧化性能取决于轧制条件和水冷条件所引起的相变、析出强化、固溶强化以及加工铁素体回复程度等材质因素的变化。尤其是轧制条件和水冷条件对相变行为的影响很大。

2.4.1.1 CCT Curve and Transformation Products of Controlled Cooling
2.4.1.1 CCT 曲线及控制冷却的转变产物

The isothermal transformation curve, also known as TTT curve, reflects the isothermal transformation law of undercooled austenite. However, in the continuous cooling transformation process, austenite in steel changes under the condition of continuous cooling. Moreover, the transformation products are different with different transformation speed. Continuous cooling transformation curve of undercooled austenite CCT curve is to cool at different cooling rates under continuous cooling conditions, measure the start point (temperature and time) and end point of undercooled austenite transformation during cooling, record them on the temperature time diagram, and then continuous cooling curve can be obtained at the start point and end point of continuous cooling.

等温转变曲线，又称 TTT 曲线，反映了过冷奥氏体等温转变的规律。但在连续冷却转变过程中，钢中的奥氏体是在不断降温的条件下发生转变的。而且转变速度不同，其转变产物也有所不同。过冷奥氏体连续冷却转变曲线——CCT 曲线就是在连续冷却条件下，以不同的冷却速度进行冷却，测定冷却时过冷奥氏体转变的开始点（温度和时间）和终点，把它们记录在温度—时间图上，连接转变开始点和终了点便可得到连续冷却曲线。

Figure 2-7 shows the *CCT* curve of eutectoid steel, on which the *TTT* curve of the steel is also drawn for comparison.

图 2-7 为共析钢的 *CCT* 曲线，在该图上也画上了该钢的 *TTT* 曲线，以做比较。

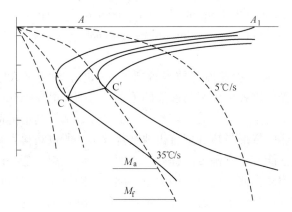

Figure 2-7　Continuous transformation curve of eutectoid steel
A—Ferrite transformation point temperature 727℃
图 2-7　共析钢连续转变曲线
A—铁素体相变点温度 727℃

When the continuous cooling rate is very small, the undercooling degree of transformation is very small, and the time of transformation beginning and ending is very long. If the cooling rate increases, the transition temperature decreases, and the time for the beginning and end of the transition shortens. And the larger the cooling rate is, the larger the temperature range of the transformation is. The CC' line in the figure is the transition stop line, which means that the transformation is not finished when the cooling curve intersects with this line, but the decomposition of austenite stops, and the rest part is supercooled to a lower temperature for martensitic transformation. The cooling curve through C and C' points is equivalent to two critical cooling rates. When the cooling rate v_c is much higher than that, austenite will all be undercooled below the point M_s and transformed into martensite.

当连续冷却速度很小时，转变的过冷度很小，转变开始和终了的时间很长。若冷却速度增大，则转变温度降低，转变开始和终了的时间缩短。并且冷却速度越大，转变所经历的温度区间也越大。图中的 CC' 线为转变中止线，表示冷却曲线与此线相交时转变并未最后完成，但奥氏体停止了分解，剩余部分被过冷到更低温度下发生马氏体转变。通过 C 和 C' 点的冷却曲线相当于两个临界冷却速度。当冷却速度很大超过 v_c 时，奥氏体将全部被过

冷到 M_s 点以下，转变为马氏体。

Therefore, when the cooling rate is lower than the lower critical cooling rate v'_c, the transformation products are all pearlite (p); when the cooling rate is higher than the upper critical cooling rate v_c, the transformation products are martensite (M) and a small amount of retained austenite; when the cooling rate is between v'_c and v_c, the transformation products are pearlite, martensite and a small amount of retained austenite.

因此，当冷却速度小于下临界冷却速度 v'_c 时转变产物全部为珠光体 p；冷却速度大于上临界冷却速度 v_c 时，转变产物为马氏体 M 及少量的残余奥氏体；冷却速度介于 v'_c 和 v_c 之间时，转变产物为珠光体、马氏体加少量的残余奥氏体。

2.4.1.2　Control Cooling Process of Threaded Steel
2.4.1.2　螺纹钢的控制冷却工艺

The controlled cooling of screw steel is also called after rolling heat quenching or heat treatment. After hot rolling, the steel bar is cooled rapidly in austenite state, the surface of the steel bar is quenched into martensite, and then the residual heat is released from the center of the steel bar for self-tempering, so as to improve the strength, plasticity, toughness and comprehensive mechanical properties. After rolling, the quenching process of steel bar is simple, energy-saving, and the surface of steel bar is beautiful and straight, which has obvious economic benefits.

螺纹钢的控制冷却又称轧后余热淬火或余热处理。利用热轧钢筋轧后在奥氏体状态下快速冷却，钢筋表面淬成马氏体，随后由其心部放出余热进行自回火，以提高强度和塑性，改善韧性，得到良好的综合力学性能。钢筋轧后淬火工艺简单，节约能源，且钢筋表面美观、条形直，有明显的经济效益。

The comprehensive mechanical and technological properties of steel bars, such as yield limit, reverse bending, impact toughness, fatigue strength and welding performance, are related to the final microstructure of steel bars. The structure obtained depends on the chemical composition of the steel, the diameter of the steel bar, the deformation condition, the final rolling temperature, the cooling condition after rolling and the self-tempering temperature.

钢筋的综合力学性能和工艺性能，如屈服极限、反弯、冲击韧性、疲劳强度和焊接性能等，同钢筋的最后组织状态有关。而获得何种组织则取决于钢的化学成分、钢筋直径、变形条件、终轧温度、轧后冷却条件、自回火温度等。

In order to obtain the required properties of steel bars, the controlled cooling process after rolling should be selected reasonably.

合理地选择轧后控制冷却工艺以便获得所要求的钢筋性能。

According to the recrystallization state of deformed austenite before quick cooling after rolling, the strengthening effect of cooling after rolling can be divided into two categories. One is that the deformed austenite has been crystallized completely, the dislocation or substructure strengthening caused by deformation has been eliminated, and the deformation strengthening effect has been weakened or disappeared, so the strengthening mainly depends on the transformation, the comprehensive mechanical properties have not been improved much, but the stress stability is high. The other is that austenite does not recrystallize or only partially recrystallized before rapid cooling after rolling. In this way, the strengthening effect of retained or partially retained deformation on Austenite in deformed austenite can be improved by adding the effect of deformation strengthening and transformation strengthening, but the tendency of stress corrosion cracking is larger.

根据钢筋在轧后快冷前变形奥氏体的再结晶状态，钢筋轧后冷却的强化效果可分为两类。一类是变形的奥氏体已经完全结晶，变形引起的位错或亚结构强化作用已经消除，变形强化效果减弱或消失，因而强化主要靠相变完成，综合力学性能提高不多，但是应力稳定性较高。另一类是轧后快冷之前，奥氏体未发生再结晶或者仅发生部分再结晶，这样，在变形奥氏体中保留或部分保留变形对奥氏体的强化作用，变形强化和相变强化效果相加，可以提高钢筋的综合力学性能，但应力腐蚀开裂倾向较大。

The control cooling method of rolled steel bar is generally divided into the following two kinds：

轧后钢筋控制冷却方法一般分为以下两种：

One is to cool down immediately after rolling, quickly cool to the specified temperature in the cooling medium, or after cooling to a certain time in the cooling device, stop the fast cooling, then air cooling and self-tempering. When the steel bar is quenched after being rolled out from the last rolling mill, the surface layer metal becomes quenched structure due to rapid cooling. However, due to the large cross-section size, its center still retains a higher temperature. After a period of time after water cooling, the heat in the core of the steel bar propagates to the surface layer, which makes it reach a new equilibrium temperature. In this way, the surface layer of the steel bar is tempered, so that it has a good toughness of the tempering structure. Due to the controlled cooling, the mechanical properties of steel have been improved obviously. For example, the yield strength of low carbon steel with carbon content (mass fraction) of 0.20%~0.26% in rolling state is $370N/mm^2$. When the temperature is reduced to 600℃ by water cooling, the yield strength can be increased to $540N/mm^2$, while the toughness remains unchanged.

一种是轧后立即冷却，在冷却介质中快冷到规定的温度，或者在冷却装置中冷却到一定的时间后，停止快冷，随后空冷，进行自回火。当钢筋从最后一架轧机轧出后采用急冷时，其表面层金属因迅速冷却而成为淬火组织。但因断面尺寸较大，其心部仍保留有较高的温度。水冷后经过一段时间，钢筋心部的热量向表面层传播，结果使它又达到一个新的均衡温度。这样一来，钢筋的表面层发生了回火，使之具有良好韧性的调质组织。由于钢材经受了控制冷却，其力学性能也有了明显的改善。例如，碳含量（质量分数）为0.20%~0.26%的低碳钢，在轧制状态下的屈服强度为$370N/mm^2$。经过水冷使之温度降到600℃时，其屈服强度可提高到$540N/mm^2$，而韧性保持不变。

Another cooling method is segmented cooling, that is, in the high-speed cooling device, in a short time, the surface of the steel bar is supercooled to below the martensite transformation point to form martensite, and the quick cooling is interrupted immediately. After a period of air cooling, the martensite in the surface layer is tempered to below seven temperatures to form tempered sorbite, and then the quick cooling is interrupted for a certain time, and the quick cooling is interrupted again for air cooling, so that sorbite, bainite and ferrite in the core were obtained. This process is called two-stage cooling. The tensile strength and yield limit of the steel bar obtained by this method are slightly lower, the elongation is almost the same, and the corrosion stability is good. At the same time, for large section steel, the internal and external temperature difference can be reduced.

另一种冷却方法是分段冷却,即先在高速冷却装置中在很短的时间内,将钢筋表面过冷到马氏体转变点以下形成马氏体,并立即中断快冷,空冷一段时间,使表层的马氏体回火到7℃以下温度,形成回火索氏体,然后再快冷一定时间,再次中断快冷进行空冷,使心部获得索氏体组织、贝氏体及铁素体组织。这种工艺叫二段冷却。采用该种方法获得的钢筋,抗拉强度及屈服极限略低,伸长率几乎相同,而腐蚀稳定性能好,同时,对大断面钢材来说,可以减小其内外温差。

2.4.1.3 Factors Affecting Control Cooling Performance
2.4.1.3 影响控制冷却性能的因素

The factors influencing the cooling performance are as follows:
影响控制冷却性能的因素有:

(1) Heating temperature. The heating temperature affects the original austenite grain size before rolling, the rolling temperature and finishing rolling temperature of each pass, the recrystallization degree and grain size of austenite between passes and after finishing rolling. When other deformation conditions are fixed, the performance of the steel bar after controlled cooling is improved obviously with the decrease of heating temperature. If it is not necessary to reduce the heating temperature of the billet and the final rolling temperature, the quick cooling device can be set before the finishing rolling to reduce the steel temperature before the final rolling.

(1) 加热温度。加热温度影响轧前钢坯的原始奥氏体晶粒大小、各道次的轧制温度及终轧温度,影响道次之间及终轧后的奥氏体再结晶程度及晶粒大小。当其他变形条件一定时,随着加热温度的降低,控制冷却后的钢筋性能明显提高。如果不降低坯料的加热温度,又需要降低终轧温度,则可以在精轧前设置快冷装置,降低终轧前的钢温。

(2) Deformation. It is very important to control the deformation of several passes before finishing rolling, and to match the deformation of passes with rolling temperature well, to obtain uniform austenite structure before rapid cooling, to prevent the generation of individual coarse grains and to cause mixed grains, and to obtain uniform structure after water cooling.

(2) 变形量。控制终轧前几道次的变形量,并将道次变形量与轧制温度较好地配合,对钢筋快冷以前获得均匀的奥氏体组织、防止产生个别粗大晶粒以及造成混晶有重要作用,水冷之后可以得到均匀组织。

(3) Finish rolling temperature. The recrystallization degree of austenite is determined by the final rolling temperature. When the cooling condition is fixed, the finishing rolling temperature directly affects the quenching condition and self-tempering condition. When the finishing rolling temperature is different, it is necessary to change the cooling process parameters to ensure that the self-tempering temperature of the steel is the same. Generally, the strengthening effect of steel is

better when the finishing temperature is lower.

（3）终轧温度。终轧温度的高低决定了奥氏体的再结晶程度。当冷却条件一定时，终轧温度直接影响淬火条件和自回火条件。终轧温度不同时，必须通过改变冷却工艺参数来保证钢的自回火温度相同。一般情况下，终轧温度较低时钢的强化效果好。

（4）The interval from finishing rolling to quick cooling. This interval mainly affects the degree of recrystallization of austenite. If the rolled steel bar is in the condition of complete recrystallization, the austenite grain size is easy to grow and the mechanical properties of the steel bar will be reduced due to the longer residence time at high temperature. It is better to have quick cooling immediately after rolling, and install the quick cooling device behind the finishing mill.

（4）终轧到开始快冷的间隔时间。这段间隔时间主要影响奥氏体的再结晶程度。如果轧后钢筋处在完全再结晶条件下，由于高温下停留时间加长，奥氏体晶粒度容易长大，使钢筋的力学性能降低。最好轧后立即快冷，将快冷装置安装在精轧机后。

（5）Cooling rate. Cooling rate is one of the important parameters for controlling cooling after rolling. Increasing the cooling speed can shorten the length of the cooler and ensure the martensite structure of the surface layer of the reinforcement. If the cooling rate is low, generally in order to achieve the required cooling temperature. The length of the cooler can be extended.

（5）冷却速度。冷却速度是钢筋轧后控制冷却的重要参数之一。提高冷却速度可以缩短冷却器的长度，保证得到钢筋表面层的马氏体组织。如果冷却速度较低，一般为了达到所需要的冷却温度。可以加长冷却器的长度。

（6）Start temperature, finish rolling temperature and self-tempering temperature of rapid cooling. The start temperature and finish rolling temperature of rapid cooling directly affect the self-tempering temperature of steel bars. The different self-tempering temperature will affect the microstructure of each point on the steel bar section after transformation, resulting in different properties of the steel bar. The final rolling temperature of fast cooling is controlled by changing cooling parameters, such as adjusting water pressure, water volume or length of cooler.

（6）快冷的开始温度、终轧温度和自回火温度。快冷的开始温度和终轧温度都直接影响钢筋的自回火温度。自回火温度不同则影响相变后钢筋截面上各点的组织状态，导致钢筋性能不同。快冷的终轧温度用改变冷却参数来控制，如调节水压、水量或冷却器的长度。

The final rolling temperature of steel bar directly affects the self-tempering temperature of steel bar. The self-tempering temperature generally decreases with the increase of the total flow of cooling water, and the larger the specification of general reinforcement is, the more cooling water will be.

钢筋的终轧温度直接影响钢筋的自回火温度。自回火温度一般随着冷却水的总流量增多而降低，一般钢筋的规格越大，冷却水量也越多。

The temperature of cooling water has an obvious effect on the cooling effect of steel bars. The higher the temperature of cooling water, the worse the cooling effect. Generally, the temperature of cooling water shall not exceed 30℃.

冷却水的温度对钢筋的冷却效果有明显的影响。冷却水的温度越高，冷却效果越差。一般冷却水的温度不超过30℃。

2.4.2 Controlled Cooling Process
2.4.2 控制冷却工艺

With the rapid development of bar mill, the control cooling technology of bar is becoming more and more perfect. In the aspect of technology, it perfectly combines the metal plastic deformation processing and heat treatment technology in the control rolling process; in the aspect of control, it also develops to realize the automatic control of the electronic computer according to the steel grade, finishing rolling temperature, etc.

随着棒材轧机的迅速发展，棒材的控制冷却技术也日趋完善。在工艺方面，把控制轧制过程金属塑性变形加工和热处理工艺完美结合起来；在控制方面也发展到能根据钢种、终轧温度等实现电子计算机的自动控制。

2.4.2.1 Technological Purpose and Advantages of Controlled Cooling Technology
2.4.2.1 控制冷却技术的工艺目的及优点

The purpose of controlled cooling technology is to improve the mechanical properties of screw steel rapidly, especially the yield strength of screw steel with poor chemical composition. It is usually used in low-carbon steel to make the mechanical properties of products exceed that of micro alloyed steel or low-alloy steel at low cost.

控制冷却技术的工艺目的是快速提高螺纹钢的力学性能，特别是化学成分差的螺纹钢的屈服强度。它通常用于低碳钢，以便在低成本条件下，使产品的力学性能超过微合金钢或低合金钢。

The advantages of the controlled cooling process are:
控制冷却工艺的优点是：

(1) The yield strength of the product can be improved without the following methods:
(1) 不采用下列方法即可提高产品的屈服强度：

1) Add expensive alloy elements (if common cooling method is adopted after rolling, it must be added);

1) 添加昂贵的合金元素（如果轧后采用普通冷却方式则必须添加）；

2) Increase the carbon content (increase the carbon content will affect the welding performance).

2) 提高碳含量（提高碳含量会影响焊接性能）。

(2) Low carbon content. Low carbon means:
(2) 碳含量低。碳含量低意味着：

1) It has good bending property without surface crack;

1) 具有良好的弯曲性能，而又不产生表面裂纹；

2) After heat treatment, the surface layer has high plasticity, that is to say, it has good anti fatigue load capacity, so the treated deformed steel can be used for dynamic load structural parts;

2) 表层经过热处理后具有高塑性，也就是说具有良好的抗疲劳载荷能力，因此可将处理过的螺纹钢用于动载结构件；

3) Good welding performance is even better than that of materials containing micro alloyed elements;

3) 良好的焊接性能甚至优于含有微合金元素的材料；

4) Compared with the products produced under normal cooling conditions, the surface will generate less oxide scale.

4) 与普通冷却条件下生产的产品相比，表面将生成更少的氧化铁皮。

(3) The thermal stability is good. Even if it is heated, its performance is better than that of ordinary screw steel.

(3) 热稳定性能好，即使加热，它的性能也比普通螺纹钢要好。

(4) The production cost is reduced by 18% compared with micro alloyed steel and 8% compared with low alloy steel.

(4) 降低了生产成本，与微合金钢相比节约成本 18%，与低合金钢相比节约成本 8%。

2.4.2.2　Metallography Principle

2.4.2.2　金属学原理

The process of controlled cooling is to quench the surface of bar and complete the heat treatment of bar by self-tempering, which is directly completed by rolling heat. When the bar leaves the last mill, there is a special heat treatment cycle, which includes three stages (As seen in Figure 2-8).

控冷工艺是指对棒材表面进行淬火，并通过自回火来完成对棒材的热处理，自回火过程直接由轧制热来完成。当棒材离开最后一架轧机时有一个特殊的热处理周期，它包括三个阶段（见图 2-8）。

Figure 2-8　Three stages of controlled cooling process

图 2-8　控冷工艺的三个阶段

First stage: surface hardening stage.

第一阶段：表面淬火阶段。

Immediately after the last mill, the bar passes through the water cooling system to achieve a short-term high-density surface cooling. Because the temperature decreasing speed is higher than the critical speed of martensite, the surface layer of screw steel is transformed into a kind of hard structure of martensite, i.e. initial martensite.

紧接最后一架轧机之后,棒材穿过水冷系统,以达到一个短时高密度的表面冷却。由于温度下降的速度高于马氏体的临界速度,因此,螺纹钢的表面层转化为一种马氏体的硬质结构,即初始马氏体。

In the first stage, the core temperature of the bar is maintained within the temperature range of austenite, so as to obtain the subsequent ferrite pearlite transformation (in the second and third stages).

在第一阶段棒材的心部温度维持在均是奥氏体的温度范围内,以便得到后来的铁素体珠光体相变(在第二阶段和第三阶段)。

At the end of this stage, the microstructure of the bar changed from the initial austenite to the following three-layer structure:

在这个阶段末,棒材的显微结构由最初的奥氏体变为如下的三层结构:

(1) The depth of surface layer is initial martensite;

(1) 表层的一定深度为初始马氏体;

(2) The middle ring region consists of a mixture of austenite, bainite and some martensite, and the content of martensite decreases from the surface to the center;

(2) 中间环形区的组成为奥氏体、贝氏体和一些马氏体的混合物,并且马氏体的含量由表面到心部逐渐减少;

(3) The core is still austenite structure.

(3) 心部仍是奥氏体结构。

The duration of the first stage depends on the depth of the martensite layer, which is the key parameter of the process. In fact, the deeper the martensite layer, the better the mechanical properties of the product.

第一阶段的持续时间依据马氏体层的深度而定,这个马氏体层深度是工艺的关键参数。实际上,马氏体层越深,则产品的力学性能越好。

The second stage: self-tempering stage.

第二阶段:自回火阶段。

When the bar leaves the water-cooling equipment and is exposed to the air, the quenched surface layer is reheated by the heat of the heat conduction center, so as to complete the self-tempering of the surface martensite, so as to ensure that the bar has enough toughness under high yield strength.

棒材离开水冷设备,暴露在空气中,通过热传导心部的热量对淬火的表层再次进行加热,从而完成表层马氏体的自回火,以保证在高屈服强度下棒材具有足够的韧性。

In the second stage, the austenite which has not been transformed in the surface layer becomes bainite, and the austenite structure in the center is still austenite.

在第二阶段,表层尚未相变的奥氏体变为贝氏体,心部仍为奥氏体结构,中间环形区到回火马氏体层之间奥氏体相变为贝氏体。

At the end of this stage, the microstructure of the bar becomes:

在这个阶段末,棒材的显微结构变为:

(1) The surface layer is tempered martensite with a certain depth;

(1) 表层为一定深度的回火马氏体；

(2) The composition of the intermediate ring zone is a mixture of bainite, austenite and some tempered martensite;

(2) 中间环形区的组成为贝氏体、奥氏体和一些回火马氏体的混合物；

(3) Austenite in the center begins to change.

(3) 心部的奥氏体开始相变。

The duration of the second stage is based on the water cooling process and bar diameter used in the first stage.

第二阶段的持续时间是依据第一阶段采用的水冷工艺和棒材直径确定的。

The third stage: final cooling stage.

第三阶段：最终冷却阶段。

The third stage occurs during the time when the bar enters the cooling bed. It consists of the isothermal transformation of austenite which has not yet been transformed in the bar. According to the chemical composition, bar diameter, finish rolling temperature, first stage water cooling efficiency and duration, the composition of phase transformation may be a mixture of ferrite and pearlite, or a mixture of ferrite, pearlite and bainite.

第三阶段发生在棒材进入冷床上的这段时间里，它由棒材内尚未相变的奥氏体的等温相变组成。根据化学成分、棒材直径、精轧温度以及第一阶段水冷效率和持续时间，相变的成分可能是铁素体和珠光体的混合物，也可能是铁素体、珠光体、贝氏体的混合物。

The physical phenomena of the three stages described above can be described in the following three forms:

以上所描述的三个阶段的物理现象可以用下列三种形式说明：

(1) Heat exchange between bar surface and cooling medium;

(1) 棒材表面和冷却介质之间的热交换；

(2) Heat transfer in bars;

(2) 棒材内的热传导；

(3) Metalology phenomenon.

(3) 金属学现象。

2.4.2.3 Mechanical Properties of Surface Hardened Bar
2.4.2.3 表面淬火棒材的力学性能

From the point of view of steel rolling production, only three of the key parameters of all processes are considered as independent control variables, they are: finishing temperature, quenching time, water flow.

从轧钢生产这个着眼点来看，在所有工艺的关键参数当中只有三个参数被认为是独立控制变量，它们是：精轧温度、淬水时间、水流量。

The microstructure and properties of the bar are changing from the surface to the center of the bar. However, it is also possible to consider the bars treated by the controlled cooling process to be approximately composed of two different structures:

在采用控冷工艺处理棒材时,从棒材表面到心部,它的显微组织和性能在不断变化。尽管如此,也可以将控冷工艺处理过的棒材考虑成近似由两种不同的结构组成:

(1) The surface layer is tempered martensite;

(1) 表层为回火马氏体;

(2) The core consists of ferrite and spherulite.

(2) 心部由铁素体和球光体组成。

The technical properties of the bar, especially the tensile properties, are determined according to the following three properties:

棒材的技术性能,特别是拉伸性能根据以下三个性能确定:

(1) Volume fraction of martensite;

(1) 马氏体的体积分数;

(2) Tensile properties of martensite;

(2) 马氏体的拉伸性能;

(3) Tensile properties of ferrite pearlite structure at the center.

(3) 心部铁素体-珠光体结构的拉伸性能。

The volume fraction of martensite depends on the initial temperature of martensitic transformation, which is a function of the chemical composition of the bar and the temperature distribution of the bar section when the bar leaves the quenching water tank.

马氏体的体积分数取决于马氏体相变的起始温度,它是棒材化学成分和当棒材离开淬水箱时棒材截面温度分布的函数。

The yield strength of tempered martensite on the surface of bar is related to chemical composition and tempering temperature. In fact, the lower the tempering temperature, the higher the yield strength and the lower the toughness. Tempering temperature is the highest temperature of the bar surface at the end of the second stage of the process, which directly depends on the quenching process adopted in the first stage. The longer the first stage is, the deeper the martensite layer is, the lower the bar temperature at the end of the first stage is, and the lower the tempering temperature is.

棒材表层的回火马氏体的屈服强度与化学成分、回火温度有关。实际上回火温度越低,屈服强度越高,韧性也越低。回火温度是工艺第二阶段末棒材表面所达到的最高温度,它直接取决于第一阶段所采用的淬火工艺。第一阶段时间越长,马氏体层越深,第一阶段末的棒材温度也就越低,则回火温度越低。

Therefore, if the chemical composition is given in the controlled cooling process, the key factor determining the mechanical properties of the bar is the temperature diagram in the quenching stage.

因此,在控冷工艺中,如果给出了化学成分,那么决定棒材力学性能的关键因素就是淬火阶段的温度简图。

When the bar diameter is given, the temperature diagram of the controlled cooling process system can change with the following factors:

当给出了棒材直径时，控冷工艺系统的温度简图能随下列因素的改变而改变：

(1) Finish rolling temperature;

(1) 精轧温度；

(2) Duration of water quenching stage;

(2) 淬水阶段的持续时间；

(3) Heat released from the surface of the bar by cooling water during the quenching stage.

(3) 淬水阶段通过冷却水释放的棒材表面热量。

The thermal conductivity between the bar surface and water cooling is one of the key parameters of the controlled cooling process. It is a function of the bar surface temperature and the design basis of the cooling equipment, cooling water pressure, water flow and temperature.

棒材表面和水冷之间的导热系数是控冷工艺的关键参数之一，它是棒材表面温度的函数，也是冷却设备、冷却水压力、水流量及温度的设计依据。

2.4.2.4 Process Control
2.4.2.4 工艺控制

The process control of the controlled cooling process is mainly completed by the control of water quantity, time and temperature. The details are as follows:

控冷工艺的工艺控制主要通过水量、时间和温度控制来完成，具体说明如下：

(1) Water quantity control. The total amount of water is regulated by water regulating valves FCV_1 and FCV_2. This control relies on a closed loop with a feedback signal from the flow meter and the operator's preset values.

(1) 水量控制。水的总量通过水调节阀 FCV_1 和 FCV_2 来调节。此控制借助于带有反馈信号的闭合回路，而反馈信号来自流量表和操作员的预设值。

(2) Time control. The length of quenching time will produce a special tempering temperature, which is directly related to the yield strength of the product. The quenching time can be controlled by adjusting the finishing speed and the number of coolers.

(2) 时间控制。淬水时间的长短会产生一个特殊的回火温度，而回火温度直接关系到产品的屈服强度。淬水时间可以通过调节终轧速度、冷却器的数量等来控制。

(3) Temperature control. The main purpose is to measure the temperature before and after the quenching line so as to obtain the accurate tempering temperature. In this way, it is very important to locate the pyrometer to measure the temperature. A pyrometer is installed in front of frame 13 to measure the temperature of the bar transported from the inner side. In addition, a pyrometer is installed 60m downstream of the quenching water line to measure the tempering temperature of the bar.

(3) 温度控制。主要测量淬水线前后的温度，以便得到准确的回火温度。这样一来，测量温度的高温计的定位就显得非常重要。一个高温计安装在 13 号机架前，用来测量输送来的棒材温度。另外，在淬水线下游 60m 处安装高温计，以便测定棒材的回火温度。

2.4.3 Process Equipment of Controlled Cooling Process
2.4.3 控冷工艺的工艺设备

2.4.3.1 Equipment Layout
2.4.3.1 设备布置

The quench line (shown in Figure 2-9) is located between the exit of the finished mill and the flying shear. It is used to quench the bars leaving the finishing mill so as to obtain the required properties.

淬水线（见图 2-9）位于成品轧机出口和倍尺飞剪之间，用于对离开精轧机的棒材进行淬火，以便得到所需要的性能。

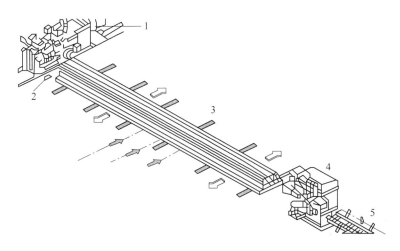

Figure 2-9 Schematic diagram of quenching water line
1—Finished rolling mill; 2—Pyrometer; 3—Quenching box; 4—Times of flying shear; 5—Lifting apron
图 2-9 淬水线示意图
1—成品轧机；2—高温计；3—淬火箱；4—倍尺飞剪；5—升降裙板

2.4.3.2 Design Points of Quenching Line Equipment
2.4.3.2 淬水线设备设计要点

According to the different final properties of the required steel, there are many types of quench line equipment. However, they all have some common points, that is, all quench water equipment are tube structure and axisymmetric, for example, the annular nozzle between the bar and the cooler is symmetrical.

根据所要求钢材最终性能不同，淬水线设备的形式也分为多种，但无论怎样，它们都有一些共同点，即所有淬水设备都是管结构的，并且是轴对称的，如棒材和冷却器之间的环形喷嘴就是对称的。

The following points shall be considered for quenching line equipment:
淬水线设备应考虑以下几点：

(1) The water flow per unit length is designed correctly to ensure high quenching efficiency.

(1) 正确地设计单位长度的水流量,以确保高的淬火效率。

(2) Number of coolers. The quenching line equipment consists of some coolers. The number of coolers is determined according to finishing temperature, finishing speed, bar diameter and tempering temperature.

(2) 冷却器的数量。淬水线设备由一些冷却器组成,冷却器的数量根据精轧温度、精轧速度、棒材直径和回火温度等几项参数来确定。

(3) The utilization rate of cooler shall be high to ensure efficient water spray.

(3) 冷却器的利用率要高,以确保高效喷水。

(4) The ratio between the inner diameter of the cooler and the cross section area of the bar shall be set correctly to ensure the proper "water injection rate".

(4) 正确地设置冷却器内径与棒材横断面面积之间的比率,以保证适当的"注水率"。

(5) Resistance of high pressure water to bars in the cooler.

(5) 在冷却器内高压水对棒材的阻力。

(6) Stability of bars in quench line.

(6) 淬水线内棒材的稳定性。

(7) At the end of the process, the straightness of the treated bar is obtained.

(7) 该工艺结束时,已处理棒材的平直度。

2.4.3.3 Technical Parameters

2.4.3.3 技术参数

The technical parameters are as follows:
技术参数如下:

The external dimension of quenching line trolley	3819mm×18600mm×2194mm;
淬水线小车外观尺寸	3819mm×18600mm×2194mm;
Weight	24500kg;
重量	24500kg;
The range of heat-treated bar	$\phi M_0 \sim 40$mm deformed steel;
热处理棒材范围	$\phi M_0 \sim 40$mm 螺纹钢;
The stroke of quenching line trolley	1700mm;
淬水线小车行程	1700mm;
Hydraulic cylinder stroke	1700mm;
液压缸行程	1700mm;
The external dimension of valve platform	12300mm×750mm×2300mm;
阀台的外观尺寸	12300mm×4750mm×2300mm;
Water system pressure (max/min)	1.2MPa/0.8MPa;
水系统压力(max/min)	1.2MPa/0.8MPa;

Flow of control valve (FCV$_1$, FCV$_2$)	40~200m³/h;
控制阀 (FCV$_1$、FCV$_2$) 流量	40~200m³/h;
Compressed air flow (under standard state)	90m³/h (for dryer);
压缩空气流量 (标准状态下)	200m³/h (for instrument);
	90m³/h (干燥器用);
	200m³/h (仪器用);
Compressed air pressure	0.5MPa;
压缩空气压力	0.5MPa;
Number of booster pumps 增压泵数量	3;
	3;
Number of bypass roller table:	20;
旁路辊道数量	20;
Diameter	φ188mm.
直径	φ188mm。

2.4.3.4　Equipment Composition
2.4.3.4　设备组成

The process equipment composition of the controlled cooling process is as follows (Shown in Figure 2-10 and Figure 2-11):

控冷工艺的工艺设备组成如下 (见图2-10、图2-11):

(1) Quenching line trolley. The trolley is equipped with two water tanks and a bypass roller table, among which the No.1 water tank has two water-cooling lines for quenching the cut products and φ10~40mm deformed steel; the No.2 water tank has only one water-cooling line for quenching φ20~40mm deformed steel.

(1) 淬水线小车。小车上布有两个水箱和一条旁路辊道,其中1号水箱有两条水冷线,用于对切分产品和φ10~40mm螺纹钢的淬火;2号水箱只有一条水冷线,用于对φ20~40mm螺纹钢的淬火。

(2) Bypass roller table. The bypass roller table is composed of 20 rollers and driven by AC motor. When the finished product does not need quenching, the bypass roller table is used.

(2) 旁路辊道。旁路辊道由20个辊子组成,由交流电机驱动,当成品不需要淬火时,使用旁路辊道。

(3) Hydraulic cylinder. The quenching line trolley is driven by two sets of hydraulic cylinders, which makes the trolley move horizontally in the vertical direction of the rolling line.

(3) 液压缸。淬水线小车由两套液压缸驱动,使小车在轧线的垂直方向水平横移。

(4) Pyrometer. One pyrometer is used to measure the population temperature of the bar in the quenching water line, and the other pyrometer is used to measure the outlet temperature of the bar in the quenching water line.

（4）高温计。一个高温计用于测量棒材在淬水线的入口温度，另外的高温计用于测量棒材在淬水线的出口温度。

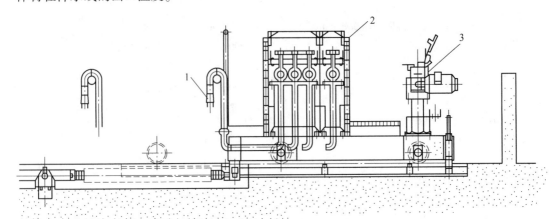

Figure 2-10　Schematic diagram Ⅰ of controlled cooling process equipment
1—Valve platform；2—Quenching line trolley；3—Bypass roller table
图 2-10　控冷工艺设备示意图一
1—阀台；2—淬水线小车；3—旁路辊道

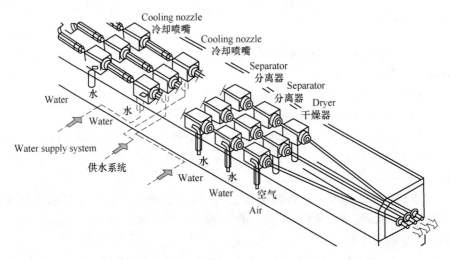

Figure 2-11　Schematic diagram Ⅱ of controlled cooling process equipment
图 2-11　控冷工艺设备示意图二

（5）Control valve. The control valve is used to control the air and water flow. Among them, FCV_1 controls line 1 of No. 1 water tank, FCV_2 controls line 2 of No. 1 water tank and No. 2 water tank.

（5）控制阀。控制阀用于控制空气和水流量。其中 FCV_1 控制 1 号水箱 1 线，FCV_2 控制 1 号水箱 2 线和 2 号水箱。

（6）Water system. The water system consists of hose, ball valve and water pipe. The ejected water is collected in the water tank and then returned to the main system of the equipment through the special pipe. The water supply system is equipped with a control valve station, which is equipped with an air flow control valve with an electric air converter, an electromagnetic flowmeter, a pressure sensor, a pressure gauge, a butterfly valve and some auxiliary facilities for operation.

（6）水系统。水系统由软管、球阀和水管组成。喷射出的水汇集在水箱里，然后经过专用管返回到设备主系统里。供水系统装有控制阀台，该阀台装备有：带有电—气转换器的空气流量控制阀、电磁流量计、压力传感器、压力计、碟阀和一些操作上使用的辅助设施。

（7）Compressed air system. The system is connected with the main air supply equipment, equipped with servo valve, filter, pressure regulator, pressure gauge and other auxiliary facilities to ensure the correct operation of the system.

（7）压缩空气系统。该系统与总供气设备相连，装备有伺服阀、过滤器、压力调节器、压力计和其他一些确保系统正确运行的辅助设施。

（8）Hydraulic system. The hydraulic system is used to drive the hydraulic cylinder installed on the quenching line, and the hydraulic oil is supplied by the hydraulic center device.

（8）液压系统。液压系统用于驱动安装在淬水线上的液压缸，液压油由液压中心装置供给。

（9）Dry oil system. Bearings, quenching line wheels, etc. for roller table shall be manually lubricated by dry nozzles.

（9）干油系统。辊道用轴承、淬水线轮子等干油嘴手动润滑。

2.4.3.5　Water Cooling Line
2.4.3.5　水冷线

The length of the whole water-cooling line is 18.6m, which can ensure that all specifications of screw steel are properly cooled to meet the required performance requirements.

整个水冷线的长度 18.6m 是可以选择的，这样能保证所有规格的螺纹钢进行适当的冷却，使其达到所需要的性能要求。

Each water-cooling line has some coolers. When the bar passes through the cooler, it is quickly surrounded by water, and the whole surface is evenly cooled.

每条水冷线都有一些冷却器，棒材通过冷却器时被水迅速包围，整个表面被均匀冷却。

In the cooler, the direction of water flow is the same as that of bar movement, which is beneficial to guide and guide the bar and reduce the movement resistance.

在冷却器里水流方向与棒材的运动方向一致，有利于棒材的导入、导出，减少了运动阻力。

The water flow rate depends on the section and speed of the bar. The cooler is mainly composed of conduit, inner ring sleeve, outer ring sleeve, gasket, bolt, etc. There is a water inlet under the outer ring sleeve, which is connected with the water supply pipe. When the outer ring sleeve is combined with the inner ring sleeve, a gap is formed (Shown in Figure 2-12). High pressure water is sprayed to the bar at this position.

水流速度根据棒材的断面和速度而定。冷却器主要由导管、内环套、外环套、垫片、螺栓等组成。外环套下面有一进水孔，与供水管相连，外环套与内环套组合时形成一个环行间隙"GAP"（见图2-12），在此处向棒材喷射高压水。

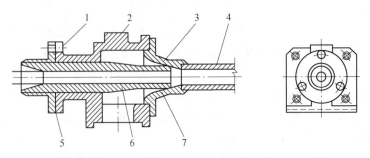

Figure 2-12　Structure of cooler
1—Bolt；2—Body；3—Outer ring sleeve；4—Conduit；5—Piece；6—Inner ring sleeve；7—GAP
图 2-12　冷却器结构示意图
1—螺栓；2—机体；3—外环套；4—导管；5—垫片；6—内环套；7—间隙

The size of "GAP" can be adjusted by the thickness of gasket, for example, "GAP" is 0.44mm, and the thickness of gasket is 2mm. Once the gap is determined, the thickness of the gasket can be selected and then fixed with bolts. Different products have different number of coolers, and bypass pipes can be used instead of coolers when they are not needed.

"GAP"的大小可通过垫片的厚度来调节，例如，"GAP"为0.44mm，垫片的厚度为2mm。一旦"GAP"确定，垫片的厚度便可选定，然后用螺栓固定。不同的产品，冷却器的数量不同，不需要冷却器时可用旁路管代替。

In the whole product range, it is not enough to have only one set of coolers. Different products require different "water injection rate". Three sets of coolers with different inner diameters are needed in the product range of $\phi 10 \sim 40$mm. Only one set of coolers can be arranged for each quenching line, so when a new product is produced, the operator will arrange the set of coolers for the product on the water-cooling line, and align the water-cooling line with the main rolling line through the quenching trolley. Obviously, the operator will also place the cooler of the next product on the remaining water-cooling line, which can save time.

在整个产品范围内，仅有一套冷却器是不够的，不同的产品要求的"注水率"不同，在小 $\phi 10 \sim 40$mm 的产品范围内需要有三套内径不同的冷却器。每条淬水线只能布置一套冷却器，因此当生产一种新产品时，操作员将该产品用的这套冷却器布置在水冷线上，并通过淬水小车使该水冷线与主轧线对齐。显而易见，操作员也将下一种产品用的那一套冷却器布置在剩余的水冷线上，这样能节约时间。

The population of the water cooling line has an exhaust bubble device, which is used to remove bubbles on the surface of the bar, so as to evenly cool the bar surface.

水冷线的入口有一个排气泡装置，用于除去棒材表面的气泡，以便对棒材表面均匀冷却。

There are two reverse separators and a reverse dryer at the outlet of the water cooling line. The structure of separator and dryer is the same, but the medium of separator is high-pressure water, and the medium of dryer is compressed air. The direction of water flow and compressed air is opposite to the moving direction of bar. The purpose of the separator is to cool the bar, at the same time, completely remove the scale and water on the surface of the bar to avoid affecting the microstructure of the bar. The function of the dryer is to remove water and residual water vapor from the surface of the bar and prevent water or water vapor from escaping from the water-cooled line. Similarly, the amount of water and compressed air in the separator and dryer is adjusted by the thickness of gasket just like that of the cooler.

在水冷线的出口有两个反向分离器和一个反向干燥器。分离器和干燥器的结构相同，只是分离器输入的介质是高压水，而干燥器输入的介质是压缩空气。水流方向、压缩空气方向与棒材的运动方向相反。分离器的作用是冷却棒材，同时，完全清除棒材表面的氧化铁皮和水，避免影响棒材的微观结构。而干燥器的作用是去除棒材表面的水和残余水蒸气，并防止水或水蒸气从水冷线逸出。同样，分离器、干燥器内的水量和压缩空气量的大小如同冷却器一样由垫片的厚度来调节。

2.4.3.6　Controlled Cooling Process Control Equipment
2.4.3.6　控冷工艺控制设备

In addition to the inlet/outlet pyrometer, the quenching line is also equipped with an independent cooling regulation system to accurately control the quenching of bars. The quench line is controlled by the following systems:

淬水线除了装有进/出口高温计之外，还装备有独立的冷却调节系统，以便精确地控制棒材的淬火。淬水线是通过下列系统进行控制的：

(1) Manual operation system;
(1) 手动操作系统；
(2) An automatic control system controlled by a dedicated PC.
(2) 由专用PC控制的自动控制系统。

The quenching line of the controlled cooling process is controlled by the full automatic control system without any manual interference. This method can avoid human error and increase the stability of products. The cooling program of each product is stored in the memory of PC. The cooling program is set in advance by the technical personnel, and can be modified or copied at any time as required. In the production process, the cooling program can be transmitted to the microprocessor to operate the NO/OFF valve and flow amplification valve of the quench water line valve station. The following parameters are stored in the cooling program:

通过全自动控制系统控制控冷工艺淬水线，不需要人工干涉。用此方法可以避免人工失误，增加产品的稳定性。在PC的存储器里储存着每种产品的冷却程序，冷却程序由技术人员提前设置好，并可根据需要随时进行修改或复制。在生产过程中冷却程序可传送到微处理机上，以操作淬水线阀台的ON/OFF阀和流量放大阀。在冷却程序里储存着下列参数：

(1) Product specification and rolling speed;
(1) 产品规格和轧制速度；
(2) Steel grade;

(2) 钢种；

(3) Population temperature of billet at mill 13；

(3) 钢坯在 13 号轧机的入口温度；

(4) The inlet temperature of the bar in the quenching line；

(4) 棒材在淬水线的入口温度；

(5) The population temperature (tempering temperature) of the bar in the cooling bed；

(5) 棒材在冷床的入口温度（回火温度）；

(6) Setting of cooler；

(6) 冷却器的设置；

(7) Setting of water flow amplification valve；

(7) 水流量放大阀的设置；

(8) Transfer switch from water line 2 to water line 2.

(8) 2 号水线向 2 号水线的转换开关。

When the product changes, the corresponding cooling program can be extracted from the PC memory, and the microprocessor can set the quenching water line control components according to this program, such as water flow amplification valve placed on the most actual water flow position, and prepare for production. The alarm is connected with each temperature setting point. When the measured temperature is inconsistent with the preset temperature of the set point, the alarm bell of the operation console will ring. At this time, the temperature of the set point can be adjusted according to the specific situation.

当产品改变时，可以将相应的冷却程序从 PC 存储器中提取出来，微处理机可根据此程序设置好淬水线控制部件，如水流量放大阀放置在实际水流量的位置上，并做好生产的准备。警报器与每一个温度放置点相连，当所测温度与设置点预先设定温度不一致时，操作台的警报铃就响了，这时可根据具体情况调节设置点温度。

2.4.3.7 Maintenance of Process Equipment
2.4.3.7 工艺设备的维护

The maintenance requirements of process equipment are as follows：

工艺设备的维护要求如下：

(1) Check whether the coolers, separators, dryers and bypass pipes of each water-cooling line are blocked to keep them unblocked.

(1) 检查各水冷线的冷却器、分离器、干燥器、旁路管有无堵塞物，以保持其畅通。

(2) Check the abrasion of cooling elements of water cooling lines regularly, and replace them in time after abrasion.

(2) 定期检查各水冷线冷却元件的磨损情况，磨损后及时更换。

(3) Regularly check gap of cooler, separator and dryer. If necessary, increase or decrease gasket according to process requirements.

(3) 定期检查冷却器、分离器、干燥器的间隙"GAP"，如需要，根据工艺要求增减垫片。

Task 2.5　Finishing of Threaded Steel
任务 2.5　螺纹钢的精整

Mission objectives
任务目标

(1) Familiar with the composition, structure and principle of the finishing line equipment of the screw steel;

(1) 熟悉螺纹钢精整线设备的组成、构造及原理;

(2) Be able to operate the cooling bed, shearing, transportation, weighing, bundling, packaging and other equipment in the finishing area of threaded steel.

(2) 能够对螺纹钢精整区的冷床、剪切、运输、称重、打捆、包装等设备进行操作。

2.5.1　Overview
2.5.1　概述

The finishing process of rebar mainly includes: cooling, cold cutting, surface quality inspection, automatic collection, bundling, weighing, marking, etc.

螺纹钢棒材的精整工序主要包括：冷却、冷剪切、表面质量检查、自动收集以及打捆、称重、标记等。

In order to realize the finishing process, corresponding equipment is equipped respectively to form the finishing system production line.

为了实现精整工艺，分别配置有相应的设备，形成精整系统生产线。

2.5.2　Shear
2.5.2　剪切

2.5.2.1　Steel Shearing Process and Selection of Shearing Machine
2.5.2.1　钢材剪切工艺与剪机的选择

Generally, the thread reinforcement is cut by shear. The types of shears used for section steel cutting can be divided into flat blade shears and various flying shears from the form, and hot shears and cold shears from the state of cutting metal.

螺纹钢筋一般采用剪机剪切。用于型钢剪切的剪机种类从形式区分有平刃剪及各类飞剪，以剪切金属的状态分有热剪切和冷剪切两类。

After finishing rolling, the head and section of the rolled piece shall be cut by flying shear, and then it shall be cooled by human cooling bed; the steel with the steel temperature below

100℃ shall be cold cut and cut into fixed length.

轧件在精轧后进行切头、分段都采用飞剪，之后进入冷床进行冷却；钢温在 100℃ 以下的钢材进行冷剪切，切成定尺长度。

The shears used for transverse cutting in rolling are called flying shears. In the production line of continuous small bar mill, flying shear is used for cutting head, section and accident shear. A set of small continuous rolling mill adopts 2~4 flying shears.

在轧件运动中进行横向剪切的剪机称作飞剪。在连续式小型棒材轧机生产线中采用飞剪作为切头、分段和事故剪之用。一套小型连轧机采用 2~4 台飞剪。

(1) The shear speed and shear capacity of each area shall match with the exit speed and rolling variety of the rolling mill. The section size that the flying shear can cut must include all varieties and specifications rolled by the rolling mill.

(1) 各区域的飞剪剪切速度及剪切能力应与轧机的出口速度和轧制品种相匹配。飞剪所能剪切的断面尺寸，必须包括轧机所轧出的全部品种和规格。

(2) In case of accident of rear rolling mill or equipment, the front flying shear can break the rolling piece in time, which is convenient to deal with the accident.

(2) 当后部轧机或设备发生事故时前面的飞剪可以及时将轧件碎断，便于处理事故。

(3) It is required to have a good shear section, especially the section shear before the cooling bed is more important, so as to eliminate the cutting head of a row of rolled pieces after the cooling bed, reduce the cutting head amount and improve the yield.

(3) 要求剪切断面好，特别是冷床前的分段剪更为重要，以省去冷床之后对一排轧件的切头，减少切头量，提高成材率。

(4) When there is a controlled cooling production line after rolling, the flying shear before the cooling bed should be able to adapt to the reduction of temperature.

(4) 当具有轧后控制冷却生产线时，冷床前的飞剪应能适应温度降低轧件的剪切。

(5) The shearing accuracy is small, the error is small, and it can be repeated to improve the yield.

(5) 剪切精度高，误差小，并且有可重复性，以提高成材率。

(6) The flying shear must be reliable, easy to maintain, simple in structure and easy to improve the productivity of the rolling line.

(6) 飞剪工作必须可靠，维修方便，结构尽可能简单，便于提高轧制作业线的生产率。

The flying shear used in modern small continuous rolling mill mainly has the following four structural forms:

现代小型连轧机采用的飞剪主要有下列 4 种结构形式：

(1) In the crank connecting rod flying shear, the cutting edge moves almost parallel in the shear area. This kind of flying shear has good quality of shear section, which is suitable for cutting large section sections and can bear large shear force. Generally, after it is used in rough rolling and medium rolling mill, the speed of sheared rolling piece is not easy to be too high, and the speed of general rolling piece is below 10m/s.

（1）曲柄连杆式飞剪，在剪切区域剪刃几乎是平行移动。这种飞剪剪切断面质量好，适合剪切大断面型材，能承受大的剪切力，一般用于粗轧、中轧机组之后，剪切的轧件速度不易过高，一般轧件速度在 10m/s 以下。

（2）Rotary flying shears or double arm flying shears, the blades of the shears rotate. This kind of flying shear is suitable for the rolling piece with high shear moving speed. The speed of rolling piece is between 10~22m/s. It is mainly used for steel cutting after medium rolling, finishing rolling and before cooling bed.

（2）回转式飞剪或称双臂杆式飞剪，剪刃作回转运动。这种飞剪适合剪切移动速度较高的轧件，轧件速度在 10~22m/s，多用于中轧、精轧机组之后和冷床之前的钢材剪切。

（3）Convertible flying shear. It is a combination of the first two flying shears, which can be transformed from one form to another by using the quick change mechanism. The same flying shear can not only cut large section steel in the form of crank at low speed (cut $\phi 70mm$ deformed steel at the bar moving speed of 1.5m/s), but also cut small section steel in the form of rotation at high speed (cut $\phi 10$ mm deformed steel at the speed of 20m/s). One shear can cover all rolling specifications, so as to replace the two movable shears set in the general rolling production line.

（3）可转换型飞剪。它是前两者飞剪的结合，能从一种形式利用快速更换机构转换成另一种形式。同一个飞剪既可以在低速状态下以曲柄形式剪切大断面轧件（在 1.5m/s 的棒材移动速度下剪切 $\phi 70mm$ 螺纹钢），也可以在高速状态下以回转形式剪切小断面轧件（在 20m/s 的速度下剪切 $\phi 10mm$ 螺纹钢）。一台剪机可以覆盖所有轧制规格范围，从而替代一般轧制生产线上设置的两台可移动剪机。

（4）The shear speed of the cold flying shear combined with the straightener is relatively low, generally only about 2~4m/s. It can cut several pieces at the same time, and the width of rolling arrangement can reach about 1200mm. The flying shear is of crank connecting rod type, which can cut 6~24m to determine the length with an error of ±10mm.

（4）与矫直机联合的多条冷飞剪剪切速度比较低，一般只有 2~4m/s。可以同时剪切数根，轧件排列的宽度可达 1200mm。飞剪为曲柄连杆式，可剪切 6~24m 定尺，误差在 ±10mm 以内。

According to its driving form, the shearing machine can be divided into the following two types：

剪机根据其驱动形式又分为以下两种：

（1）Clutch brake type, driven by a continuous DC motor driven flywheel. The control system is simple, but the slip and separation of the clutch are not good, resulting in poor shearing accuracy, and increasing the amount of metal cutting.

（1）离合器制动器型，由一台连续运转的直流电机带动飞轮驱动。控制系统简单，但由于离合器打滑和分离不爽，造成剪切精度差，而增加金属切削量。

（2）Direct drive type, driven by high power and low inertia motor. Each cut completes a start and stop action. It is also called start stop working system.

（2）直接驱动型，采用高功率低惯量电机驱动。每次剪切均完成一次启动、停止动作。也称作启停工作制型。

Most of the modern continuous rolling profile production lines adopt the direct drive type, which has the following advantages:

现代连续轧制型材生产线大部分采用直接驱动型，其优点是：

(1) Simple mechanical structure, less maintenance.

(1) 机械结构简单，维修量少。

(2) High shearing accuracy and repeatability can improve the yield.

(2) 剪切精度高，重复性好，可以提高成材率。

(3) The cycle time is short, and a shearing cycle is completed within 0.5s, which is beneficial to the setting of the length of the upper cooling bed.

(3) 周期时间短，大约在 0.5s 之内完成一个剪切周期，有利于上冷床轧件长度的设定。

However, the control system of the direct drive shear is relatively complex and expensive.

但是，直接驱动型剪机的控制系统比较复杂，造价较高。

In order to cut the head of high-speed rolling piece, Italy bomini company designed a set of special mechanism, called high-speed cutting shear. A vertical swing steel divider can move materials to a double channel guide slot installed at the section shear outlet. The guide and guard of the guide slot are overlapped. When the rolled piece is close to the shear, the steel divider is on the upper position. After the shear action is completed, the lower cutting edge will lift the rolled piece into the upper guide slot, and then the steel divider will fall back, and the material head will be sent to the cutting head collection slot, while the material tail (due to its possible long length) will be send people to break the scissors.

意大利波米尼公司为了剪切高速轧件的头部，设计了一套特殊机构，称作高速切头剪。一个垂直摆动分钢器，可以将来料拨到安装在分段剪出口的一个双通路导槽中，导槽的导卫重叠布置，当轧件接近剪机时，分钢器处于上位，剪切动作完成后下剪刃将把轧件提升进入上导槽，然后分钢器回落，将料头送入切头收集槽，而料尾（由于可能比较长）被送入碎断剪。

It can accelerate and decelerate quickly without stopping to shorten the shearing period, reaching 0.5s. In this way, the hot segmented shear can be used to cut directly. When the shear length is 6m, the rolling speed can reach 12m/s. In this way, it is not necessary to carry out cold shear on the rolled piece at the outlet of the cooling bed.

在不停车的情况下快速地加减速以缩短剪切周期，达到 0.5s。这样就可以采用热分段剪直接进行定尺剪切。剪切长度为 6m 时，轧件速度可达 12m/s。这样就不需要在冷床出口处对轧件进行冷剪。

In order to improve the accuracy of hot shear tolerance, it is necessary to accurately set the speed of the workpiece at the exit of finishing mill. The rolling speed is calculated by two photocells by recording the time that the end of the bar passes a fixed distance. The shear accuracy can only be guaranteed if the speed of the rolled piece is kept constant, even after the end of the rolled piece leaves the finishing mill. Therefore, the hot section shear is equipped with a pair of pinch rolls, and the opening can be adjusted to adapt to different shapes, sections and sizes. For larger size rolling pieces, it is sometimes necessary to install a steel drawing equipment before the shear, so as to facilitate the movement speed of the tail of the rolling piece leaving the finishing mill to be stable.

为了提高热剪切公差的精度，必须精确设定轧件在精轧机出口处的速度。轧件速度由两个光电管，通过记录棒材端部经过一段固定距离的时间来计算。只有轧件速度保持恒定才能保证剪切精度，即使在轧件尾部离开精轧机之后也是如此。因此，热分段剪均配有一对夹送辊，开口度可以调整以适应不同的形状、断面和尺寸大小。对于更大尺寸的轧件，有时必须在剪机前安装一个拉钢设备，以利于轧件尾部离开精轧机的移动速度稳定。

2.5.2.2 Optimized Shear System of Sectional Flying Shear after Finishing Rolling
2.5.2.2 精轧后分段飞剪的优化剪切系统

In production, the change of billet quality results in the difference of total length of rolled piece, the error of cutting head and tail and the tolerance of unit mass of rolled product, which makes the length of rolled piece on the cooling bed not all the same. Therefore, the last section of segmented shear will appear short length feeding into the cooling bed. If it is not removed from the cooling bed, it will become the accident source of any automatic cutting system.

在生产中坯料质量的变化导致轧件的总长度不同，以及轧制过程中切头、切尾的误差和所轧成品单位质量的公差，造成上冷床的轧件长度不可能全部相同。因此，分段剪切的最后一段会出现短尺料进入冷床，如果不在冷床上剔除，将会成为任何自动定尺剪切系统的事故根源。

The optimized cutting system of segmented shear proposed by Italy bomini company and Danieli company can ensure that the length of all the rolled pieces on the cooling bed is an integral multiple of the fixed length. The shear procedure of the segmented flying shear is as follows:

意大利的波米尼公司和达涅利公司所提出的分段剪优化剪切系统都可以保证所有上冷床的轧件长度为定尺长度的整数倍。其分段飞剪的剪切程序如下：

First, most of the rolled pieces are cut according to the whole length of the cooling bed, then the rest are cut into short times of the length, and the action cycle of the lifting skirt of the cooling bed is adjusted to match. The program can avoid any problems caused by the irregular length of short ruler.

首先将绝大部分轧件按冷床的全长进行剪切，然后将剩下部分剪切成短倍尺，同时调整冷床提升裙板的动作周期以相匹配。这种剪切程序可以避免由于短尺长度的无规律所带来的任何问题。

If the cut length is shorter than the required length of the upper cooling bed, adjust the length of the penultimate section of steel, and reduce it according to the finished product length, so as to increase the section length of the last section to meet the required length of the upper cooling bed. At this time, the action cycle of the lifting skirt should also be matched accordingly.

如果剪成短倍尺小于上冷床的要求长度时，则调整倒数第二段钢材长度，按成品倍尺减小，以增加最后一段的分段长度，达到上冷床的要求长度。这时提升裙板的动作周期也应相应配合。

The optimized shear system of hot segmented flying shear includes the following functions:
热分段飞剪的优化剪切系统包括以下功能:

(1) Cutting head and tail;

(1) 切头、切尾;

(2) The recovery length of the rolled piece is shorter than the fixed length;

(2) 回收长度短于定尺的轧件;

(3) Break the rolled piece shorter than the preset length;

(3) 对短于预设定长度的轧件进行碎断;

(4) In case of an accident, the shear is broken.

(4) 出现事故时进行碎断剪切。

There are two ways to solve the problem of short gauge material after cutting. The first way is to break the tail shorter than the fixed length, so that the short gauge material can not be put on the cooling bed; the second way is to set up short gauge material collection device before or after the cooling bed, such as Tangshan steel bar continuous rolling mill and Fushun Steel gear steel bar rolling mill. The first method is often used in small-scale continuous rolling mill of carbon steel, while the short gauge collection device is often used in small-scale continuous rolling mill of alloy steel due to the high price of raw materials. Generally, the short ruler material is collected above 3m and broken below 3m.

解决剪后短尺料的方法有两种,第一种方法是将短于定尺长度的尾部进行碎断,使短尺料不上冷床;第二种方法是在冷床前或冷床后设置短尺料收集装置,例如唐钢棒材连轧机、抚顺钢厂齿轮钢棒材轧机。碳素钢小型连轧机组多采用第一种方法,而合金钢小型连轧机组由于原材料价格贵,多采用短尺料收集装置。一般是短尺料在3m以上收集,3m以下碎断。

When the on-line heat treatment system after rolling is adopted, the self-tempering process of the martensite layer on the surface has not been completed when the steel bars passing through the forced water cooler and being quenched on the surface arrive at the section shear before the cooling bed, and the quenched layer is very hard. When cutting this kind of varieties, the flying shear with greater shear capacity must be selected to ensure the working reliability and repeatability of the flying shear.

当采用轧后在线热处理系统时,从强制水冷器中穿水冷却,经表面淬火的螺纹钢到达冷床前分段剪时还没有完成表面马氏体层的自回火过程,淬火层非常硬,当剪切这类品种时,必须选择剪切能力更大的飞剪,以保证飞剪的工作可靠性和可重复性。

2.5.3 Cooling
2.5.3 冷却

The structure, properties and section shape of hot-rolled profiles are directly affected by different cooling systems. The uneven cooling of each part of section steel will cause different structure changes. Different cooling rate and transformation time lead to different microstructure and thickness. At the same time, uneven cooling is easy to cause deformation and distortion of section steel, especially the section steel with special section.

热轧后的型材采用不同的冷却制度对其组织、性能和断面形状有直接影响。型钢的各部位冷却不均将引起不同的组织变化。冷却速度不同，相变时间不同，所得组织、粗细程度都有差别。同时，冷却不均易引起型钢，特别是异形断面型钢的变形扭曲。

In order to improve the mechanical properties of section steel and prevent the distortion caused by uneven deformation, according to the characteristics of steel type, shape and size, different cooling methods and process systems are adopted in the three cooling stages of primary cooling, two cooling and three cooling after rolling.

为了提高型钢的力学性能，防止不均匀变形导致扭曲，根据钢材的钢种、形状、尺寸大小等特点，在轧后的一次冷却、二次冷却和三次冷却的三个冷却阶段分别采用不同的冷却方法和工艺制度。

The different stages of cooling after rolling of section steel are divided as follows: the first stage is from the final rolling temperature to the temperature at which phase transformation occurs; the second stage is the temperature range at which phase transformation occurs; and the third stage is the cooling after phase transformation. Controlled cooling after rolling is to control the opening and final cooling temperature, cooling speed and cooling time in each stage to obtain the required structure and properties.

型钢轧后冷却的不同阶段是这样划分的：从钢材的终轧温度到开始发生相变的温度为第一阶段；发生相变的温度范围为第二阶段；相变后的冷却为第三阶段。钢材轧后控制冷却就是在每个阶段中控制其开冷和终冷温度、冷却速度和冷却时间，以获得所需的组织和性能。

2.5.3.1　Cooling of Deformed Steel Bar on Cooling Bed
2.5.3.1　螺纹钢棒材在冷床上的冷却

(1) Brake of rolling piece.

(1) 轧件的制动。

After the rolling piece is cut in sections, it enters into the cold bed conveying roller table separately driven by the motor through the steel separator and separation baffle.

轧件在轧后分段剪切之后经分钢器和分离挡板进入由电机单独传动的冷床输入辊道。

There are two braking methods for the workpiece entering the cooling bed and entering the roller table:

进入冷床输入辊道的轧件的制动方法有两种：

1) Natural braking by friction.

1) 通过摩擦自然制动。

2) This method can only be used to brake the ribbed steel bars with surface quenching and self-tempering after rolling.

2) 通过磁性制动板强制制动，这种方法仅用于进行轧后余热表面淬火和自回火的带肋钢筋的制动。

When the moving speed of the rolled piece is more than 16m/s, the rolling piece is sent to the cooling bed by the roller table through the separation baffle plate and a set of brake sliding plate.

当轧件的移动速度大于 16m/s 时轧件由辊道送入冷床是通过分离挡板和一套制动滑板来完成的。

The brake procedure using the release damper is as follows:

采用分离挡板的制动程序如下:

1) After the hot-rolled piece is sheared in sections, the rolling piece accelerates on the roller table, so as to quickly separate from the later rolling piece and enter the entrance roller table.

1) 热轧件经分段剪剪切之后,轧件在辊道上加速,从而与后面在轧制中的轧件快速脱离,进入入口辊道。

2) When the brake slide plate is lowered to the low position, the first rolling piece slides down, and the natural friction brake is started until it stops completely.

2) 当制动滑板降到低位时,第一根轧件滑下,开始自然摩擦制动直到完全停止。

3) At the same time, the head of the next rolling piece enters into the upper area of the inclined roller table, and the separation baffle keeps it in this position until the brake sliding plate returns to the middle position.

3) 与此同时,下一根轧件的头部进入倾斜辊道的上区,并由分离挡板将其保持在这一位置,直到制动滑板回升到中位。

4) At this time, the release baffle can be lifted and opened up to make the rolling piece slide to the side wall of the brake sliding plate.

4) 此时分离挡板可以向上抬起、打开,使轧件滑到制动滑板的侧壁。

5) The first rolling piece slides into the first tooth of the cooling bed from the brake slide plate lifted to the upper position.

5) 第一根轧件由提升到上位的制动滑板滑到冷床的第一个齿内。

6) Put down the separation baffle to start the braking cycle of the third rolling piece, and at the same time, the moving rack of the cooling bed will send the first rolling piece into the next tooth.

6) 放下分离挡板,可以开始第三根轧件的制动周期,同时冷床的动齿条将第一根轧件送入下一齿内。

When the ribbed steel bar is quenched by waste heat after rolling, it is ferromagnetic because the temperature of the rolled piece is lower than the Curie point, so the magnetic brake device is used to brake the rolled piece in front of the cooling bed at the population of the cooling bed. The magnetic brake device is installed on the brake slide plate of the cold bed population. The magnetic brake device can ensure that the rolled piece is basically aligned on the rack of the cooling bed.

轧制带肋钢筋采用轧后余热淬火工艺时,由于轧件温度低于居里点而具有铁磁性,因而在冷床入口处,使用磁性制动装置将轧件制动在冷床前。磁性制动装置安装在冷床入口处的制动滑板上。使用这种磁性制动装置能保证轧件在冷床齿条上基本对齐。

The magnetic brake device can shorten the width of the cooling bed, or keep the width of the cooling bed unchanged to improve the final rolling speed of the rolled piece, so as to improve the overall production capacity.

使用磁性制动装置可以缩短冷床宽度，或者保持冷床宽度不变而提高轧件的终轧速度，从而提高总体生产能力。

In order to adapt to the high-speed rolling of ribbed steel bars and deformed steel bars, HSD (high-speed conveying system in upper cooling bed) has been developed by German Symantec Company. The system is composed of pinch roll, high-speed flying shear, brake pinch roll, double rotating groove and synchronous device. It can carry out high-speed bar conveying with the help of fast acceleration tilting cooling bed. It can meet the requirements of 36m/s workpiece speed and 40m/s maximum workpiece speed on the conveying roller table of the cooling bed. The product specification range is $\phi 6 \sim 32mm$.

为适应带肋钢筋和螺纹钢的高速轧制，德国西马克公司研制出上冷床高速输送系统（HSD）。该系统由夹送辊、高速度飞剪、制动夹送辊、双旋转槽和同步装置组成，并配合快速加速度倾斜式冷床，能进行高速度的棒材输送，可以满足冷床输入辊道上轧件速度达到36m/s的要求，最高轧件速度可以达到40m/s。产品规格范围为$\phi 6 \sim 32mm$。

The high-speed conveying system of the upper cooling bed is divided into two groups and arranged in parallel.

这套上冷床高速输送系统分为两组并列布置。

The high-speed flying shear can cut in sections according to the optimized cutting process, which can be matched with single line, double line or split rolling process. The brake pinch roll is used to act on the end of the bar and send the bar to the double rotating groove. The synchronous device is used to send the rolled piece to the fast moving cooling bed, and the rolled piece moves fast on the cooling bed.

高速飞剪根据最优化剪切工艺进行分段剪切，可配合单线、双线或切分轧制工艺。利用制动夹送辊作用于棒材尾部并将棒材送入双旋转槽，利用同步装置将轧件送上快速移动冷床，轧件在冷床上快速移动。

（2）Alignment of the rolled piece on the alignment roller table.

（2）轧件在齐头辊道上的齐头。

In the process of cooling on the cooling bed, the end trimming is to prepare for group cold cutting of multiple pieces, so as to reduce the cutting head and save metal.

轧件在冷床上冷却过程中齐头是为成组进行多根冷剪做准备，以减少切头，节约金属。

According to the running direction of the rolling piece, i.e. the bar coming down from the brake sliding plate will determine its flush direction on the cooling bed. When the outlet direction of the cooling bed is opposite to the inlet direction, the end of the optimized length of the rolled piece will be used as the reference for unloading the cooling bed rack. The movement period of the rack determines the minimum length of the optimized rolled piece without any problem in the cooling of the rolled piece. Because the distribution of rolled pieces on the cooling bed is not very scattered, it is easy to carry out tail end alignment on the end alignment roller table.

依据轧件的运行方向，即从制动滑板上下来的棒材将要在冷床上决定其齐头方向。当冷床的出口方向与入口方向相反时，优化长度的轧件将以其尾部作为参照卸入冷床齿条。齿条的运动周期，决定了轧件冷却中不出现问题的优化轧件的最小长度。由于轧件在冷床上分布并不很分散，所以在齐头辊道上就很容易进行尾部齐头。

When the outlet direction of the cooling bed is the same as the direction of population, in the unloading system mentioned above, in order to ensure a good end trimming effect, a certain number of holes and grooves need to be carved on the end trimming roller, and at the same time, the problem caused by the head of the previous rolling piece should be prevented when discharging to the cooling bed. Therefore, when the rolling speed of the rolled piece exceeds 16m/s, the separation baffle shall be used to isolate the next incoming material without affecting the lifting brake slide plate.

当冷床的出口方向与入口方向相同时，前面所说的卸料系统中，为保证良好的齐头效果，在齐头辊上需要刻一定数量的孔槽同时要防止在卸料到冷床上时前一根轧件头部造成的问题。因此，在轧件轧制速度超过16m/s时要采用分离挡板，隔离下一根来料，不影响提升制动滑板。

Special attention should be paid to the fact that when long cooling bed (110m or longer) is used, it is difficult or almost impossible for small-diameter screw steel (or small flat steel) to align on the fixed baffle plate, because the punch force caused by the alignment is enough to bend the workpiece and generate a compressive stress, so that the workpiece can jump out of the tooth slot. In this way, the mill needs to be stopped to clean the cooling bed.

应特别注意的是采用长冷床（110m 或更长）时，小直径螺纹钢（或小扁钢）将很难或几乎不可能在固定挡板上进行齐头，因为此时齐头所造成的轧件冲力足以使轧件弯曲并产生一个压应力，而使轧件跳出齿槽。这样轧机就需要停车以清理冷床。

A good solution is to use two rows of single hole groove roller table, the first row (high transmission speed) will transport the rolled piece to the near end position as far as possible, and then the second row (adjustable at low speed) will make the bar accurately close to the baffle, and stop in front of the baffle. A set of corresponding induction system can make the guide roller stop at the contact of the rolling piece. The flush effect is within 25mm.

一种好的解决办法是采用两排单孔槽齐头辊道，第一排（传动速度高）将轧件尽量运到近终位置，然后第二排（低速可调）使棒材准确地接近挡板，停在挡板之前。一套相应的感应系统可以使导辊在轧件一接触时即行停车。齐头效果在25mm之内。

(3) Cooling of steel on cooling bed.

(3) 钢材在冷床上的冷却。

When the steel is cooled on the cooling bed, in order to ensure the smooth progress of the subsequent process, it is necessary to control the end quality of the rolled piece well and keep the straightness of the rolled piece as much as possible to prevent the warping and side bending, which will seriously affect its smooth entry into the straightening machine.

钢材在冷床上冷却时，为保证后步工序的顺利进行，必须很好地控制轧件的端部质量，并尽可能地保持轧件的平直度，以防止出现翘头和侧弯，这些将严重影响其顺利进入矫直机。

In the process of cooling, the most important thing is to prevent the distortion caused by uneven cooling or different phase transformation, especially when the cooling section is not symmetrical. In order to solve this problem, modern cooling beds adopt special size parameters, keep straightness, and design the best rack shape and tooth length to compensate for the uneven cooling caused by different cooling curves.

在冷却过程中最重要的是防止轧件由于不均匀冷却或不同时相变而引起的扭曲变形，特别是在冷却断面不是均匀对称的型钢条件下更容易发生扭曲。为了解决这个问题，当今现代冷床均采用特殊的尺寸参数，保持平直度，设计最佳的齿条形状和齿形长度，以补偿由于不同的冷却曲线所产生的不均匀冷却。

In order to control the structure and properties of the bar, according to the different steel grades and sizes, the cooling speed, the open cooling temperature, the final cooling temperature and the different cooling methods of the steel on the cooling bed are controlled.

为了控制棒材的组织和性能，根据钢种、尺寸大小的不同，控制钢材在冷床上的冷却速度、开冷、终冷温度，以及不同的冷却方式。

In the cooling bed, the rolling piece can be cooled by natural air, ventilation or forced water cooling. At this time, the water spray pipe is placed at the lower part of the rolling piece. The cooling bed can also be equipped with a set of water-cooled regulating system to control and reduce the distortion, especially when the angle steel is cooled. The special nozzle arranged at the required position sprays water to the hot metal surface for cooling. According to the different cross-section shape, the spray position and spray intensity are also different, and can be adjusted at any time.

轧件在冷床上既可以采用自然风冷方式，也可以采用通风冷却，也有的采用强制水冷方式，此时喷水管置于轧件下部。还可以在冷床上装备一套水冷调节系统，以控制和降低扭曲，特别是冷却角钢时。在所需位置布置的特殊喷嘴，向热金属表面喷水冷却，根据断面形状的不同，喷水的部位及喷水强度也有所不同，并且随时可调。

The temperature of the cooling bed under the rolled piece should be kept below 100℃, because it is conducive to the straightening and cold shearing of the rolled piece. When rolling special steel, such as bearing steel or spring steel, the cooling bed can be equipped with movable heat preservation covers at the upper and lower sides of its initial section to reduce the cooling speed, so as to realize the delayed cooling process. The dynamic and static racks of the cooling bed are inclined in the horizontal direction, so as to change the contact point and make the screw steel rotate continuously during the moving process of the rolling piece. On the contrary, the rack of the cooling bed can be designed into a special shape for the steel requiring rapid cooling on the cooling bed, such as austenitic stainless steel, so that the rolled piece can be cooled in water for a certain period of time after coming down from the conveying roller table of the cooling bed, and then the water can be discharged into the cooling bed for air cooling.

轧件下冷床的温度应保持在100℃以下，因为这有利于轧件的矫直和冷剪切。轧制特殊钢如轴承钢或弹簧钢时，冷床可以在其初始段上下两侧安装可移动保温罩来降低冷却速度，从而实现延迟冷却工艺。冷床的动静齿条在水平方向上均有一定倾斜，以便在移动轧件过程中不断改变接触点并使螺纹钢旋转。相反，对于需要在冷床上进行快速冷却的钢种如奥氏体不锈钢，冷床的齿条可以设计成特殊的形状，以使轧件从冷床输入辊道上下来后立即侵入水中冷却一定时间，再出水进入冷床空冷。

2.5.3.2 Structural Characteristics and Technical Performance Parameter Selection of Cooling Bed

2.5.3.2 冷床的结构特点与技术性能参数选择

(1) Structure and characteristics of a cooling bed.

(1) 冷床的结构形式和特点。

The common structure of the cooling bed mainly includes step rack type cooling bed, swing type cooling bed, inclined roller type cooling bed, chain type cooling bed, etc. In recent years, with the progress of small-scale continuous rolling technology, the quality requirements of steel products are higher and higher. It is hoped that there is no relative sliding friction between the rolled piece and the bed surface on the cooling bed, so as to avoid scratching the surface of the rolled piece. At the same time, it is hoped that the rolled piece can be cooled evenly and straightened in the cooling process. Step rack type cooling bed is widely used because of its uniform cooling and straightening in the process of cooling. Generally, the straightness can reach 4mm/m, the best can reach 2mm/m and the surface scratch is small.

冷床的结构形式常见的主要有步进式齿条型冷床、摇摆式冷床、斜辊式冷床、链式冷床等。近年来随着小型连轧技术的进步，对钢材产品质量要求也越来越高，希望轧件在冷床上与床面没有相对滑动摩擦，以免划伤轧件表面，同时希望轧件在冷却过程中冷却均匀并得到矫直。步进式齿条型冷床，因其轧件冷却均匀，并在冷却过程中得到矫直，一般平直度可达 4mm/m，最好可达到 2mm/m 而且表面擦伤小，因而得到越来越广泛的应用。

In addition to the conventional brake slide plate mechanism, the cooling bed mechanism on the bar also has a tail clamp, a brake chute, a switch, etc.

棒材上冷床机构除了常规的制动滑板机构外，还有夹尾器、制动落料槽、转辙器等。

The rolled piece enters into the chute or switch of cast iron at the speed of 17~18m/s or higher. The tail of the rolled piece is clamped by the tail of the population clamp in a very short time, and it is braked to about 3m/s, then the chute or switch is opened, and the rolled piece is placed on the cooling bed. This mechanism is only suitable for the production of carbon steel rolling pieces of small bar mill.

轧件以 17~18m/s 或更高的速度进入铸铁的滑槽或转辙器，入口夹尾器在很短的时间内将轧件的尾部夹住，制动至 3m/s 左右，然后打开滑槽或转辙器，将轧件放在冷床上。这种机构只适用于生产碳素钢轧件的小型棒材轧机。

(2) Selection of technical performance parameters of cooling bed.

(2) 冷床技术性能参数的选择。

1) Selection of cooling bed width.

1) 冷床宽度的选择。

The width of the cooling bed depends on the length of the rolled piece and the section shear length after the machine, and is larger than the maximum section length. For example, if the length of the finished rolled piece is l_1, and the maximum length of the flying shear section is L, the width of the cooling bed B is usually taken as:

冷床宽度根据轧件长度及机后分段剪切长度而定,并且比最大的分段长度更大些,例如轧件成品长度为 l_1,飞剪分段最大长度为 L,则冷床的宽度 B 通常取为:

$$B = L + l_1$$

As long as the cooling capacity is enough, it is not appropriate to pursue the width of the cooling bed excessively. The width of the cooling bed can also be determined by multiplying the maximum rolling speed of the finished product by 4 and adding buffer length, that is:

只要冷却能力足够,过分地追求冷床宽度是不合适的。冷床宽度也可以用成品最高轧制速度乘以 4 后加缓冲长度确定,即

$$B = v_{\text{Maximum rolling}} \cdot 4 + l_1$$
$$B = v_{轧最大} \times 4 + l_1$$

Where　　　B——Cooling bed width, m;

$v_{\text{Maximum rolling}}$——Maximum rolling speed of finished rolling piece, m/s;

l_1——Buffer length, generally the length of finished rolled piece, m.

式中　　　B——冷床宽度, m;

$v_{轧最大}$——成品轧件最高轧制速度, m/s;

l_1——缓冲长度,一般为轧件成品长度, m。

If the width of the cooling bed is too wide, the total mass of the equipment will be increased and the investment will be increased due to the increase of the cooling bed transportation, output roller table and brake steel paddle. If the width of the cooling bed is too narrow, the length of the cooling bed will be lengthened and the span of the workshop will be enlarged. At the same time, because of the short operation cycle time of the section shear and the too fast action frequency of the cooling bed, it is difficult for the mechanical and electrical equipment of the section flying shear and the cooling bed to adapt.

冷床宽度过宽将由于冷床输入、输出辊道以及制动拨钢器随之增长,而使设备总质量增加,投资加大。冷床宽度过窄,则使冷床长度加长,造成厂房跨度加大,同时,由于分段剪操作周期时间短和冷床动作频率过快,使分段飞剪及冷床的机电设备难以适应。

2) Selection of operation cycle of cooling bed.

2) 冷床动作周期的选择。

The action cycle of the cooling bed refers to the time required for one step of the cooling bed. The stepping motion of the cooling bed is mostly realized by cam mechanism, so the action cycle of the cooling bed is actually the time when the cam mechanism rotates for one cycle. This time should be less than the interval time between two shears in the section behind the finishing mill. This time t is determined by the moving speed of the rolled piece and the length of the segment, that is

冷床的动作周期是指冷床步进一次所需的时间。冷床的步进运动多以凸轮机构来实现,因此,冷床的动作周期实际上也就是凸轮机构旋转一周的时间。这个时间要小于精轧机后面分段剪两次剪切之间的间隔时间。这一时间 t 是由轧件移动速度和分段长度决定的,即

$$t = L/v$$

Where t——Action cycle of cooling bed, s;

  L——The minimum length of the piece to be segmented, m;

  v——Rolling speed (i.e. maximum rolling speed), m/s.

式中 t——冷床的动作周期，s；

  L——轧件被分段的最小长度，m；

  v——轧件速度（即最大轧制速度），m/s。

Considering the time required by the feeding and unloading mechanism of the cooling bed, the operation cycle of the cooling bed should be shortened by about 1~2s.

再考虑到冷床上、下料机构所需的时间，冷床的动作周期应再缩短1~2s。

3) Selection of rack spacing.

3) 齿条间距的选择。

The weight of the rack occupies a large proportion in the weight of the rack type cooling bed. If the distance between the racks is too small, the number of racks will inevitably increase, thus increasing the weight of the cooling bed. However, if the rack spacing is too large, the steel will deflect on the cooling bed and affect the product quality. Proper rack spacing shall ensure that the deflection of steel due to self-weight on the cooling bed does not exceed the standard bending degree (less than 6mm/m) issued by the Ministry, and at the same time minimize the number of racks. The steel deflection on the cooling bed is affected by two factors: ①the size of the steel section, when the rack spacing is the same, if the steel section is small, the deflection will be large; if the section is large, the deflection will be small; ②the higher the temperature is, the greater the deflection will be.

齿条的重量在齿条型冷床的重量中占有很大的比例，齿条间距选择过小，则必然增加齿条数量，从而增加冷床重量。但齿条间距过大，又会使钢材在冷床上产生挠度而影响产品质量。适当的齿条间距应是保证钢材在冷床上因自重而产生的挠度不超过部颁标准弯曲度（小于6mm/m）的同时尽量减少齿条数量。冷床上钢材挠度受两个因素的影响：①钢材断面大小的影响，在齿条间距相同的情况下，钢材断面小则挠度大，断面大则挠度小；②钢材温度的影响，温度越高，挠度越大。

When determining the distance between the racks of the cooling bed, the parameters of the cooling bed can be determined by analogy, or by calculating the deflection of the steel between the racks or the cantilever wall at the head and tail.

具体确定冷床齿条间距时可以对比已有的冷床参数用类比法确定，也可以通过计算钢材在齿条间或头、尾悬臂墙的挠度来确定。

4) Determination of rack pitch and profile angle.

4) 齿条齿距、齿形角的确定。

Rack pitch and profile angle are related to steel section. For the screw steel, there is a certain proportion between the pitch T and the diameter of the screw steel.

齿条齿距、齿形角与钢材断面有关。对螺纹钢来说，齿距T与螺纹钢直径有一定的比例关系。

There are two kinds of tooth form angle in China. One is (a) 30°/60° tooth shape, as shown in the Figure 2-13. The advantage of this kind of tooth shape is that it is easier to roll the steel when the steel is stepped, which is beneficial to the straightening of the steel in the cooling process. At the same time, this kind of tooth shape is also more suitable for the cooling of flat steel, angle steel and other special-shaped materials. The other rack tooth angle is 45°, as shown in (b) in Figure 2-13. The tooth shape is suitable for cooling the threaded steel, so that the straightening plate on the feeding side can be made of equal angle steel, which simplifies the processing of the equipment and reduces the weight of the equipment.

齿形角国内较常见的有两种。一种是 30°/60°齿形，如图 2-13 (a) 所示。这种齿形的好处是在钢材步进时，较容易实现钢材的滚动，有利于钢材在冷却过程中的矫直，同时，这种齿形也较适合冷却扁钢、角钢等异形材。另一种齿条齿形角为 45°，如图 2-13 (b) 所示。这种齿形适合于冷却螺纹钢，从而进料侧的矫直板可以采用等边角钢制造，简化了设备加工，设备重量也较轻。

Figure 2-13　Schematic diagram of tooth pitch and tooth shape angle of step rack type cooling bed
(a) 30° and 60° tooth shape; (b) 45° tooth shape
图 2-13　步进式齿条型冷床齿距、齿形角示意图
(a) 30°、60°齿形；(b) 45°齿形

5) Determination of length of cooling bed.

5) 冷床长度的确定。

The length of the cooling bed depends on the time required to cool the steel.

冷床长度取决于钢材冷却需要的时间。

The relationship between the heat transferred by the rolled piece and the temperature reduction of the rolled piece is as follows:

轧件所传递的热量与轧件温度降低的关系为：

$$dQ = KCdT_w$$

The relationship between heat transfer and cooling time is as follows:

轧件所传递的热量与冷却时间的关系为：

$$dQ = a(T_0 - T_w)dt$$

The above two formulas are the heat transferred by the rolling piece to each $1m^2$ cooling area, so the two formulas are equal, so it is concluded that:

以上两式均为轧件对每 $1m^2$ 冷却面积所传递的热量，因而两式相等，故得出：

$$\alpha(T_0 - T_w)dt = KCdT_m$$

By integrating this formula in the range of 0~d time and temperature, we can get:

将此式在 0~d 时间和温度范围内积分，可得：
$$M = \ln[(T_{WA} - T_0)/(T_w - T_0)]$$

Where K——Weight of rolled piece per square meter of surface area, kg/m²;
  C——Mass heat capacity, J/(kg·℃);
  T_0——Ambient temperature, ℃;
  T_w——Rolling piece temperature, ℃;
  T_{WA}——Temperature at the beginning of rolling cooling, ℃;
  T_{WC}——Temperature at the end of rolling piece cooling, ℃;
  t——Cooling time of rolled piece, min;
  α——Heat transfer coefficient, J/(m²·h·℃).

式中 K——每平方米表面积的轧件重量，kg/m²;
  C——质量热容，J/(kg·℃);
  T_0——周围环境温度，℃;
  T_w——轧件温度，℃;
  T_{WA}——轧件冷却开始时的温度，℃;
  T_{WC}——轧件冷却终止时的温度，℃;
  t——轧件冷却时间，min;
  α——传热系数，J/(m²·h·℃)。

The heat transfer coefficient α α consists of two parts, one is the radiation heat transfer coefficient α_s, the other is the contact heat transfer coefficient α_b, and there are:
传热系数 α 由两部分组成，一为辐射传热系数 α_s，一为接触传热系数 α_b，而且有：
$$\alpha = \alpha_s + \alpha_b$$

For deformed and rectangular section steels, α_s and α_b can be based on the average temperature T_m of the steel. Find out from the relevant curve.
对于螺纹钢和矩形断面钢材，α_s 和 α_b 可根据钢材的平均温度 T_m 确定。由有关曲线查出。

The average temperature of the rolled piece can be calculated as follows:
轧件的平均温度可按下式计算：
$$T_m = T_0 + (T_{WA} - T_{WC})/M$$

According to the cooling time, the length of the cooling bed L_k can be calculated as:
根据冷却时间可求出冷床长度 L_k 为：
$$L_k \geq (n+1)T$$

Where T——Rack pitch of a cold storage, mm;
  n——The total number of steel bars steel, and there is: $n = 6m/t_k$;
  t_k——Rolling cycle of each steel, s.

式中 T——冷库齿条齿距，mm;
  n——在 t 时间内需上冷床的钢材总根数，并有：$n = 6m/t_k$;
  t_k——每根钢材的轧制周期，s。

According to the hourly output of rolling mill and cooling time of rolling piece, the length of cooling bed can also be calculated according to the following formula:

根据轧机的小时产量和轧件的冷却时间,也可按下式计算冷床的长度:

$$L_k = Qat/G$$

Where L_k——Distance between gifts on the cold bed, m;
Q——Maximum hourly output of rolling mill, t/h;
G——Weight of rolled piece, t/piece;
a——Distance between gifts on the cold bed, m;
t——Cooling time of rolled piece, h.

式中 L_k——冷床长度,m;
Q——轧机最高小时产量,t/h;
G——根轧件的重量,t/根;
a——轧件在冷床上的间距,m;
t——轧件的冷却时间,h。

Considering the on-line straightening requirements of the rolled piece, the rolled piece shall be cooled to about 100℃.

考虑到轧件的在线矫直要求,轧件应冷到约100℃。

6) Determination of cam eccentricity.

6) 凸轮偏心距的确定。

The eccentricity e of the cooling bed cam can be taken as $e=T/2$ (T is rack pitch), $e<T/2$ for 30°/60° profile, and $e>T/2$ for 45° profile. $e<T/2$ or $e>T/2$ can change the contact point between the steel and the rack during the cooling process on the cooling bed. At the same time, the steel has a certain rolling in the step process to achieve the purpose of straightening the steel. However, in addition to bars, there are also profiles in the production, so it is better to take $e=T/2$. The advantage is that the steel step is stable, and the profiles will not have sliding friction with the cooling bed. In recent years, the static rack is inclined to the step direction at an angle, so as to avoid the defect that the contact point between steel and rack is always a fixed position.

冷床凸轮偏心距 e 可以取成 $e=T/2$（T 为齿条齿距）,用于30°/60°齿形可以取 $e<T/2$,而用于45°齿形可以取 $e>T/2$。$e<T/2$ 或 $e>T/2$ 都可以改变钢材在冷床上冷却过程中与齿条的接触点,同时在步进过程中钢材有一定的滚动,以达到矫直钢材的目的。但是,在生产的品种中除了棒材之外还有型材,则最好取 $e=T/2$,其优点是钢材步进平稳,型材不会与冷床产生滑动摩擦。近年来采用静齿条与步进方向倾斜一个角度,从而避免了钢材与齿条接触点始终是一个固定位置的缺点。

2.5.4 On Line Sizing and Shearing of Bar
2.5.4 棒材在线定尺剪切

In the traditional rolling production line, after the workpiece is cooled on the cooling bed, there is a manual fixed type cold shear at the output roller table of the cooling bed. The bar is cut into a fixed length by the positioning baffle. There are two types of fixed cold shear: one is open type (or C-frame type), the other is closed type. Their selection depends on the shear force and the width of the blade. Both shears are equipped with two cutting edges, one fixed, and the other movable cutting edge installed on the sliding plate, operated by the pneumatic clutch or start stop DC motor. The shearing machine is equipped with a set of bar pressing device to clamp the bar during the shearing process. The movable baffle is used to change the length of the fixed length by adjusting its position, and the minimum cutting head is realized to ensure the shearing accuracy of the bar. The sheared bar is unloaded by the vertical transport chain or mobile trolley.

在传统的轧制生产线上，轧件在冷床上冷却后，在冷床的输出辊道处有一台手动固定式冷剪，棒材由定位挡板齐头，将轧件切成定尺。这种固定式冷剪有两种形式：一种为开口式（或称作C形框架式），另一种为闭口式。它们的选择取决于剪切力和剪刃的宽度。两种剪机均装有两个剪刃，一个固定，另一个活动剪刃安装在滑板上，由气动离合器或启停式直流电机操作。剪机上装有一套棒材压紧装置，在剪切过程中卡住棒材。采用可移动挡板，通过调整其位置来改变定尺长度，并实现最小的切头，保证棒材的剪切精度。剪切后的棒材，由垂直方向的运输链或移动小车卸料。

After the steel is cooled, several pieces of cold flying shear are directly used to cut into fixed length. The crank type cold flying shear is driven by DC motor, the shear speed can be changed from 1.5m/s to 2.5m/s, and the shear length is 4~18m. For example, Tanggang bar factory adopts pendulum type cold flying shear, with a shearing capacity of 350t, which can be cut into 6~12m fixed length bar, with a maximum shearing speed of 2.5m/s, which is higher than other forms of cold shearing machine, so that each row of materials can be arranged less, without affecting the production capacity of flying shear. Reducing the width of cutting material can improve the shearing accuracy, reduce the amount of cutting head and reduce the accident rate.

钢材冷却后直接采用冷飞剪多条剪切成定尺。曲柄式冷飞剪采用直流电机传动，剪切速度可以从1.5m/s变到2.5m/s，剪切定尺长度为4~18m。例如唐钢棒材厂采用摆式冷飞剪，剪切能力350t，可剪切成6~12m的定尺棒材，最高剪切速度达2.5m/s，已经高于其他形式的冷剪机，因此可以使每排料排得少一些，而不影响飞剪的生产能力。减小剪切料的排料宽度有利于提高剪切精度，减小切头量，减小事故率。

2.5.5 Steel Packaging (Bundling)
2.5.5 钢材的包装（打捆）

Steel packaging is the back process of steel rolling production, and it is an essential and important process. The complete steel packaging shall have multiple functions such as steel input, bundling forming, and bundling, bundling output, weighing and marking. The key equipment to complete all the above functions is the main binding machine. The core part of the binding machine is the head part, which is universal. The auxiliary machines are matched with different packaging varieties and different process conditions of various factories to form various types of units.

钢材包装是轧钢生产的后部工序，是必不可少的重要工序。完整的钢材包装应具有钢材输入、捆包成型、捆扎、捆包的输出及称重和标记等多种功能。为完成以上全部功能的关键设备是主机捆扎机。捆扎机的核心部件是机头部分，是通用的。辅机则随着包装品种的不同和各工厂现场工艺条件的差异而配套，形成各种形式的机组。

The binding machine of the main machine can be divided into steel belt binding machine (binding machine) and wire binding machine according to the binding materials used. If it is not a special requirement of the user, it would be cheaper to bundle with wire. According to the classification of transmission mode, there are three types of strapping machine: pneumatic, hydraulic and mechanical.

主机捆扎机按所用的捆扎材料分类有钢带捆扎机（打捆机）和线材捆扎机两种。如果不是用户的特殊要求，用线材打捆要便宜些。按传动方式分类，捆扎机有气动、液压和机械三种形式。

The binding materials are selected by the binding machine of the main machine according to the variety of the packed steel. Generally, steel strips are used for binding small sections, steel pipes and packing boxes, and steel wires (wire rods) are used for binding large sections, deformed steel bars, wire rods and small bundles. Steel banding does not damage the steel surface, with high tensile strength and neat and beautiful packing. Wire binding cost is low, wire source is convenient, not easy to break. The transmission mode is selected according to the production site conditions, site size, production efficiency, environmental temperature, etc.

主机捆扎机根据包装钢材品种选定捆扎材料，一般小型型材、钢管、包装箱多采用钢带捆扎，大型型材、螺纹钢、线材、小型捆包多采用钢丝（线材）捆扎。钢带捆扎不破坏钢材表面，抗拉强度高，捆包规整美观。线材捆扎成本低，钢丝来源方便，不易崩断。传动方式则根据生产现场条件、场地大小、生产效率高低、环境温度等来选定。

The matching of auxiliary machines is completely determined according to the binding object, process content, production process, site conditions, productivity and operation level.

辅机的配套则完全根据捆扎对象、工艺内容、生产流程、现场条件、生产率和操作水平来确定。

2.5.5.1 Bomini Company's Baler
2.5.5.1 波米尼公司的打捆机

The baler designed by Italian company Bomini can bind any section of steel. It is fully automatic hydraulic operation. The binding material used is $\phi 5.5 \sim 6.5mm$ wire rod. An important feature of the utility model is that the knotting position is always on the top of the bundle, so that the bound profile can easily run along the roller table. The baler can be mobile or stationary. The binding action is used to feed the tightened packing line directly along the outline of the bundle, so as to avoid scratching the surface of the product. The packing line is only bent at the corner of the bale and does not produce friction. The binding process is shown in Figure 2-14.

由意大利波米尼公司设计的打捆机，能够捆扎任何断面的型钢。它完全是自动化的液压操作。所用捆扎材料为 $\phi 5.5 \sim 6.5mm$ 线材。它的一个重要特点是，打结的位置总是在捆的上面，使捆好的型材容易沿辊道运行。打捆机可以是移动式的，也可以是固定式的。打捆动作为沿着捆的外轮廓把拉紧的打包线直接喂入，这样可避免划伤产品的表面。打包线只在捆的角部弯曲，并不产生摩擦。捆扎的动作过程如图 2-14 所示。

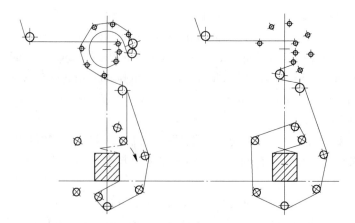

Figure 2-14 Binding action of baler
图 2-14 打捆机捆扎动作

This way of binding can ensure that the corners of the bales are well bound. This is very important for making a square or rectangular bale, which prevents the bale from becoming a semicircle.

这种捆扎方式可以保证对捆垛的角部进行良好捆扎。这对于将捆垛打成方形或矩形捆是非常重要的,它可以防止捆形变成半圆形。

Three sets of counting devices have been reserved in the transverse moving area behind the conveying roller table in the bar binding area according to the production process requirements, and a set of electronic weighing device has been equipped.

在棒材捆扎区域的输入辊道后的横移区,按生产工艺要求预留了三组计数装置,已经配备了一套电子称重装置。

2.5.5.2 Binding Unit of Xi'an Steel Factory
2.5.5.2 西安钢厂的捆扎机组

The working process of the baling mill is: conveying roller table.

该捆轧机组的工作过程为:输入辊道。

Send the bar to the front of the traverse position, the short ruler separator will suck up the fixed length material, the input lifting baffle will fall, the roller table will start, the short ruler material will be sent out to the short ruler collection tank, the roller table will stop, the short ruler separator will put down the fixed length material, move it to the step chain road through the lifting traverse machine, and then the separation conveyor will go through the counter, and the points will be sent to the collection arm of the four in one binding machine, and the collection arm will transport the bar to the V-shaped roller table, after forming and binding, it is output to the fixed baffle in the direction of end alignment, and then it is sent to the aggregate bench by the lifting and traverse trolley, and lifted out in a centralized way.

将棒材送至横移位置前面,短尺分离机将定尺料吸起,输入升降挡板落下,辊道启动,将短尺料送出至短尺收集槽前,辊道停止,短尺分离机将定尺料放下,经升降横移机移至步进链道上，再由分离输送机经计数机，点数后送到四合一捆扎机的收集臂中，收集臂将棒材传送到 V 形辊道上，经成形、捆扎后，向齐头方向输出至固定挡板，再由升降横移小车送到集料台架上，集中吊出。

The binding unit can bind bar with the size of $\phi 12\sim 40$mm and the length of 9m and 12m. The bar temperature is required to be below 300℃. The annual output is 350000t, the hourly output is 75t, and the bale weight is less than 3t. The binding material is 08F wire rod of $\phi 5.5$mm. Three operation modes of online full-automatic, semi-automatic and manual are adopted, with the bundling speed of 45s/lane and the bundling diameter of no more than 450mm.

该捆扎机组可捆扎棒材尺寸为 $\phi 12\sim 40$mm，长度为 9m、12m。棒材温度要求在 300℃ 以下。年产 350000t，小时产量为 75t，捆重小于 3t。捆扎材料为小 $\phi 5.5$mm 的 08F 线材。采用在线全自动、半自动和手动三种操作方式，捆扎速度 45s/道，捆包直径不大于 450mm。

The counting speed of the counter is $3\sim 5$ pieces/s. The operating conditions of the binding machine are: the ambient temperature of the mechanical hydraulic part is $-5\sim +50$℃; the ambient temperature of the electrical part is $-5\sim +50$℃; The relative temperature is less than 83%, without frost.

计数机的点数速度为 3~5 根/秒。捆扎机设备的使用条件为：机械液压部分环境温度 $-5\sim +50$℃；电气部分环境温度 $-5\sim +50$℃；相对温度小于 83%，不结霜。

2.5.6 Typical Finishing Process of Continuous Rolling Bar
2.5.6 连轧棒材典型精整工艺

After heating, rough rolling, medium rolling and finishing rolling, the continuous casting billet is rolled into ribbed steel bar, and then into the finishing stage. Some varieties or specifications need controlled cooling or surface residual heat quenching of ribbed bars after rolling.

连铸方坯经加热、粗轧、中轧和精轧之后轧成带肋钢筋，而后进入精整阶段。某些品种或规格轧后需要进行控制冷却或带肋钢筋的表面余热淬火。

When the rolled piece enters the downstream of the rolling mill, the double length flying shear after the water quenching line will be used for double length cutting, and it will go through the cooling, fixed length cutting and bundle finishing processes on the cooling bed. The following is an example of Tangshan Iron and Steel Co., Ltd. for the finishing process.

轧件进入轧机下游淬水线之后的倍尺飞剪进行倍尺剪切，并经过冷床上的冷却、定尺剪切和成捆等精整工序。下面以唐钢棒材连轧机为例，对精整工序分别介绍。

2.5.6.1 Multiple Shear of Bar
2.5.6.1 棒材的倍尺剪切

After finishing rolling, the rolled piece is first sent to the pinch roll before the multiple length shear. The pinch roll is double drive type, its function is to clamp the rolled piece into the flying shear for multiple length cutting and tail cutting; in addition, when the cooling system fails, the rolled piece is clamped and sent to the breaking shear for breaking and the tail steel of large-size rolled piece is broken. If it is necessary to collect short gauge steel, when the short gauge steel collecting bed is short, in order to make the short gauge steel in high-speed operation enter the collecting bed smoothly, the pinch roll can also slow down the rolling piece.

终轧后的轧件首先送入倍尺剪前的夹送辊。夹送辊为双驱动型,其作用是夹持轧件进入飞剪进行倍尺剪切和切尾;另外,当冷却系统出现故障时将轧件夹持送入碎断剪进行碎断和大规格轧件尾钢的碎断。如果需要收集短尺钢材,当短尺钢材收集床较短时,为使高速运行的短尺材顺利进入收集床,夹送辊还可以对轧件进行减速。

The pinch roll before the double length shear is two roll horizontal type. The upper roller is movable and driven by the air cylinder. When the air cylinder starts, the upper roller is pressed down so that the rolled piece can be clamped. The maximum moving speed of the clamped workpiece is 18m/s. The clamping force of the rolled piece is 5737N. There are grooves on the pinch roll, which can be replaced according to the cross section of different products.

倍尺剪前的夹送辊为二辊水平式。上辊可动,由气缸驱动,当气缸开动时,上辊压下,以便能夹持轧件。夹持的轧件移动速度最大达 18m/s。对轧件的夹持力为 5737N。夹送辊上有槽,可根据不同产品的断面进行更换。

Guide device is also installed at the inlet and outlet of pinch roll, mainly a bell mouth conduit, which is fixed on the bracket, and the height of conduit bracket can be adjusted to align with the rolling line. According to different rolling specifications, choose different specifications of conduit.

在夹送辊入口和出口处还装有导向装置,主要是一个喇叭口导管,它固定在支架上,导管支架高度可以调整,以便对准轧制线。根据不同的轧制规格,选用不同规格的导管。

The double length shear of the factory is a start/stop rotary flying shear driven by a DC motor. The performance of the shear is as follows:

该厂的倍尺剪为启/停回转式飞剪,由一台直流电机驱动。剪机的性能如下:

The maximum moving speed of shear rolled piece is 18m/s;

剪切轧件的最大移动速度 18m/s;

Cross section area of shear rolled piece is 2000mm^2;

剪切轧件断面面积 2000mm^2;

The temperature of shear rolled piece is 550℃;

剪切轧件温度 550℃;

Number of cutting edges is 2;

剪刃数 2 片;

The cutting speed is 2.7~19.8m/s.

剪刃速度 2.7~19.8m/s。

The cutting speed of the shear is adjustable, which is proportional to the rolling speed of the bar. In the cutting position, the cutting edge speed is ahead or behind the rolling speed. The leading or lagging speed can reach 10% of the rolling speed. The shear rate, lead or lag rate will be set according to the rolling procedure or by the operator.

该剪的剪刃速度是可调的,与棒材轧制速度成正比。在剪切位置,剪刃速度比轧件速度超前或滞后。超前或滞后的速度可以达到轧件速度的10%。剪切速度、超前或滞后速度将根据轧制程序或由操作人员设定。

It is an effective measure to increase the yield and make rational use of the cooling bed area.
倍尺优化剪切工艺是提高成材率、合理利用冷床面积的有效措施。

After the billet is rolled into bar by continuous rolling mill, the length of the finished rolled piece is much longer than the length acceptable by the cooling bed, so it must be cut into the length acceptable by the hot flying shear. In order to avoid short length steel and improve the yield, the length of steel on the cooling bed is usually cut into multiple of fixed length.

钢坯经连轧机轧成棒材之后，成品轧件的长度远大于冷床所能接受的长度，因此，必须经热飞剪剪切成冷床所能接受的长度。为了避免在定尺剪切时产生短尺钢，提高成材率，一般将上冷床的钢的长度剪切成定尺的倍数。

In the actual production, the length of billet, tolerance of finished product size and the number of cutting head in the rolling process fluctuate within a certain allowable range, resulting in the continuous change of the length of the steel rolled out by the finishing mill, which may make the length of the last section of steel after multiple length cutting less than the minimum length acceptable by the cooling bed. In this case, the steel tail whose length is less than the minimum length that the cooling bed can accept is usually cut by the end of multiple length steel. If the length of the last section of steel at this time is only slightly less than the minimum length acceptable for the cooling bed, but greater than the fixed length, in this case, the broken tail steel will cause the reduction of the yield.

在实际生产中，钢坯的长度、成品尺寸的公差、轧制过程中切头的多少，都在一定的允许范围内波动，导致精轧机轧出钢材的长度也在不断变化，这就可能使倍尺剪切后所得到的最后一段钢的长度小于冷床所能接受的最小长度。这种情况下，通常是在倍尺钢剪切结束后，把长度小于冷床所能接受最小长度的钢尾由碎断剪碎断。如果这时最后一段钢的长度只是稍小于冷床所能接受的最小长度，而大于定尺长度，这种情况下碎断尾钢将会造成成材率的降低。

In order to solve this practical problem, the optimal shearing process of multiple length steel is formed. Its purpose is to get the maximum number of finished bars from the given finished bar length, reduce the short length and improve the yield.

为了解决这一实际问题，形成了倍尺钢优化剪切工艺。其目的是从给定的成品棒长中得到最大数量的成品长度的棒材，减少短尺，提高成材率。

The essence of the optimized shearing process is that when the length of the tail steel is less than the minimum length acceptable to the cooling bed and greater than the fixed length of the finished product, a part of the length of the double length steel before the tail steel (i. e. the double length steel of the second last upper cooling bed) is reserved for the tail steel, so that the length of the tail steel can reach the length acceptable to the cooling bed, while the length of the second last one is reduced, but it can still meet the requirements of upper cooling bed. If the tail steel length is too small, after optimization with the penultimate steel, both of them can not meet the length requirements of the upper cooling bed, the optimization starts from the penultimate steel. The final optimization result is that the last three steels can reach the minimum length of the upper cooling bed, and the tail steel whose length is less than the fixed length is received by the short length collecting bed. If the length is less than the minimum length acceptable for the short bed, the guide shear after shearing shall be used for crushing.

优化剪切工艺的实质是当尾钢长度小于冷床所能接受的最小长度而大于成品定尺长度时,就把尾钢之前的一段倍尺钢(即倒数第二根上冷床的倍尺钢)的长度留一部分(相当于定尺长度或倍数)给尾钢,使尾钢长度达到冷床所能接受的长度,而倒数第二根长度有所减小,但仍能满足上冷床要求。如果尾钢长度太小,在与倒数第二根钢优化后,二者都达不到上冷床的长度要求时,则优化从倒数第三根倍尺钢开始。最终优化结果是使最后三根钢都能达到最小上冷床长度,最后长度小于定尺长的尾钢由短尺收集床接收。如果其长度还小于短尺床所能接受的最小长度,则经剪后的导向器导向碎断剪进行碎断。

The guide is a guide plate driven by an electric motor. The distance between the head of guide plate and flying shear is 650mm. The position of the guide is controlled by positioning the holding brake. The action of the guider is controlled by the encoder of the flying shear, so as to ensure that the action of the guider is synchronous with the flying shear. The breaking shear is located behind the guide and is used for breaking or accident cutting of short steel.

导向器是一个由电机带动的导板。导板头部距倍尺飞剪650mm。用定位抱闸来控制导向器的位置。导向器动作由倍尺飞剪编码器控制,以保证导向器的动作与飞剪同步。碎断剪位于导向器之后,用于短尺钢的碎断或事故剪切。

The technical parameters of breaking shear are as follows:
碎断剪的技术参数如下:
Rolling speed   18m/s;
轧件速度   18m/s;
The shear area is $1257mm^2$ in hot state and $800mm^2$ in cold state;
剪切面积热态$1257mm^2$,冷态$800mm^2$;
The temperature is 500℃ when the maximum section of rolled piece is $800mm^2$;
轧件温度最大断面$800mm^2$时500℃;
Breaking length   580mm;
碎断长度   580mm;
Blade speed   3.8~20m/s.
刃片速度   3.8~20m/s。

The short ruler collecting bed is located next to the conveying roller table after crushing and shearing, which is used to collect the short ruler tail steel after optimized shearing.

短尺收集床位于碎断剪后、输送辊道的旁边,用于收集优化剪切后所得到的短尺尾钢。

The length of the short length collecting bed is 18m, and the acceptable bar length: the shortest is 4m, the longest is 12m.

短尺收集床长度为18m,可接收的棒材长度:最短为4m,最长为12m。

2.5.6.2 Cooling of Bar on Cooling Bed
2.5.6.2 棒材在冷床上的冷却

(1) Bar input and brake.

(1) 棒材的输入与制动。

The steel cut by multiple length shear is transported and braked by roller with brake skirt and reaches the cooling bed, as shown in Figure 2-15.

经倍尺剪剪切后的钢材经过带制动裙板的辊道输送和制动并达到冷床上，如图2-15所示。

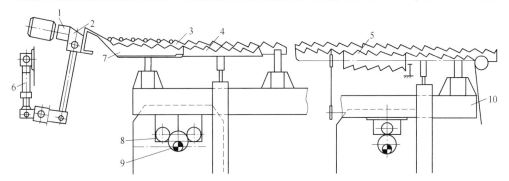

Figure 2-15 Schematic diagram of roller table with apron and step cooling bed
1—Cooling bed roller table; 2—Lifting apron; 3—Moving rack; 4—Fixed (static) rack; 5—Alignment roller table;
6—Hydraulic cylinder; 7—Leveling plate; 8—Eccentric wheel; 9—Transmission shaft; 10—Moving beam

图2-15 带裙板辊道和步进式冷床示意图
1—冷床入口辊道；2—升降裙板；3—动齿条；4—固定（静）齿条；5—对齐辊道；
6—液压缸；7—矫直板；8—偏心轮；9—传动轴；10—动梁

The total length of cooling bed roller table is about 230m, including 186m roller table with apron and the rest without apron.

冷床入口辊道总长约230m，其中带裙板辊道186m，其余为不带裙板辊道。

There are 186 rollers in the transportation roller table, and each roller is driven separately by adjustable speed AC motor. The control of the roller table is divided into three sections. The speed of each section is controlled separately. In order to realize the correct brake of the steel and the separation of the front and rear steel heads and tails, the speed of the roller table can be changed within 15% of the rolling mill speed in production. Generally, No. 1 roller speed is set to be +5% ahead of the rolling mill speed, No. 2 roller speed is set to be +10% ahead of the rolling mill speed, and No. 3 roller speed is set to be +5% ahead of the rolling mill speed. All the rollers keep a fixed angle of 12° towards the cooling bed to facilitate the bar to slide into the skirt.

运输辊道共有186个辊，每个辊由可调速的交流电机单独驱动。辊道的控制分为三段，各段速度单独控制，为实现钢的正确制动以及前后倍尺钢头尾的分离，在生产中辊道速度可在大于轧机速度的15%范围内变化。通常设定1号辊道速度超前轧机速度+5%，2号辊道超前+10%，3号辊道超前+5%。所有辊子均保持一固定向冷床方向的12°倾角，以利于棒材滑入裙板。

The brake apron is a series of plates which can move up and down in the vertical direction on one side of the transport roller table. The steel is braked by the friction resistance between the plate and the steel, and the steel is sent to the cooling bed straightening plate by the lifting movement. The skirt has three positions in the vertical direction, as shown in Figure 2-16.

制动裙板是位于运输辊道一侧的一系列可在垂直方向上下运动的板，利用板与钢材之间的摩擦阻力使钢材制动，并通过提升运动把钢材送入冷床矫直板。裙板在垂直方向有三个位置，如图2-16所示。

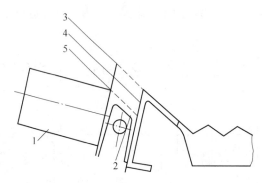

Figure 2-16 Diagram of three positions of brake skirt
1—Roller table; 2—Apron; 3—Off position; 4—Middle position; 5—Low position
图2-16 制动裙板三个位置示意图
1—辊道；2—裙板；3—离位；4—中位；5—低位

The brake skirt is divided into two parts in the rolling forward direction.
制动裙板在轧件前进方向上共分成两部分。
The first part is located in front of the cooling bed, and the skirts are separated/combined by 18 hydraulic adapters, so that the first part of the skirt is divided into two sections: fixed section (the skirt is often in the high position in production) and movable section (moving together with the skirt on the cooling bed), so as to meet the requirements of accurate braking and transportation of different specifications and varieties. The second part of the skirt is located on the cooling bed, which is the same width as the cooling bed and is 132m long. It is composed of multiple skirts, which can only be separated from each other.

第一部分位于冷床之前，裙板之间通过18个液压接手进行分离/结合，使第一部分裙板分成固定段（裙板在生产中经常处在高位）和活动段（与冷床上裙板一起运动）两段，以满足不同规格品种准确制动运输的要求。第二部分裙板位于冷床上，与冷床宽度相同，长132m，由多块裙板构成一个整体，单块裙板之间只能分开。

The total length of roller table with lifting apron is 186m, including 132m cooling bed section and 54m in front of the cooling bed.
带升降裙板的辊道总长186m，其中包括132m的冷床段及冷床前的54m。

There are five electromagnetic skirts distributed in the brake skirt, which generate electromagnetic force when electrified. For the threaded steel bars cooled by water penetration, the braking force can be enhanced to make the steel bars brake rapidly. The magnetic force can be adjusted according to the size of the specification. The UN quenched water and large diameter bars do not need to be braked by the electromagnetic skirt.

在制动裙板中分布有5块电磁裙板，通电时产生电磁，对于经过穿水冷却的螺纹钢筋可增强其制动力，使钢筋快速制动。磁力大小可根据规格大小进行调节，未淬水的及大直径的棒材不需要用电磁裙板来制动。

When the current one time steel enters into the skirt for braking, the skirt will be lowered to the lowest position, and the next one will be prevented from entering into the skirt by pneumatic steel puller. The pneumatic steel puller is installed at the entrance of the skirt plate of the movable section. Its function is to lift the 1.2m long movable steel puller by the air cylinder when the tail of the current double length bar is sliding and braking, so as to prevent the head of the next steel from entering the skirt plate, but still running along the roller table. When the steel puller is raised, the movable skirt is raised to the middle position, and the latter steel is braked along the side of the skirt. The movement of the steel paddle is synchronized with the movement of the movable skirt and the head position of the next steel. When the bar running speed is lower than 10m/s, the steel shifter is not used.

当前一根倍尺钢进入裙板进行制动时，裙板降至最低位置，通过气动拨钢器来阻止下一根倍尺钢头部进入裙板。气动拨钢器安置在活动段裙板的入口处，其作用是当前一根倍尺棒材的尾部正滑和制动时，由气缸将1.2m长的可移动拨钢器抬起，防止下一根钢的头部进入裙板，而仍沿辊道运行。在拨钢器抬起的同时，活动裙板升至中位，后一根钢材沿裙板侧面制动运行。拨钢器的运动与活动裙板的运动和下一根钢的头部位置同步。在棒材运行速度低于10m/s时，不采用拨钢器。

The position of the steel shifter is determined according to the different specifications and varieties of rolled products, as well as the time or distance of braking, and is manually positioned at the corresponding position.

拨钢器的位置是根据轧制产品规格、品种的不同，以及需要制动的时间或距离来决定的，由人工定位在相应位置上。

(2) Brake and steel separation process of apron roller table.

(2) 裙板辊道制动及分钢工艺过程。

The brake process of bar is realized by the friction resistance between steel and brake plate. The friction force is determined by the friction coefficient between steel and brake plate and the number (or total length) of brake plate.

棒材的制动过程是由钢材与制动板间的摩擦阻力来实现的。摩擦力的大小由钢材与制动板间摩擦系数和制动板的数量（或总长度）所决定。

In order to achieve the accurate brake positioning and timely delivery of the double length steel to the cooling bed after shearing, and to make the steel fall on the position close to the head of the cooling bed as much as possible, so as to reduce the working time of the end trimming roller table in the subsequent work; in addition, to ensure the brake time, and the interval time of the last two steel tail heads matches the action cycle of the skirt plate, so as to achieve the steel splitting action smoothly, it is necessary to accurately control the steel enters the starting position P_1 of the brake plate, the braking distance S and the position P_2 that stops on the cooling bed after braking.

为顺利实现剪后倍尺钢的制动定位准确和向冷床的及时输送,并尽可能使钢材落在靠近冷床头部的位置上,以减少后序工作中齐头辊道的工作时间;另外,要保证制动时间,并且后两根钢材尾头的间隔时间与裙板动作周期相匹配,顺利实现分钢动作,则需要准确控制钢材进入制动板开始制动位置 P_1、制动距离 S 与制动后停在冷床上的位置 P_2。

After the bar is sheared by multiple length shear, the tail reaches P_1 and enters the brake plate to start braking. After time t, it reaches P_2 position. t is called the braking time. The braking distance S and the braking time t mainly depend on the speed v of the rolling piece and the friction coefficient f. In production, the friction coefficient f is considered as a constant, so the braking distance S and braking time t mainly depend on the moving speed v of the rolling piece.

棒材经倍尺剪剪切后,尾部到达 P_1 并进入制动板开始制动,经时间 t 后,到达 P_2 位置, t 称为制动时间。制动距离 S 和制动时间 t 主要取决于轧件的速度 v 和摩擦系数 f。在生产中,摩擦系数 f 认为是一个常数,所以制动距离 S 和制动时间 t 主要取决于轧件的运动速度 v。

In production, in order to ensure that the rolled piece with the highest transfer speed can be positioned at P_2, there must be enough braking distance, which requires that the brake plate in front of the cooling bed has enough length. According to the calculation and the experience of bar rolling mill of Tangshan Iron and Steel Co., Ltd., if the length of the front apron of the cooling bed is 54m, it can meet the braking requirements when the maximum rolling speed is 18m/s. When rolling bars of other specifications, the rolling speed is different, but it is lower than $v = 18$m/s, so the braking distance S is less than 54m. In order to ensure the correct positioning of the rolled piece on the cooling bed, the position P_1 of the starting brake on the roller table in front of the cooling bed is different for bars of different specifications. Therefore, the length of the movable section in the front brake skirt of the cooling bed is variable, and the length of the movable section in the skirt can be determined according to the brake distance of the produced bar. The skirt is connected by 18 hydraulic adapters. When one of them is disconnected, the skirt is divided into fixed section and movable section.

在生产中,为保证移送速度最高的轧件能够定位于 P_2,必须有足够的制动距离,这就要求冷床前制动板有足够的长度。根据计算和唐钢棒材轧机的经验,冷床前裙板长度为 54m 就能满足最大轧件速度≤18m/s 时的制动要求。当轧制其他规格棒材时,其轧件速度各不相同,但都低于 $v=18$m/s,所以制动距离 S 都小于 54m。为了保证轧件在冷床上的正确定位,对不同规格的棒材,在冷床前辊道上开始制动的位置 P_1 是不同的。所以在冷床前段制动裙板中活动段的长度是可变的,并且可根据所生产棒材的制动距离,确定裙板中活动段的长度。裙板由 18 个液压接手联接,当其中一个断开时,裙板就从此处分为固定段和活动段。

Separate the tail and head of the front and rear double length steel to ensure the smooth braking of the former steel. The head and tail of the two steels do not interfere with each other. The time required for sequential steel division shall match the movement cycle of the skirt. Skirt action cycle refers to the time required for skirt to complete an action cycle. The action cycle of skirt plate and the position of front and rear rolling pieces are different as shown in Figure 2-17.

将前后两支倍尺钢材尾部和头部分开，保证前一根钢材顺利制动，两根钢材头尾互不干扰，要求顺序分钢所需时间要与裙板运动周期相匹配。裙板动作周期是指裙板完成一个动作循环所需的时间。裙板动作周期与前后轧件所处位置如图 2-17 所示。

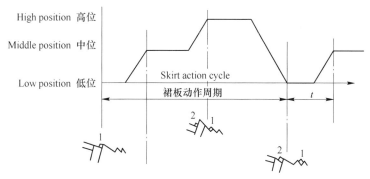

Figure 2-17 Action cycle of skirt and position
1—The former steel with multiple length; 2—The latter steel with multiple length
图 2-17 裙板动作周期与前后轧件的位置
1—前一根倍尺钢；2—后一根倍尺钢

In order to realize the smooth steel separation, when the former steel leaves the roller table and starts to brake, a distance ΔS must be set apart from the latter steel to ensure the time from the time when the former steel leaves the roller table to the time when the next steel head reaches the P_1 position of the front end of the brake plate. The time interval Δt must meet the time required for the brake plate to rise from the lowest position to the middle position, so as to prevent the next steel head from being at the brake plate enter the brake skirt at low position. It can be seen from the action cycle diagram of the brake skirt that the time t required for the brake plate to move from the low position to the middle position requires:

为了实现顺利分钢，前一根钢离开辊道开始制动时，必须与后一根钢拉开一段距离 ΔS，以保证前一根钢离开辊道的时间到下一根钢头部到达制动板前端 P_1 位置时间，其时间间隔 Δt 必须满足制动板从最低位置升到中间位置所需的时间，以防止下一根钢头部在制动板处于低位时进入制动裙板。从制动裙板动作周期图可知，制动板从低位到中位所需时间 t 要求：

$$t \leqslant \Delta t$$

The time interval Δt is achieved by the velocity difference between the front and back bars. The roller table is divided into three sections, and the speed of each section is higher than the rolling speed. The former steel is accelerated after being sheared by multiple length shear, so that the distance between the two steels is increased. When the rolling speed v is more than $10 m/s$, due to the high rolling speed, the time interval Δt generated by the acceleration of the roller table can not meet the corresponding action time of the skirt. In order to ensure the separation of the front and rear steel heads and tails, the pneumatic steel puller is used to prevent the lower steel head from entering the apron.

时间间隔 Δt 是靠前后两根棒材的速度差来实现的。辊道分为三段，每段速度均高于轧件速度。前一根钢经倍尺剪剪切后加速，使两支钢间距加大。当轧制速度 $v>10m/s$ 时，由于轧制速度快，辊道加速产生的时间间隔 Δt 不能满足裙板相应动作所需时间。为保证前后钢头尾分开，利用气动拨钢器，阻止下根钢头部进入裙板。

（3）Bar straightening and cooling.

（3）棒材的矫直与冷却。

The cooling bed used in the plant is a start stop walking beam rack cooling bed, which is located between the apron roller table and the layer forming equipment in the cold shear area. Cold bed is used for air cooling, straightening and end trimming of hot bar after rolling. After cooling, the bar is transported to the layered chain and trolley, transferred to the exit roller table of the cooling bed before the cold shear, and then sent to the pendulum type cold flying shear, and cut into a fixed length.

该厂采用的冷床为启停式步进梁齿条冷床，位于裙板辊道与冷剪区成层设备之间。冷床对轧后热状态的棒材进行空冷、矫直、齐头。棒材冷却后输送到成层链和小车上，移送到冷剪前冷床出口辊道上，然后送至摆式冷飞剪，剪成定尺。

The cooling bed is 132m wide and 12.5m long (from the center of the cooling bed population roller table to the center of the cooling bed outlet roller table).

冷床宽 132m，长 12.5m（从冷床入口辊道中心到冷床出口辊道中心距离）。

The cooling bed consists of straightening plate, moving rack, static rack and alignment roller table. The moving rack and the static rack have their own length. Because the bar is easy to deform when it just enters the cooling bed, the racks are closely arranged at the population end of the cooling bed, and the spacing between the adjacent dynamic and static racks is 150mm. However, due to the low steel temperature at the outlet end of the cooling bed, the spacing between the adjacent dynamic and static racks is 300mm.

冷床由矫直板、动齿条、静齿条和对齐辊道等组成。而动齿条和静齿条又各有长短之分。因为棒材在刚进入冷床时温度高容易变形，因而在冷床入口端，齿条的排列紧密，相邻动静齿条间距为 150mm，而在冷床出口端由于钢温较低，不易变形弯曲，则相邻的动静齿条间距变成 300mm。

The straightening plate is located at the population side of the cooling bed, and the bar from the population roller of the cooling bed first falls on the straightening plate. Each straightening plate is 1.15m long and 250mm wide, with 10 teeth on it. Only the moving rack movement gap is left between the blocks. The straightening plate replaces the static rack here. The bar is transferred forward one by one from the moving rack until it moves out of the straightening plate. The function of the straightening plate is to maintain the maximum straightness of the bar.

矫直板位于冷床的入口侧，由冷床入口辊道上下来的棒材首先落到矫直板上。矫直板每块长 1.15m，宽 250mm，上有 10 个齿形，块与块之间只留出动齿条移动的间隙。矫直板在此处代替了静齿条。棒材由动齿条一个槽一个槽地向前传递，直到移出矫直板。矫直板的作用是保持棒材的最大平直度。

The bar passes through the straightening plate and enters into the main body of the cooling bed. It is a walking cooling bed composed of a static rack and a moving rack arranged alternately in length. The moving rack is installed on the moving beam. There are 22 moving beams in the cooling bed, and 20 moving racks are installed on each moving beam. There are wheels under each moving beam, and the moving beam is placed on the eccentric wheel. The eccentric wheel drives the beam to move up, down, front and back of the moving beam, so as to transmit the bar on the rack tooth by tooth.

棒材经过矫直板而进入冷床的本体，它是由长短交替布置的静齿条与动齿条组成的步进式冷床。动齿条安装在动梁上。冷床共有 22 个动梁，而每个动梁上安装有 20 个动齿条。每个动梁下有轮子，动梁平放在偏心轮上。偏心轮转动带动此梁，使动梁上、下、前、后移动，从而将齿条上面的棒材一齿一齿地向前传送。

When the bar on the moving rack is transported to the alignment roller table, the alignment roller table can send the bar to the baffle plate for alignment, so that the bar can enter the fixed length cold flying shear in a neat and consistent manner. The alignment roller table has 8 teeth with the same profile as the rack, and the position is higher than the fixed rack. The alignment roller table is driven by the chain driven by the gear on the constant speed AC motor.

当动齿条上的棒材被传送至对齐辊道时，可由对齐辊道将棒材送到挡板处对齐，使棒材能整齐一致地进入定尺冷飞剪。对齐辊道上有 8 个与齿条完全相同的齿形，且位置比定齿条要高。对齐辊道是由恒速交流电机上的齿轮带动链子单独驱动的。

The 132m long alignment roller table is divided into 8 sections, each section has 12 rollers, which are controlled separately, and which sections can be operated according to the multiple length of the bar.

132m 长的对齐辊道分为 8 段，每段 12 个辊子，分别控制，可根据棒材的倍尺长短来决定哪些段运行。

（4）Marshalling and translation of bars.

（4）棒材的编组与平移。

The bar straightened, cooled and leveled by the cooling bed is transferred by the moving rack to the collection chain or marshalling chain close to the cooling bed. The total width of the collection chain is 126m and the working length is nearly 2m. The action of the collection chain is intermittent, that is to say, a bar is placed on the collection chain for each direction of the moving rack, and the collection chain acts once, so that the bar forms a layer, which is convenient for the next transportation of the trolley. At the same time, the collection chain also plays a role in adjusting the production rhythm. If there is a small failure of a later equipment, the collection chain can also collect some bars to play a buffer role.

经过冷床调直、冷却和齐头后的棒材，由动齿条传递到紧靠冷床的收集链或称编组链上。收集链的总宽度达 126m，工作长度近 2m。收集链的动作是间断性的，即动齿条每向收集链上放一根棒材，收集链动作一次，使棒材形成层状，便于下一步运输小车的运输。同时，收集链也起到调整生产节奏的作用，如果后面的某个设备出现小故障，收集链还可以收集一些棒材，起到缓冲的作用。

Use the translation device (or transport trolley) to lift the laminated bar on the collection chain, translate it to the top of the output roller table, put the laminated bar on the roller table, keep the layered bar, and send it to the fixed length cold flying shear for shearing.

利用平移装置（或称运输小车）将收集链上的成层棒材托起，平移到输出辊道上方，将成层棒材放到辊道上，保持棒材的层状，由辊道送到定尺冷飞剪进行剪切。

When the cold cut fixed length flying shear uses the slotted shear blade to cut the bar, it is arranged in parallel with the collection chain, close to one end of the cold shear, and the special layered trolley (about 6m wide) is set at the head of the cooling bed to replace the collection chain. When receiving the bar from the moving rack of the cold bed, use the teeth of the upper gear plate of the layered trolley to separate the bar, so that the bar on the trolley can form a certain number of layers, and transport the layer of bar to the output roller table through the transport trolley, and then through the separation of the upper gear plate of the layered trolley, so as to maintain a certain distance between the bars, so that the bar can be fed smoothly into the grooved edge of the flying shear with cold fixed length。

当冷切定尺飞剪采用带槽剪刃剪切棒材时，采用与收集链并列布置，靠近冷剪一端，冷床头部处设置的专用成层小车（宽度约 6m）代替收集链。在接受从冷床动齿条上下来的棒材时，利用成层小车上齿板的齿将棒材分开，使小车上的棒材按一定数量形成层，并将这一层棒材通过运输小车运至输出辊道，再经过成层小车上齿板的分离，使棒材之间保持一定的距离，以便棒材能顺利喂入冷定尺飞剪的带槽刃中。

The output roller table of the cooling bed operates at the speed of the cold shear line, and an electromagnetic roller is equipped on the roller table to ensure the positioning and head alignment of the bar and the smooth feeding of the head of the bar to the cutting edge of the cold shear, especially in the production of large-scale bar.

冷床输出辊道按冷剪剪切线速度运行，并在辊道上备有电磁辊，以确保棒材的定位及头部对齐，确保棒材头部顺利喂入冷定尺剪的剪刃，尤其在生产大规格棒材时这一点更为重要。

2.5.6.3　Cutting to Length of Bar after Cooling
2.5.6.3　冷却后棒材的定尺剪切

When the bar is naturally cooled to below 100℃ in a cold bed, it is cut to a certain length on a cold flying shear.

棒材经冷床自然冷却到 100℃ 以下时，在冷飞剪机上进行定尺剪切。

Continuous shear line (CCL) is used in the factory for sizing shear. The pendulum type cold shear can cut the bar in operation. The bar starts from the last slot of the cooling bed and is cut into layers according to the predetermined number of shear pieces. The conveying system then sends the bar layer to the cold flying shear for fixed length cutting until it is sent to the collection area by the cold shear output roller table. The whole process can be completed continuously and automatically.

该厂采用连续剪切线（CCL）进行定尺剪切。所用摆式冷剪可剪切运行中的棒材。棒材从离开冷床的最后一个槽开始，按预定剪切根数成层，传送系统再将棒材层送入冷飞剪进行定尺剪切,直到由冷剪输出辊道送到收集区。整个工艺过程可全部连续、自动地完成。

(1) Magnetic conveying bar.

(1) 磁力输送棒材。

Three magnetic rollers and magnetic conveyor (also known as magnetic chain) are installed between the output roller table and the cold shear of the cooling bed to ensure the alignment of the bar layer head and the fixation of the bar spacing, to ensure that the bar enters the cold shear guide slot continuously and accurately, and to prevent the bar from slipping on the roller table.

在冷床输出辊道和冷剪之间安装有3个磁力辊道和磁力输送机(也称磁性链),以保证棒层头对齐和棒材间距固定,保证棒材连续、准确地进入冷剪导槽,防止棒材在辊道上打滑。

The magnetic conveyor is driven by a variable AC motor, and its working system is continuous, variable speed and no inversion. The magnetic conveyor is synchronized with the output roller table of the cooling bed.

磁力输送机由一个可变交流电机驱动,工作制度为连续、可变速、无反转,该运输机上装有永磁板。磁力输送机与冷床输出辊道同步。

(2) Cutting to length of bar layer.

(2) 棒材层的定尺剪切。

The bar layer is sent to the swing type cold flying shear by the cold bed output roller table and magnetic conveyor. The shear can be used to cut the moving or static bar layer vertically, and the multiple length bar of about 100~132m can be cut into the fixed length bar required by users.

由冷床输出辊道和磁力输送机将棒材层送入摆动式冷飞剪。该剪可以对运动或静止的棒材层进行垂直剪切,将长约100~132m的倍尺棒材剪切成用户需要的定尺棒材。

The technical properties of cold shear with fixed length in a bar mill are as follows:

某棒材厂的定尺冷剪技术性能如下:

Mode　　　　　　　　　Swing crank flying shear;

形式　　　　　　　　　摆动曲柄式飞剪;

Shear capacity　　　　350t;

剪切能力　　　　　　350t;

Blade stroke　　Stroke of upper blade + stroke of lower blade-150mm+20mm=170mm;

剪刃行程　　上剪刃行程+下剪刃行程-150mm + 20mm = 170mm;

The maximum rotation speed of shear spindle　　　　161r/min;

剪切主轴转速最大　　　　　　　　　　　　　　　161r/min;

The maximum shear speed　　　　1.5m/s;

最大剪切速度　　　　　　　　　1.5m/s;

The maximum shearing period　　　　2.8s/time;

最大剪切周期　　　　　　　　　2.8s/次;

Cutting edge width　　　　　　　800mm;

剪刃宽度　　　　　　　　　　　800mm;

Shear accuracy	±15mm;
剪切精度	±15mm;
Power of two DC speed regulating main motors drive one reducer	430kW+430kW;
两台直流调速主电机带动一台减速机传动，其功率为	430kW+430kW;
Speed regulation range:	0~800r/min;
调速范围	0~800r/min;

The swinging arm of the cold shear is 1500mm long and the swinging angle is 11.23°.

冷剪的摆动臂长 1500mm，摆动角 11.23°。

A turnover plate controlled by the air cylinder is arranged at the cold shear inlet to exclude the end material head of the bar after shearing. In addition, an upper pressing plate controlled by the air cylinder is provided to ensure that the head of the bar can enter the guide groove of the cutting edge smoothly. The position of the pressing plate is preset automatically according to the diameter of the sheared bar. When the bar diameter exceeds 12mm, the pressing plate rises. For the bar diameter less than or equal to 12mm, the pressing plate remains in the down position.

在冷剪入口处设有一个由气缸控制的翻板，用来排除剪切后的棒材尾端料头。另外还设有一个由气缸控制的上压板，用来保证棒材头部顺利进入剪刃导槽。压板的位置根据所剪切棒材的直径预先自动设定，棒材直径超过 12mm 时压板升起，对于直径小于或等于 12mm 的棒材，压板保持在下降位置。

In order to ensure the quality of shear steel and improve the shear capacity, small-sized bars are cut with double inclined flat shear blades, and large-sized bars are cut with slotted shear blades. 7 kinds of cutting edges are required for cutting all specifications of bars with a diameter of 12~50mm. The type, specification, number of slots, slot spacing and shape of each cutting edge are clearly specified.

为了保证剪切钢材质量和提高剪切能力，小规格棒材采用双斜度平剪刃剪切，大规格的棒材采用带槽剪刃剪切。剪切直径为 12~50mm 的所有规格棒材时共需 7 种剪刃，每种剪刃的剪切品种、规格、开槽数量、开槽间距以及剪刃形状都有明确规定。

(3) Conveying, counting and collecting of fixed length bar.

(3) 定尺棒材的输送、计数与收集。

After the bar is cold cut to a fixed length, it enters into a bar layer transfer device composed of a double roller table and a chain conveyor, a bar counter, a bar splitting device and a movable collecting basket. The chain conveyor has two bed bodies in total, the one close to the cold shear is called bed B, and the one far away from the cold shear is called bed A. the rear process of each bed body is equipped with rod splitting device, basket transport device, roller table, baler and chain bale unloading bed, each forming a production line. Double roller table is the boundary between the two production lines. The maximum fixed length of bar that can be placed in each bed is 12m.

棒材经冷剪切成定尺后进入由双辊道及链式运输机、棒材计数器、分棒装置及可动收集筐组成的棒层移送装置。链式运输机共有两个床体，靠近冷剪一侧的称为 B 床，远离冷剪的称为 A 床，每个床体的后部工序均设有分棒装置、料筐运输装置、辊道、打捆机和链式卸捆床，各成一条生产线。双辊道即是两条生产线的分界之处。每个床体可放置棒材最大定尺长度为 12m。

The first cutting bar of cold shear enters the first row of roller table of bed B first along the exit roller table after cold shear, and then continues to move forward and enters into the first row of roller table of bed A, at this time, the second shear bar has entered the first row of roller table of bed B, when the head of bar reaches 1~2m before the end of roller table, the cover plate of small roller table under the two roller tables of bed body is supported by two trolleys, so that the bar stops moving on the small cover plate, at this time, the third shear and the fourth shear bar can enter the first row of roller table. The trolley moves horizontally to the direction of the chain conveyor to make the bar reach the position of the second row of roller table, and the trolley lowers to make the upper plane height of the cover plate lower than the roller table, so the first shear and the second shear bar are placed on the second row of roller table of bed A and bed B respectively. The two groups of roller tables rotate to both sides respectively, and the bars collide with the baffles of A and B beds respectively to align the bar heads. At this time, the trolley returns to the bottom of the first row of roller tables at the low position, ready to support the third shear and the fourth shear bars, and at the same time, the bars of the first shear and the second shear are also lifted and moved to the top of the chain bed of the first section of chain conveyor. The two rows of small cover plates of roller table are one car body, the car body traverse travel is 1250mm, the lifting travel is 363.7mm, and the operation cycle is synchronous with the production rhythm of cold shear. In order to prevent the bar from slipping on the first row of roller table, 5 electromagnetic roller tables are installed on the roller table.

冷剪第一剪棒材沿冷剪后出口辊道首先进入 B 床第一排辊道，然后继续前进，进入 A 床第一排辊道，此时第二剪棒材已进入 B 床第一排辊道，在棒材头部到达辊道端部前 1~2m 时，两个床体辊道下边的小辊道盖板由两个小车托起，这样棒材在小盖板上停止运动，此时第三剪、第四剪棒材已经可以进入第一排辊道。小车向入口链式输送机方向横移，使棒材到达第二排辊道位置，小车下降使盖板上平面高度低于辊道上面，便将第一剪和第二剪棒材分别放在 A 床和 B 床的第二排辊道上。两组辊道分别向两侧旋转，棒材分别撞到 A、B 床挡板，使棒材头部对齐，此时小车在低位返回至第一排辊道下面，准备托起第三剪、第四剪棒材，托起并横移的同时，也把第一剪、第二剪的棒材托起并横移至第一段链式运输机的链床上面。两排辊道小盖板为一个车体，车体横移行程为 1250mm，升降行程为 363.7mm，动作周期与冷剪的生产节奏同步。为防止棒材在第一排辊道上打滑，辊道上安装有 5 个电磁辊道。

If any defect is found or the length is too short, the fixed length bar cut from the cold shear shall be manually removed from the roller table and placed in the collection tank.

由冷剪过来的定尺棒材，如发现有缺陷或定尺超短时，由人工在辊道上剔出，并放置于收集槽内。

The bar is transferred to the bar counter through the inlet chain conveyor, the intermediate chain conveyor and the outlet chain conveyor, and the bar is counted according to the bundling requirements. The bar counter consists of a bar separation screw, a counting wheel and a counting photocell.

棒材经入口链式输送机、中间链式输送机和出口链式输送机被移送到棒材计数器处，按打捆要求对棒材计数。棒材计数器由一个棒材分离丝杠、计数轮和一个计数光电管组成。

The maximum production capacity of each counter is 9 pieces/s, and the accuracy is 99.9% (more than 1000 bars).

每个计数器的最大生产能力为 9 支/s，精度为 99.9%（1000 支以上棒材）。

The screw thread and the tooth shape of the counting wheel of the screw are changed according to the product specification. When the bar is larger than $\phi 32mm$, the screw is not used for counting, but the counting wheel and the photoelectric counter are directly used for counting.

螺旋丝杠的螺纹和计数轮的齿形根据产品规格而变化，大于 $\phi 32mm$ 的棒材计数时不用螺旋丝杠，而是直接由计数轮和光电计数器完成计数。

This counter is only used in the case of Baling according to the number of bars. This device is not used when baling in layers.

该计数器只用于按棒材根数打捆的情况。当按层数打捆时，不使用该装置。

The bars counted according to the number of bars or the number of layers are transmitted to the collection basket through the rod splitting device. The main equipment of the rod splitting device includes: rotating guide plate, baffle plate, scraper wheel, holding plate, and end trimming device, discharge plate, etc.

按棒材根数或按层数计数后的棒材经分棒装置传送到收集筐中。分棒装置的主要设备有：转动导板、挡板、刮轮、保持板、齐头装置和卸料板等。

If the diameter of the bar is less than 20mm, the bar through the counter is separated by a group of rotating guide plates and transmitted to the retainer plate along the baffle plate. After the end trimming device is aligned, the retainer plate unloads the bar into the discharge plate, and then the discharge plate places the bar into the movable collection basket.

当按根数收集时，若棒材直径小于 20mm，则通过计数器的棒材由一组转动导板分离并沿着挡板传送到保持板中，在齐头装置齐头后由保持板将棒材卸入卸料板中，然后由卸料板将棒材放入可移动收集筐中。

If the bar diameter is greater than or equal to 20mm, the counted bar is directly transmitted to the movable collection basket by a rotating guide plate (different from the above shape) on one side of the counter and a group of scraper wheels. At this time, other rotating guide plate, baffle plate, end trimming device, retainer plate and discharge plate are all out of the production line.

若棒材直径大于或等于 20mm，则计数后的棒材由位于计数器一侧的一个转动导板（形状与上述不同）与一组刮轮配合，直接传送到可动收集筐中。此时，其他转动导板、挡板、齐头装置、保持板和卸料板均退出生产线。

（4）Transfer, weighing and collection of bar bundle.

（4）棒捆的传送、称量与收集。

In combination with bed A and B, there are two corresponding collection stations A and B, the bars unloaded by the collection basket are transported to the baling station by the conveying roller table to be baled. After the transfer roller bed completes the baling positioning and baling of the bar bundle, the bundled bar is transferred to the collection conveyor chain. Transfer the bars with different length to the weighing station.

与 A、B 床相配合,设有 A、B 两个相对应的收集站,由收集筐卸下的棒材由传送辊道传送到打捆站打成棒捆。传送辊道完成棒捆的打捆定位、打捆后,向收集传送链传送打好捆的棒材。将打好捆的不同定尺长度的棒材传送到称重站。

Bundle collection stations A and B can be divided into two completely independent parts: A_1, A_2 and B_1, B_2. For the bundle with a length of 6m, the two parts A_1 and A_2 or B_1 and B_2 of the collection station act independently of each other. The working sequence is that the first bale is sent to A_1 or B_1, the second bale is sent to A_2 or B_2, and then to and fro. For bundles longer than 6m, A_1 and A_2 or B_1 and B_2 will work at the same time to form a collection station, which can only receive one bundle at a time.

棒捆收集站 A 和 B 均可分成两个完全独立的 A_1、A_2 和 B_1、B_2 两部分。对棒长 6m 的捆,收集站的两部分 A_1 和 A_2 或 B_1 和 B_2 相互独立动作。工作顺序为第一捆送到 A_1 或 B_1,第二捆送到 A_2 或 B_2,以后往复循环。对于大于 6m 长的棒捆,A_1 和 A_2 或 B_1 和 B_2 将同时工作,形成一个收集站,每次只能接收一个棒捆。

Each bale collection station consists of two collection chains and a weighing system. The operation mode of the collection chain (single/double) has been pre-selected according to the fixed length during the operation of the conveying roller table.

每个棒捆收集站都由两段收集链和一个称重系统组成。收集链的运行方式(单/双)已在传送辊道运行过程中根据定尺长度预先选定。

When the transfer roller table transfers the bundle to a predetermined position, the liftable chain rises and transfers the bundle forward to the scale. Then, the first segment of the liftable chain descends and begins to weigh.

当传送辊道将棒捆传送到预定位置后,可升降的传送链升起并将棒捆向前传送到电子秤,然后,第一段可升降链下降并开始称重。

The weighing station is located in the middle of the lifting chain, which is a pressure sensing weighing system. The weighing range is 1000~5000kg, the weighing scale is 1kg, and the weighing accuracy is 0.1% of the measuring range. The weighing system also includes a display terminal, a printer and an alphabetic data keyboard, which can input people, display and print. The printing contents include: standard, specification, heat number, steel type, inspector, date, hour, bundle weight and total weight. The system can transmit the above contents to the label printer, and can store the heat number and weight of each bundle of bars on duty.

称重站位于可升降链的中部,为压力传感式称重系统,称重范围为 1000~5000kg,称重刻度 1kg,称量精度为量程的 0.1%。该称重系统还包括一台显示终端、一台打印机和一台字母数据键盘,可输入、显示、打印。打印的内容包括:标准、规格、炉号、钢种、检验者、日期、小时、捆重和总重。该系统可将上述内容传送给标牌打印机,并且能够存储当班每捆棒材的炉号和重量。

The second horizontal conveyor chain is used to collect the weighed bundle. When the electronic scale sends out the weighing end signal, the lifting chain can be raised and transferred to the horizontal chain population, and then the lifting chain can be lowered to prepare for the next cycle. At the same time, the horizontal chain moves the bundle forward according to the predetermined distance, and completes the tag out work on the horizontal chain.

第二段水平传送链用来收集称重后的棒捆。当电子秤发出称重结束信号后,可升降链升起并将棒捆传送到水平链入口,然后可升降链下降准备进行下一个循环,同时,水平链按预定距离向前移动棒捆,并在水平传送链上完成挂牌工作。

According to the national standard, at least two labels shall be hung for each bundle of delivered steel. The labels shall have the supplier's name (or factory logo), brand number, furnace batch number, specification, weight and other marks. Tanggang Bar Factory uses two kinds of signs, one for domestic delivery, and the other for export of steel.

按国家标准规定,成捆交货的钢材每捆至少要挂两个标牌,标牌上应有供方名称(或厂标)、牌号、炉批号、规格、重量等印记。唐钢棒材厂采用两种标牌,一种用于国内交货,另一种用于出口的钢材。

Task 2.6　Product Defects and Quality Control
任务2.6　产品缺陷及质量控制

Mission objectives

任务目标

(1) Master the production characteristics and quality requirements of screw steel products;

(1) 掌握螺纹钢产品的生产特点和质量要求;

(2) Be able to identify and inspect the defects of screw steel products, analyze the causes of defects and deal with them.

(2) 能够识别并检验螺纹钢产品缺陷,分析缺陷形成原因并进行处理。

2.6.1　Quality Requirements for Use of Threaded Steel
2.6.1　螺纹钢使用的质量要求

With the rapid development of the national economy, as the main users of the screw steel, the construction of industrial and civil buildings, water conservancy projects, roads and bridges has more and stricter requirements on the variety and quality of the screw steel, especially on the strength level and comprehensive performance. The consumption proportion of the high-strength screw steel (HRB400) has increased year by year. This trend promotes the development of screw steel production technology. HRB500 or higher level screw steel production has been listed in the development plan of production enterprises.

随着国民经济的高速发展,作为螺纹钢主要用户的工业及民用建筑、水利工程、道路桥梁的建设对螺纹钢的品种及质量要求越来越严格，特别是对强度级别、综合性能的要求越来越高，较高强度级别螺纹钢（HRB400）的消费比例逐年提高。这种趋势促进了螺纹钢生产技术的发展，HRB500甚至更高级别的螺纹钢生产已列入生产企业的发展规划。

Screw steel products are divided into two categories according to national standards：

螺纹钢产品按国家标准分为两类：

（1）Hot rolled ribbed bars for reinforced concrete refer to the bars with round cross section formed by hot rolling and naturally cooled, and two longitudinal ribs and cross ribs evenly distributed along the length direction on the surface. The longitudinal section of the transverse rib is crescent shaped and does not intersect with the longitudinal rib. The product specification is usually $\phi 6 \sim 50mm$, and the brand is HRB335, HRB400 and HRB500.

（1）钢筋混凝土用热轧带肋钢筋，系指经热轧成形并自然冷却的横截面通常为圆形，且表面通常带有两条纵肋和沿长度方向均匀分布的横肋的钢筋。其横肋的纵截面呈月牙形，且与纵肋不相交。产品规格通常为$\phi 6 \sim 50mm$，牌号为HRB335、HRB400和HRB500。

（2）The residual heat treatment steel bar for reinforced concrete refers to the ribbed steel bar formed by hot rolling and treated by residual heat. Its shape is the same as that of the hot rolled ribbed steel bar. The product specification is usually $8 \sim 40mm$, and the brand is KL400.

（2）钢筋混凝土用余热处理钢筋，系指经热轧成形并余热处理的带肋钢筋，其外形与热轧带肋钢筋相同。产品规格通常为$8 \sim 40mm$，牌号为KL400。

Due to the needs of production process and user use, screw steel products $\phi 6 \sim 10mm$ are generally delivered in the form of wire rod, while $\phi 12 \sim 50mm$ are delivered in the form of straight rod.

由于生产工艺及用户使用的需要，螺纹钢产品$\phi 6 \sim 10mm$规格一般以盘条状态交货，$\phi 12 \sim 50mm$规格的则以直条状态交货。

The requirement for the quality of screw thread steel is to ensure the safety performance of building components, and its performance requirements are mainly as follows：

对螺纹钢质量的要求是必须保证建筑构件的安全性能，其使用性能要求主要为：

（1）Fatigue strength. It is shown that the fatigue strength under the repeated action of low load is one of the properties that must be examined and evaluated before the research and design stage and the formulation of design specifications. The main factors that affect the fatigue strength are stress concentration, inhomogeneity of structure and environmental conditions. The steel bars with smooth surface have better fatigue resistance, and the steel bars with large surface shape change are easy to be induced by stress concentration at the sudden change of shape fatigue failure.

（1）疲劳强度。显示较低载荷反复作用下的疲劳强度是钢筋研发阶段和制订设计规范前必须考核并做出评价的性能之一，影响疲劳强度的主要因素有应力集中、组织不均匀性以及环境条件，表面平滑的钢筋抗疲劳性能较好，表面形状变化较大的钢筋易在形状突变处应力集中而诱发疲劳破坏。

（2）Stress relaxation property. The phenomenon of stress relaxation of steel bars under long-term stress will increase structural deformation and reduce structural durability, which is essentially caused by the dissipation of dislocations in the steel and the dissolution of interstitial atoms.

（2）应力松弛性能。钢筋在长时受力下应力松弛的现象，将增大结构变形、降低结构耐久性，本质上是由于钢材内部位错的消散和间隙原子的脱溶引起的。

（3）Low temperature performance. With the decrease of ambient temperature, the change of tensile properties and impact toughness of steel bars, especially the adaptability to welding and the deterioration of the properties of welded joints, will seriously affect the stability and durability of reinforced concrete structures.

（3）低温性能。随着环境温度下降，钢筋的拉伸性能、冲击韧性的变化，尤其是对焊接的适应性及焊接接头性能的变坏，将严重影响钢筋混凝土结构的稳定性和耐久性。

（4）Corrosion resistance. The corrosion of steel bars caused by concrete mixing with water will eventually lead to structural damage. For structures under special environment, such as quays, piers, undersea buildings, etc., the leading opinion of the design department is to galvanize the steel bars or use stainless steel bars, and the design life is extended from 30 to 100 years.

（4）耐蚀性。因混凝土掺水而引起钢筋的锈蚀，最终将导致结构的损毁。对于特殊环境下的结构，如码头、桥墩、海底建筑等，设计部门的主导意见是对钢筋进行镀锌处理，或采用不锈钢钢筋，设计寿命则由30年延至100年。

（5）Durability. The durability affects the working life of the structure. The steel bars with smaller diameter are more sensitive to corrosion. The main factors affecting corrosion are environment, concrete protective layer and steel bar surface state (whether there is protective layer). Harbor Engineering, hydraulic engineering, chemical engineering and municipal engineering have higher requirements for durability.

（5）耐久性。耐久性影响结构的工作寿命，直径较细的钢筋对锈蚀比较敏感，影响锈蚀的主要因素是环境、混凝土保护层和钢筋表面状态（是否有防护层）。港工、水工、化工、市政工程对耐久性有较高要求。

（6）Delivery status. The delivery status has a great influence on the construction. And the above steel bars are delivered as straight bars, forming many joints in the structural reinforcement. Fine steel bars are generally delivered by wire rods, reducing joints, but the straightening process must be increased before use, which has certain impact on the strength.

（6）交货状态。交货状态对施工影响很大。以上的钢筋以直条交货，在结构配筋中形成许多接头。细钢筋一般以盘条交货，减少了接头，但使用前须增加调直工序，对强度有一定影响。

In order to ensure the service performance, the basic properties of the threaded steel are as follows：
为保证使用性能，螺纹钢必须具备的基本性能是：

（1）Strength is the most basic property of steel. Generally, the higher the strength of the stressed reinforcement, the better the performance, but there are certain limits. As the elastic modulus of steel is basically a constant value （£ = 2.0×10^{5}MPa）, the large deformation （elongation） caused by high stress when the strength is too high will affect the normal use （deflection, crack). Therefore, the design strength of reinforcement in concrete structure is limited to 360MPa, and too high strength is meaningless. The improvement of the strength mainly depends on the improvement of the material (alloying); the strength can also be improved by heat treatment and cold processing, but the ductility loss is too large; the area ratio of the base circle of the deformed steel bar (the ratio of the bearing cross-sectional area to the nominal area after deducting the discontinuous cross rib) also has a certain impact on the strength.

（1）强度是钢筋最基本的性能。一般受力钢筋强度越高，性能就越好，但也有一定限度。由于钢材弹性模量基本为一常值（£ = 2.0×10⁵MPa），强度过高时高应力引起的大变形（伸长）将影响正常使用（挠度、裂纹）。因此，混凝土结构中钢筋设计强度限为360MPa，太高的强度没有意义。提高强度主要靠材质改进（合金化）；也可通过热处理和冷加工提高强度，但延性损失太大；变形钢筋的基圆面积率（扣除间断横肋后承载截面积与公称面积之比）对强度也有一定影响。

(2) Ductility is the deformation ability of steel bars, which is usually expressed by the elongation measured by tensile test, and the yield strength ratio also reflects its ductility. However, due to different gauge distance, the current general elongation index (A_5、A_{10}、A_{100}) only reflects the local residual deformation in the necking area, and the measurement error of fracture splicing is large, which is difficult to truly reflect the ductility of the reinforcement. At present, the international has begun to use the total elongation (Uniform elongation A_{gt}) under the maximum tension. It is a scientific index to describe the ductility of reinforcement, as shown in Table 2-2. The ductility of steel bars is affected by the material. The increase of carbon equivalent can improve the strength, but the ductility decreases. After cold working, the value of steel bar decreases in order of magnitude (A_{gt} decreases) from more than 20% before processing to about 2% after processing, and it still develops with aging, especially brittle when the surface shrinkage is large. The seismic structure has a clear ductility requirement for the stressed reinforcement.

（2）延性是钢筋的变形能力，通常用拉伸试验测得的伸长率来表达，屈强比也反映了其延性。但目前通用的伸长率指标（A_5、A_{10}、A_{100}）因标距不同，只反映颈缩区域的局部残余变形，且断口拼接测量误差较大，难以真正反映钢筋的延性。目前，国际上已开始用最大拉力下的总伸长率（均匀伸长率A_{gt}）来描述钢筋的延性，是比较科学的指标，见表2-2。影响钢筋延性的因素是材质，碳当量加大虽能提高强度，但延性降低。钢筋冷加工后值呈数量级减小（A_{gt}由加工前超过20%降到加工后的2%左右），而且随时效仍有发展，面缩率较大时尤具脆性。抗震结构对受力钢筋有明确的延性要求。

Table 2-2　Classification of ductility requirements for steel bars in foreign countries

表2-2　国外对钢筋的延性要求分级

Index 指标	R_m/R_e	$A_{gt}/\%$	Type of reinforcement 钢筋类型
Medium ductility steel 中等延性钢	1.05	2.5	Cold worked steel bar 冷加工钢筋

Continued Table 2-2

Index 指标	R_m/R_e	$A_{gt}/\%$	Type of reinforcement 钢筋类型
High ductility steel 高延性钢	1.08	5.0	Hot rolled steel bars (heat treatment, microalloyed steel bars) 热轧钢筋(热处理、微合金化钢筋)
Aseismic steel 抗震钢	1.15	8.0	Hot rolled steel bars (heat treatment, microalloyed steel bars) 热轧钢筋(热处理、微合金化钢筋)

(3) Cold bending performance is to meet the requirements of steel processing. In case of bending, hook or repeated bending, the reinforcement shall be free from crack and fracture. The inner diameter of steel bar arc with good ductility is small and the construction adaptability is strong.

(3) 冷弯性能是为满足钢筋加工的要求。在弯折、弯钩或反复弯曲时,钢筋应避免裂纹和折断。延性好的钢筋弯弧内径小,施工适应性强。

(4) Weldability is a problem to be considered in the application of reinforcement. When the carbon content is higher than M, the weldability becomes worse, and it can not be welded when the carbon content is more than 0.55%. For the reinforcement strengthened by heat treatment and cold processing, the welding will lead to the reduction of the strength of the reinforcement in the welding area, which should be paid attention to in use.

(4) 焊接性能是钢筋应用时应考虑的问题。碳当 M 较高时焊接性能变差,超过 0.55% 时不可焊。通过热处理、冷加工而强化的钢筋,焊接会引起焊接区钢筋强度的降低,使用时应予以注意。

(5) The anchorage performance and anchorage ductility (the anchorage is still maintained in case of large slip) are the basis of the joint stress of reinforcement and concrete in the structure. The mechanical properties of plain steel bars are poor due to cementation and friction; the deformation steel bars are subject to force due to occlusion, which is related to their shape, and depends on the height of the transverse rib, the ratio of the area of the rib (the ratio of the projected area of the transverse rib to the surface area) and the shape of the concrete teeth.

(5) 锚固性能及锚固延性(大滑移时仍维持锚固)是钢筋在结构中与混凝土共受力的基础。光面钢筋靠胶结及摩擦,受力性能较差;变形钢筋以咬合作用受力,与其外形有关,取决于钢筋的横肋高度、肋面积比(横肋投影面积与表面积之比)以及混凝土咬合齿的形态。

(6) The stability of mass is very important to the steel bars. The steel products produced on a large scale generally have good homogeneity and stable quality. The cold-worked steel bars produced by small-scale workshops usually have large dispersion, unstable mechanical properties and high unqualified rate. In the case of unstable base metal and lack of management and inspection, it will be very serious, often affecting the safety and reliability of the structure.

(6) 质量的稳定性对受力钢筋十分重要。规模生产的钢筋产品一般均质性好,质量稳定。小规模作坊式生产的冷加工钢筋一般离散度大,力学性能不稳定,不合格率高。在母材不稳定和缺乏管理和检验的情况下将十分严重,往往影响结构的安全可靠性。

In order to ensure the basic properties of the screw steel, the main quality contents specified in the technical delivery conditions of the national standard are: chemical composition, mechanical properties, process performance, boundary dimension and surface quality, etc.

我国国家标准为保证螺纹钢的基本性能,在其交货技术条件中规定的主要质量内容为:化学成分、力学性能、工艺性能、外形尺寸及表面质量等。

2.6.1.1 Chemical Composition, Mechanical Properties and Process Performance
2.6.1.1 化学成分、力学性能及工艺性能

The chemical composition of steel directly affects the mechanical and technological properties of steel, which is the key point of quality control of screw steel. Chemical composition, composition segregation, surface defects, internal defects and non-metallic inclusions are the main contents of product quality inspection in steel-making and continuous casting processes. Among them, the composition control of steelmaking process is particularly important. In order to ensure the homogeneity of steel in use, the chemical composition fluctuation of the same batch is generally required to be controlled within a very small range. With the development of steel-making technology, the chemical composition of screw steel produced on a large scale can be controlled within the expected range. Especially for the diameter effect of screw steel caused by the defect of composition system of screw steel in our country, most of the production enterprises adopt the method of subdividing the composition control range and producing different specifications of screw steel with billets of different composition range to ensure the service performance of screw steel.

钢的化学成分直接影响着钢材的力学及工艺性能,是螺纹钢质量控制的重点。化学成分、成分偏析、表面缺陷、内部缺陷、非金属夹杂是螺纹钢生产检验炼钢、连铸工序产品质量的主要内容。其中,炼钢工序的成分控制尤为重要,为保证钢材在使用中的均质性,一般都要求同一批次的化学成分波动控制在很小的范围内。我国标准规定的化学成分允许范围比较大,随着炼钢技术的不断进步,目前规模生产的螺纹钢化学成分都能控制在预期的范围内。特别是对由于我国螺纹钢成分体系缺陷而产生的螺纹钢直径效应,生产企业大都采用将成分控制范围细分,以不同成分范围的坯料生产不同规格的螺纹钢的方法来保证螺纹钢的使用性能。

The mechanical property is the most important quality index for the use of screw steel, which mainly includes yield strength, tensile strength, elongation and surface shrinkage, etc. special users will also require yield strength ratio, etc.

力学性能是螺纹钢使用最重要的质量指标,其内容主要包括屈服强度、抗拉强度、伸长率及面缩率等,特殊用户还会要求屈强比等内容。

Process performance is a quality index to ensure that the threaded steel will not be damaged in the process of processing. The main contents are bending performance and reverse bending performance.

工艺性能是保证螺纹钢在用户加工过程中不被破坏的质量指标,主要内容就是弯曲性能、反弯性能。

For mechanical properties and process properties, on the premise of meeting relevant standards or user requirements, the uniformity requirements shall be met. The smaller the performance difference of the same batch of steel bars, the better.

对于力学性能和工艺性能在满足有关标准或用户要求的前提下,还要满足均匀性要求,同一批次的钢筋性能差越小越好。

2.6.1.2 Boundary Dimension and Surface Quality
2.6.1.2 外形尺寸及表面质量

The overall dimension of the rebar is mainly to ensure the mechanical properties and anchorage performance. The national standard has detailed requirements for the dimensions and deviation of various parts of the rebar, especially for the nominal section area, theoretical weight and weight deviation of each specification.

螺纹钢筋的外形尺寸主要是为满足力学性能和锚固性能提供保证,国家标准对螺纹钢外形各部位尺寸及其偏差都有详细要求,特别对各规格公称截面面积、理论重量、重量偏差做了严格规定。

In order to ensure the mechanical properties of the rebar, the surface of the rebar shall be free of cracks, scabs and folds, but it is allowed to have bumps not exceeding the height of the transverse rib and other defects with the depth and height not exceeding the allowable deviation of the size of the part.

为保证螺纹钢筋的力学性能,钢筋表面不得有裂纹、结疤和折叠,但允许有不超过横肋高度的凸块及深度和高度不大于所在部位尺寸允许偏差的其他缺陷存在。

2.6.2 Production Characteristics and Quality Control of Screw Steel
2.6.2 螺纹钢的生产特点及质量控制

2.6.2.1 Production Characteristics of Threaded Steel
2.6.2.1 螺纹钢的生产特点

In the process of using, the mechanical properties and technological properties of the screw steel mainly depend on the chemical composition, and the anchoring performance is guaranteed by its shape. In the production process, strictly according to the scope of national standards for control, product quality can basically meet the use requirements. Due to the large control range of the ingredients stipulated in Chinese standards, the mechanical properties of the final product can be ensured by adjusting the content combination of each element during the smelting process. In addition, low-temperature rolling, controlled rolling and controlled cooling can be used in the rolling process to further improve its comprehensive properties. The screw steel is a simple section steel, and its overall dimension depends on the pass design. Although the size deviation is strictly controlled, the rolling process is relatively simple, and the operation efficiency of the advanced production line is very high. The calendar operation rate of the modern production workshop is generally more than 85%, the yield and the sizing rate are also close to 100%, and the annual output is up to 1.2 million tons. It can be seen from this that the production features of screw steel are simple process, flexible adjustment, strict control and high production efficiency.

螺纹钢在使用过程中重点要求的力学性能和工艺性能取决于化学成分,而锚固性能是由其外形来保证。在生产过程中,严格按国家标准规定的范围进行控制,产品质量基本都可满足使用要求。由于我国标准规定的成分控制范围都比较大,在冶炼过程中可通过调整各元素含量组合,来确保最终产品的力学性能,另外在轧制过程中还可采用低温轧制、控轧控冷等技术来进一步改善其综合性能。螺纹钢属于简单断面型钢,外形尺寸取决于孔型设计,虽然尺寸偏差控制严格,但轧制工艺比较简单,先进的生产线作业效率非常高,现代化的生产车间其日历作业率一般都可达85%以上,成材率、定尺率也都接近100%,年产量最高已达1200000t。由此可看出,螺纹钢的生产特点是:工艺简单、调整灵活、控制严格、生产效率较高。

2.6.2.2 Quality Control in Steelmaking Process
2.6.2.2 炼钢过程中的质量控制

The metallurgical quality of billets plays a decisive role in the quality of final products. Many internal and external defects of products are caused by the poor metallurgical quality of billets. If the final mechanical properties are unqualified, it is mostly caused by the unqualified chemical composition, serious segregation, excessive inclusions or uneven morphology of the billet; if cracks, cracks, pockmarks, etc. occur on the steel surface, most of them are caused by the subcutaneous bubbles or double skin of the continuous casting billet. These defects not only affect the appearance and internal quality of the products, but also cause accidents in the rolling process, such as splitting, tearing and so on. Therefore, in order to ensure the metallurgical quality of billets, the quality control of the whole process of steelmaking process is particularly important.

坯料的冶金质量对最终产品的质量起决定性的作用,产品的许多内部和外部缺陷究其原因是由于坯料的冶金质量不良所致。如最终的力学性能不合格,多是由于坯料的化学成分不合格、偏析严重、夹杂物过多或形态不均所引起的;如发生在钢材表面的裂纹、发裂、麻点等,大多数是由连铸坯的皮下气泡或重皮造成的。这些缺陷除影响产品的外观和内在质量外,还会使轧制过程产生事故,如劈头、撕裂等会在轧制过程中引起堵钢、缠辊等事故。因此,为保证坯料的冶金质量,对炼钢工序全过程的质量控制显得尤为重要。

The quality control of smelting process includes:

冶炼过程的质量控制包括:

(1) Accurate control of liquid steel composition. The content of carbon, manganese, silicon and main alloy elements in molten steel should fluctuate less, and the harmful impurities such as sulfur, phosphorus, oxygen, hydrogen and nitrogen should be minimized in Chinese standard stipulates that the allowable fluctuation range of carbon content is 0.05%, while that of foreign products is only 0.02%. Although the converter has a certain desulfurization capacity, in order to control the sulfur content, it is necessary to control the sulfur content of the molten iron, which generally does not exceed 50% of the standard, and the physical content is usually about 1/3 of the standard.

（1）精确控制钢液成分。钢液中碳、锰、硅及主要合金元素含量波动要小，硫、磷、氧、氢、氮等有害杂质要尽量少。我国标准规定：碳含量允许波动范围为 0.05%，而国外产品碳含量的波动量仅为 0.02%。尽管转炉有一定的脱硫能力，为控制硫含量还是要控制投入铁液的硫含量，一般不超过标准规定的 50%，实物含量常在标准规定的 1/3 左右。

(2) Keep high purity. After smelting, the molten steel must be separated from the slag, and the slag shall be blocked for tapping or raked for tapping. It is not allowed to pour the molten steel and slag into the ladle at the same time, and the mixture of the slag will cause inclusion.

（2）保持高的纯净度。冶炼后的钢液必须与钢渣分离，要挡渣出钢或扒渣出钢，不允许钢液和渣同时倒入钢包中，钢渣相混会造成夹杂。

(3) The temperature fluctuation range should be small, and the superheat should be controlled at 15℃.

（3）温度波动范围要小，过热度要控制在 15℃。

The quality control in the continuous casting process includes：

连铸过程中的质量控制包括：

(1) Slag retaining tapping and protective casting. In the process of casting, the molten steel adopts the protection slag, long nozzle, inert gas protection, etc., so that the secondary oxidation caused by the contact with air can be completely avoided in the process of billet forming, so as to reduce the inclusion of chemicals inside the billet.

（1）挡渣出钢和保护浇铸。钢液在浇铸过程中采用保护渣、长水口、惰性气体保护等，使钢液在成坯的过程中完全避免和空气接触而产生二次氧化，以减少铸坯内部的化学物质夹杂。

(2) Level control. Keep the molten steel at a constant height in the mold to control the required casting speed.

（2）液面控制。使钢液在结晶器内保持恒定的高度，以控制所要求的浇铸速度。

(3) Casting temperature control. In order to ensure the normal casting and slab quality, it is necessary to keep the proper superheat of molten steel in the tundish, which is usually controlled at about 30℃.

（3）浇铸温度控制。为保证正常的浇铸和铸坯质量，要保持钢液在中间包中的适当过热度，通常过热度控制在 30℃ 左右。

(4) Surface quality control of slab. The technology of air mist cooling and multi-point straightening is used to control the surface temperature fluctuation of billet and disperse the surface deformation rate, so as to reduce the cracks caused by thermal stress and deformation stress on the surface of continuous casting slab.

（4）铸坯表面质量控制。采用气雾冷却和多点矫直技术，控制坯料表面温度波动和分散表面变形率，减少连铸坯表面由于热应力和变形应力而造成的裂纹。

2.6.2.3 Quality Control during Rolling

2.6.2.3 轧钢过程中的质量控制

Rolling process is the last and most important process in the production of screw steel. The control level of the whole process directly affects the final mechanical properties, geometric dimensions and surface quality of products. Strict temperature control, rolling process control and cooling control are the guarantee of the final product quality.

轧钢工序是螺纹钢生产的最后也是最重要的工序,对工序全过程的控制水平直接影响着产品最终的力学性能、几何尺寸和表面质量。严格的温度控制、轧制过程控制和冷却控制是产品最终质量的保证。

(1) Temperature control.

(1) 温度控制。

Heating temperature control: the heating temperature of billet actually includes the surface temperature and the temperature difference along the section, and sometimes the temperature difference along the length of billet. The final heating temperature of billet in the furnace is determined after considering the actual conditions such as rolling process, structural characteristics of rolling mill and structural characteristics of furnace. The specific time for heating to the specified temperature and section temperature difference depends on the billet size, steel type, heating method, temperature system adopted and some other conditions. In the production process of screw steel, the billet is usually heated by three-stage continuous heating furnace, the temperature of soaking section is generally controlled at 1200~1250℃, the temperature of upper heating section is controlled at 1250~1300℃, and the temperature of lower heating section is controlled at 1280~1350℃. In order to ensure the uniform heating temperature of the billet, the heating speed shall be strictly controlled to prevent the excessive temperature difference between the inside and outside of the billet caused by the excessive speed; the temperature in the furnace shall be adjusted correctly to keep the temperature at all points along the furnace width uniform, and the lower heating temperature shall be higher than the upper heating temperature in the heating section 20~30℃: reduce the temperature difference between the billet length and the billet top and bottom as far as FI is concerned; have enough heat preservation time in the soaking section to further improve the uniformity of billet heating temperature.

加热温度控制:钢坯的加热温度实际上包括表面温度和沿断面上的温度差,有时还包括沿坯料长度方向上的温度差。钢坯在炉内的最终加热温度是考虑了轧制工艺、轧机的结构特点以及炉子的结构特点等实际情况后确定的。加热到规定的温度和断面温差所需的时间,取决于坯料的尺寸、钢种、加热方式、采用的温度制度以及一些其他条件。在螺纹钢生产过程中,钢坯的加热通常采用三段连续式加热炉,均热段温度一般控制在1200~1250℃,上加热段温度控制在1250~1300℃,下加热段温度控制在1280~1350℃。为保证钢坯加热温度均匀,要严格控制加热速度,防止速度过快造成坯料内外温差过大;正确调整炉内温度使沿炉宽各点的温度保持均匀,在加热段确保下加热温度高于上加热温度20~30℃,尽量减少坯料长度上和钢坯上下面的温度差;在均热段还要有足够的保温时间,以利于进一步提高钢坯加热温度的均匀性。

Rolling temperature control: rolling temperature control includes opening temperature control and finishing temperature control. The change of rolling piece temperature is restricted by the heating quality of heating furnace, rolling process, rolling mill layout, etc. Therefore, the rolling temperature of different production lines should be determined according to their actual conditions. In order to ensure the stability of rolling process and workpiece size, the start rolling temperature is usually controlled at 1050~1150℃, and the finish rolling temperature is controlled at 850℃; in the more advanced continuous rolling production line, due to the rapid rolling speed, the temperature rise in the rolling process is greater than the temperature drop, in order to improve product performance and energy saving, the start rolling temperature can be controlled at 900~950℃.

轧制温度的控制：轧制温度控制包括开轧温度控制和终轧温度控制，轧件温度的变化受加热炉的加热质量、轧制工艺、轧机布置方式等的制约，因此，不同形式的生产线应结合各自的实际情况确定其轧制温度。为保证轧制过程和轧件尺寸的稳定，开轧温度通常控制在1050~1150℃，终轧温度则控制在850℃以上；在比较先进的全连轧生产线上，由于轧制速度较快，轧制过程的温升大于温降，为达到改善产品性能和节能的目的，可将开轧温度控制在900~950℃。

（2）Rolling process control.

（2）轧制过程控制。

Rolling is the key process to ensure the dimensional accuracy of products. The temperature, tension, groove and guide wear, mill adjustment, roll and guide processing and installation directly affect the dimensional accuracy of products.

轧制是保证产品尺寸精度的关键工序，轧制过程中的温度、张力、轧槽及导卫的磨损、轧机的调整、轧辊及导卫的加工和安装都直接影响着产品的尺寸精度。

The temperature fluctuation of rolled piece directly affects its deformation state and the size of deformation resistance, thus causing the fluctuation of rolled piece size. Therefore, the rolling temperature should be strictly controlled in the rolling process to keep the rolling temperature as consistent as possible.

轧件的温度波动直接影响其变形状态和变形抗力的大小，从而造成轧件尺寸的波动。所以在轧制过程中要严格控制轧制温度，使轧制温度尽可能保持一致。

In the continuous rolling production, tension is inevitable, especially in the rough and medium rolling mill because of the large section and short stand spacing, tension rolling can only be used. The fluctuation of tension is the most important factor to affect the dimensional accuracy. Therefore, in the rolling process, the tension fluctuation should be strictly controlled by the precise adjustment of the Telex system, so as to ensure the constant tension on the premise of realizing the micro tension rolling.

在连轧生产中，张力是不可避免的，特别在粗中轧机组由于轧件断面较大且机架间距较短，只能采用张力轧制。而张力的波动又是影响尺寸精度的最关键因素，因此在轧制过程中，应通过电传系统的精确调整，严格控制张力波动，在实现微张力轧制的前提下，确保张力的恒定。

The processing, installation, adjustment and wear of process equipment such as rolling mill, roll and guide directly affect the dimensional accuracy of rolling piece. Therefore, the processing and installation of process equipment shall be carried out in strict accordance with the requirements of process design. In the rolling process, the process specification shall be strictly implemented, and the worn process equipment shall be adjusted or replaced in time.

轧机、轧辊、导卫等工艺装备的加工、安装、调整以及在使用过程中的磨损直接影响着轧件的尺寸精度,因此,要严格按工艺设计的要求来进行工艺装备的加工和安装,在轧制过程中,严格执行工艺规程,及时调整或更换磨损的工艺装备。

(3) Cooling process control.

(3) 冷却过程控制。

The cooling after rolling of screw steel generally adopts two ways: natural air cooling and controlled cooling. As the cooling speed directly affects the performance of steel bars, uneven cooling will inevitably result in uneven performance. Therefore, no matter what cooling method is adopted, the uniformity of cooling shall be ensured, and the consistency of cooling speed between head and tail and each steel shall be maintained in the production process.

螺纹钢的轧后冷却,一般采用自然空冷和控制冷却两种方式,由于冷却速度直接影响钢筋性能,不均匀冷却必然造成性能的不均,因此,不管采用何种冷却方法,均应保证冷却的均匀性,在生产过程中保持头尾以及每支钢之间冷却速度的一致。

2.6.3 Inspection and Inspection of Product Quality
2.6.3 产品质量的检查、检验

In the inspection rules of GB 1499—1998 standard, the inspection of screw thread steel is divided into characteristic value inspection and delivery inspection. A characteristic value inspection is applicable to: (1) third party inspection; (2) inspection of product quality control by the supplier; (3) inspection proposed by the demander and agreed by the supplier and the demander. Delivery inspection is applicable to the inspection of acceptance batch of reinforcement. For batch rules, measurement methods and positions of different inspection items, sampling number, sampling methods and positions, sample inspection and test methods, etc., the standards have made detailed provisions. In addition, for the reinspection and determination, the standard has made the requirements that "the reinspection and determination of reinforcement shall conform to the provisions of GB/T 17505".

GB 1499—1998 标准的检验规则中将螺纹钢的检验分为特性值检验和交货检验。A 特性值检验适用于:(1) 第三方检验;(2) 供方对产品质量控制的检验;(3) 需方提出要求,经供需双方协议一致的检验。交货检验适用于钢筋验收批的检验。对于组批规则、不同检验项目的测量方法及位置、取样数目、取样方法及部位、试样检验试验方法等,标准中都做了详细的规定。此外,对复验与判定,标准做出了"钢筋的复验与判定应符合 GB/T 17505 的规定"的要求。

2.6.3.1 Routine Inspection
2.6.3.1 常规检验

In order to find out the waste products in time and reduce the loss of quality, the routine inspection of quality is usually set on the transport bench before the packaging of finished products. Therefore, most manufacturers call the transportation stand before packaging as the inspection stand, where the quality inspectors complete the inspection of the external dimension and surface quality of the threaded steel and the sampling of other inspection items.

在螺纹钢生产线上，为及时发现废品，减少质量损失，通常把质量的常规检验设置在成品包装前的输送台架上。因此，大多数厂家把包装前输送台架称为检验台架，质检人员在此完成螺纹钢的外形尺寸及表面质量的检验和其他检验项目的取样工作。

（1）Overall dimension.

（1）外形尺寸。

The overall dimension of the reinforcement shall be measured one by one, mainly including the inner diameter, height of longitudinal and transverse ribs, spacing between transverse ribs, maximum clearance at the end of transverse ribs, length of fixed length and curvature, etc. The measuring tools are vernier caliper, straightedge, steel tape, etc.

钢筋的外形尺寸要求逐支测量，主要测量钢筋的内径、纵横肋高度、横肋间距及横肋末端最大间隙、定尺长度及弯曲度等。测量工具为游标卡尺、直尺、钢卷尺等。

The method of measuring the average height of the cross rib center on both sides of the same section is adopted for the measurement of the height of the cross rib of the ribbed steel bar. That is to measure the maximum outer diameter of the steel bar and subtract the inner diameter. Half of the value obtained is the height of the rib, which should be accurate to 0.1mm. When the relative rib area needs to be calculated, the height of the quarter of the measured transverse rib should be increased.

带肋钢筋横肋高度的测量采用测量同一截面两侧横肋中心高度平均值的方法，即测取钢筋最大外径，减去该处内径，所得数值的一半为该处肋高，应精确到0.1mm。当需要计算相对肋面积时，应增加测量横肋1/4处高度。

The distance between transverse ribs of ribbed bars is measured by the method of average rib distance. That is to say, measure the center distance between the first and the 11th transverse rib on one side of the reinforcement. Divide this value by 10 to get the transverse rib spacing, which should be accurate to 0.1mm.

带肋钢筋横肋间距采用测量平均肋距的方法进行测量。即测取钢筋一面上第1个与第11个横肋的中心距离，该数值除以10即为横肋间距，应精确到0.1mm。

The length measurement generally adopts the method of sampling inspection, which is carried out at a certain time interval. The length deviation is ±25mm when delivering according to the fixed length. When the minimum length is required, the deviation is +50. When the maximum length is required, the deviation is −50.

长度测量一般采用抽检的方法，按一定的时间间隔进行测量，其长度偏差按定尺交货时的长度允许偏差为±25mm，当要求最小长度时，其偏差为+50，当要求最大长度时，其偏差为−50。

The bending degree is also measured by the method of regular spot check. Generally, the middle bending degree is measured by the method of stay wire. The bending degree shall not affect the normal use, and the total bending degree shall not be greater than 0.4% of the total length of the reinforcement. When obvious bending phenomenon is found, it shall be measured one by one.

弯曲度也采用定时抽检的方法进行测量,一般用拉线的方法测量弯曲度,弯曲度应不影响正常使用,总弯曲度不大于钢筋总长度的 0.4%。当发现有明显弯曲现象时应逐支测量。

(2) Surface quality.

(2) 表面质量。

The surface quality of reinforcement shall be checked one by one. Usually, visual inspection, magnifying glass low power observation and tool measurement are used. The requirement is that the end of the steel bar should be cut straight, and the local deformation should not affect the use. The surface of the reinforcement shall be free of defects affecting the service performance, and the height of the convex block on the surface shall not exceed the height of the transverse rib.

钢筋的表面质量应逐支检查。通常采用目视、放大镜低倍观察和工具测量相结合的方法来进行。其要求是钢筋端部应剪切正直,局部变形应不影响使用。钢筋表面不得有影响使用性能的缺陷,表面凸块不得超过横肋的高度。

(3) Sampling.

(3) 取样。

GB 1499—1998 requires that two tensile, two bending and one reverse bending tests shall be conducted for each batch of steel bars to test the mechanical and technological properties of the steel bars. The samples of these tests are usually also collected on the test bench. Two steel bars are selected on the bench, one tensile and one bending sample are taken on each of them, and then one reverse bending sample is taken on any steel bar. The samples of different test items in the same batch are bound firmly, and labeled with batch number and production number are sent to the laboratory for inspection.

GB 1499—1998 要求每批次钢筋应做两个拉伸、两个弯曲和一个反弯试验以检验钢筋的力学性能和工艺性能。这些检验的样品通常也在检验台架上采集,在台架上任选两支钢筋,在其上各取一个拉伸和一个弯曲试样,再在任一支钢筋上取一反弯试样,同一批次不同检验项目的试样分别捆扎牢固,贴上注明批次、生产序号的标签送试验室进行检验。

If it is necessary to inspect the chemical composition of the reinforcement, one of the above samples can be sent for the chemical composition inspection after the mechanical property inspection.

若需对钢筋的化学成分进行检验时,可在上述试样做完力学性能检验后,任选其一送去进行化学成分检验。

2.6.3.2 Quality Objection Handling
2.6.3.2 质量异议处理

Since the final quality inspection should be carried out by the project supervisor before the use of the rebar by the user, the quality objections are also generated before the final use. The quality objections of deformed steel bars can be generally divided into three categories:

由于螺纹钢筋在用户使用前都要由工程监理进行最后的质量检验,所以其质量异议也都产生在最终使用之前。螺纹钢筋的质量异议一般可分为四类：

（1）Quality objections that do not affect performance. This kind of objection is mostly caused by the difference in understanding of steel bar production standard, service performance and use method. Through direct communication with users, it can help users solve problems in use. Generally, this kind of objection will not cause quality loss.

（1）不影响使用性能的质量异议。这类异议大多是由于对钢筋生产标准、钢筋使用性能及使用方法的认识差异所造成，可通过和用户的直接沟通，帮助用户解决使用中的问题，这类异议一般不会造成质量损失。

（2）It does not affect the use performance, but may cause quality objection of increased user use cost. For example, bending, corrosion, oil pollution, etc. produced in the process of production, storage and transportation of steel bars can be treated by concession, giving certain economic compensation to users or pre-treatment for users before use, reducing the user's use cost and achieving the user's satisfaction. Such objections will cause different degrees of economic losses.

（2）不影响使用性能，但可造成用户使用成本增加的质量异议。如钢筋在生产、储存、运输过程中产生的弯曲、锈蚀、油污等，可通过让步的方法来处理，对用户给予一定的经济补偿或替用户进行使用前的预处理，降低用户的使用成本，达到使用户满意的目的，这类异议会造成不同程度的经济损失。

（3）Quality objection seriously affecting the service performance of reinforcement. The quality objection caused by the lack of quality control and inspection during the production process, which causes the unqualified products to flow into the user's hands, can be handled through the method of return and replacement. If the user's construction period is delayed, the user shall be compensated for the delay. Once such objections occur, they may cause great economic losses to producers.

（3）严重影响钢筋使用性能的质量异议。由于在生产过程中对质量控制、检验的缺失使不合格品流入用户手中而产生的质量异议，可通过退货、换货的方法来处理，如果延误了用户的工期，还要对用户进行误工补偿。这类异议一旦发生，可能会造成生产者的重大经济损失。

（4）Quality objections caused by inspection methods and equipment. Since the methods and equipment used by the producers and users of the corrugated steel bars in the inspection of the quality of the steel bars cannot be completely consistent, which will inevitably result in the difference of the inspection results, both parties shall communicate in a timely manner to find out the causes of the differences. If there is no consensus, the quality inspection agency agreed by both parties can be proposed to reclassify and cause the defects of the products.

（4）检验方法、检验设备造成的质量异议。由于螺纹钢筋的生产者和使用者在对钢筋质量进行检验时所使用的方法、设备不可能完全一致，不可避免地会造成检验结果的差异，双方应及时沟通，找出差异产生的原因，若达不成共识，可提请双方一致认可的质量检测机构重新进行检测产品缺陷分类和查找原因。

2.6.3.3 Product Defect Classification and Causes
2.6.3.3 产品缺陷分类和原因

In the whole production process of rebar, due to the continuous changes of production equipment, production environment and process parameters, there will be some quality defects in each process from smelting, continuous casting to rolling. These defects will affect the final rebar product quality in varying degrees. This section focuses on the main casting billet and rolling defects that affect the final product quality class and analysis, in order to reduce product defects and quality loss in the production process.

在螺纹钢筋的整个生产过程中,由于生产设备、生产环境、工艺参数处于不断的变化之中,从冶炼、连铸到轧制各工序都会产生一些质量缺陷,这些缺陷会不同程度对最终的螺纹钢筋产品质量产生影响,本节重点对影响最终产品质量的主要铸坯和轧钢缺陷进行分类和分析,以期在螺纹钢的生产中,尽量减少产品缺陷,降低生产过程中的质量损失。

(1) Slab defects.

(1) 铸坯缺陷。

1) Segregation.

1) 偏析。

Segregation is an important quality problem of continuous casting slab. The smaller the section of continuous casting slab, the more serious the segregation. The reason is that the solidification time of molten steel in the mold is not the same and the columnar crystal growth is not balanced, which makes the alloy elements such as carbon, sulfur and phosphorus enriched in the last part of the solidification and forms the segregation of chemical composition. This segregation is usually accompanied by porosity or even shrinkage. The segregation of continuous casting slab reduces the strength and plasticity of metal, which seriously affects the mechanical and technological properties of steel bars. The segregation defects can be effectively reduced by enlarging the section size of slab, strictly controlling the superheat of molten steel, reducing the content of phosphorus, sulfur and manganese, and adopting electromagnetic stirring.

偏析是连铸坯的一个重要的质量问题,连铸坯断面越小偏析越严重。其产生的原因是由于结晶器内钢液凝固时间不一致,柱状晶生长不均衡,使得碳等合金元素及硫、磷等富集于凝固最晚的部分,形成化学成分的偏析。这种偏析通常会伴生着疏松甚至出现缩孔。连铸坯的偏析降低了金属的强度和塑性,严重地影响着钢筋的力学和工艺性能。扩大连铸坯断面尺寸、严格控制钢液过热度、降低磷、硫、锰的含量及采用电磁搅拌可有效地减少偏析缺陷。

2) Central porosity.

2) 中心疏松。

In the process of continuous casting slab crystallization, because of the interpenetration and interdiction between the dendrites, the liquid which is rich in low melting point components is isolated between the dendrites. After this part of liquid is condensed, because there is no supplement of other liquid, many dispersed small shrinkage cavities will be formed between dendrites, thus the center of continuous casting slab will be loose. If the looseness is serious, it will affect the mechanical properties of finished steel bars.

在连铸坯结晶过程中,由于各枝晶间互相穿插和互相封锁作用,是富集着低熔点组元的液体被孤立于各枝晶之间。这部分液体在冷凝后,由于没有其他液体的补充,会在枝晶间形成许多分散的小缩孔,从而形成连铸坯的中心疏松。如果疏松严重,会影响成品钢筋的力学性能。

3) Shrinkage cavity.

3) 缩孔。

During continuous casting, the metal solidifies from the periphery to the center, and the liquid in the center solidifies at the latest, forming a closed shrinkage cavity in the center. If only the metal around and at the bottom solidifies first, an open shrinkage cavity is formed at the upper part of the slab. If the closed shrinkage cavity does not contact with the air during rolling, it can be welded. If the larger shrinkage cavity is rolled again, it may cause the jam accident. The opening shrinkage often leads to splitting and steel piling accidents.

连铸时金属由四周向心部凝固,心部液体凝固最晚,会在心部形成封闭的缩孔。如果仅四周及底部的金属先凝固,则在铸坯的上部形成开口的缩孔。封闭的缩孔在轧制时如不与空气接触可以焊合,较大的缩孔,再轧制时可能造成轧卡事故。开口缩孔往往会造成劈头、堆钢事故。

4) Crack.

4) 裂纹。

The cracks of continuous casting slab can be divided into corner crack, edge crack, middle crack and center crack. The corner crack is located at the corner of the slab, which has a certain depth from the surface and is perpendicular to the surface. In serious cases, it propagates along the diagonal to the slab. The corner crack is caused by the side depression and serious square breaking of the corner of the billet, which makes the tensile stress between the local metals greater than the intergranular bonding force. The edge cracks are distributed at the junction of equiaxed grain and columnar crystals around the slab and propagate along the columnar crystals to the interior, which is caused by the deformation of the bulging slab when it is straightened by the guide roller. The middle crack is generated in the area of columnar crystal and propagates along the columnar crystal, which is generally perpendicular to the two sides of the slab. In serious cases, the center of the slab also exists at the same time, which is caused by the thermal stress generated when the slab is forced to cool. The central crack is produced in the area of columnar crystal near the center and perpendicular to the surface of slab, and can pass through the center when serious. The reason is that the casting speed is too high, the billet is straightened by the guide roller in the liquid core state, and the bearing pressure is too large. All internal cracks that are not exposed can be welded as long as they are not in direct contact with the air during rerolling, which will not affect the product quality, but the cracks that are not welded will affect the mechanical properties of the reinforcement.

连铸坯的裂纹可分为角部裂纹、边部裂纹、中间裂纹和中心裂纹。角部裂纹在铸坯的角部,距表面有一定的深度,并与表面垂直,严重时沿对角线向铸坯内扩展。角部裂纹是由于铸坯角部的侧面凹陷及严重脱方,使局部金属间产生的拉应力大于晶间结合力所造成的。边部裂纹分布在铸坯四周的等轴晶和柱状晶交界处,沿柱状晶向内部扩展,是由鼓肚的铸坯通过导辊矫直时变形引起的。中间裂纹在柱状晶区域产生并沿柱状晶扩展,一般垂直于铸坯的两个侧面,严重时铸坯中心的四周也同时存在,是由铸坯被强制冷却时产生的热应力造成的。中心裂纹在靠近中心部位的柱状晶区域产生并垂直于铸坯表面,严重时可穿过中心。是由于铸速过高,铸坯在液体状态下通过导辊矫直,所承受的压力过大所致。凡是不暴露的内部裂纹,只要再轧制时不与空气直接接触可以焊合,不影响产品质量,但焊合不了的裂纹影响钢筋的力学性能。

(2) Rolling defects.

(2) 轧钢缺陷。

1) Roll up.

1) 结卷。

The scabs are tongue like, massive and fish scale like embedded on the surface of the steel bar. Its size and thickness are different, and its shape is closed or not closed, connected or not connected with the main body, warped or not warped, single or multiple flaky. The distribution of scabs caused by cast steel is irregular, with inclusions below.

结疤呈舌状、块状、鱼鳞状嵌在钢筋表面上。其大小厚度不一,外形有闭合或不闭合、与主体相连或不相连、翘起或不翘起、单个或多个成片状。铸钢造成的结疤分布不规则,下面有夹杂物。

Causes: ①There are residual scabs, bubbles on the surface of ingot (Billet) or the ratio of depth to width of surface cleaning is unreasonable. ②The groove is badly scored, and the meat or bonding metal of a groove before the finished hole is dropped. ③If the rolled piece slips in the pass, the metal will be piled up or the foreign metal will be brought into the slot along with the rolled piece. ④The slot is severely worn or scratched by foreign objects.

产生原因:①铸锭(坯)表面有残余的结疤、气泡或表面清理深宽比不合理。②轧槽刻痕不良,成品孔前某一轧槽掉肉或黏结金属。③轧件在孔型内打滑造成金属堆积或外来金属随轧件带入槽孔。④槽孔严重磨损或外物刮伤槽孔。

2) Crack.

2) 裂纹。

The cracks are generally straight and sometimes Y-shaped. Its direction is mostly consistent with the rolling direction, and the gap is generally perpendicular to the steel surface.

裂纹一般呈直线状,有时呈"Y"状。其方向多与轧制方向一致,缝隙一般与钢材表面垂直。

The causes are as follows: ①The air bubbles and non-metallic inclusions under the ingot (Billet) are exposed after rolling fracture or the cracks and tensile cracks of the ingot (Billet) are not removed. ②Uneven heating, low temperature, poor pass design, poor processing or improper cooling of rolled steel. ③The groove of rough rolling is seriously worn.

产生原因:①铸锭(坯)皮下气泡、非金属夹杂物经轧制破裂后暴露或铸锭(坯)

本身的裂纹、拉裂未清除。②加热不均、温度过低、孔型设计不良、加工不精或轧后钢材冷却不当。③粗轧孔槽磨损严重。

3) Fold.

3) 折叠。

Folding is a kind of defect along the rolling direction, whose shape is similar to the crack and has a certain angle with the surface of the reinforcement. In general, it is straight and zigzag, and appears on the surface of steel bar in full length or intermittently.

折叠是沿轧制方向，外形与裂纹相似，与钢筋表面呈一定斜角的缺陷。一般呈直线状，也有锯齿状，通长或断续出现在钢筋表面上。

Causes: ①The ear of a rolled piece in front of the finished hole. ②Improper pass design, serious slot wear, poor design and installation of guide and guard device, etc., cause "step" of rolled piece, improper adjustment of rolled piece or metal accumulation caused by rolling piece slipping, and folding during rolling.

产生原因：①成品孔前某道轧件出耳子。②孔型设计不当，槽孔磨损严重，导卫装置设计、安装不良等，使轧件产生"台阶"或轧件调整不当或轧件打滑产生金属堆积，再轧时造成折叠。

4) Pit.

4) 凹坑。

The pit is a strip or block on the surface, which is periodically or irregularly distributed on the surface of the reinforcement.

凹坑是表面条状或块状的凹陷，周期性或无规律地分布在钢筋表面上。

Causes: ①There are protrusions on the working surface of the rolling groove, rolling guide plate and straightening roll, and periodic pits are produced after the rolling piece passes. ②During the rolling process, the external hard metal is pressed on the surface of the rolled piece and forms after falling off. ③The ingot (Billet) stays in the furnace for a long time, which causes the oxide scale to be too thick, which is pressed on the surface of the rolled piece during rolling and forms after falling off. ④The rough rolling hole is seriously worn, the metal on the surface of the rolled piece is gnawed down, and the surface of the rolled piece is pressed again when rolling again, which is formed after falling off. ⑤The ingot (Billet) scabs and falls off. ⑥The rolled piece collides with the hard object or the steel is piled up unevenly and pressed.

产生原因：①轧槽、滚动导板、矫直辊工作面上有凸出物，轧件通过后产生周期性凹坑。②轧制过程中，外来的硬质金属压入轧件表面，脱落后形成。③铸锭（坯）在炉内停留时间过长，造成氧化铁皮过厚，轧制时压入轧件表面，脱落后形成。④粗轧孔磨损严重，啃下轧件表面金属，再轧时又压入轧件表面，脱落后形成。⑤铸锭（坯）结疤脱落。⑥轧件与硬物相碰或钢材堆放不平整压成。

5) Bump.

5) 凸块。

The convex block is a periodic bulge on the surface of the reinforcement except the transverse rib.

凸块是钢筋除横肋外表面上周期性的凸起。

Causes: there are sand holes, falling blocks or cracks in the finished hole or front hole groove of the finished product.

产生原因：成品孔或成品前孔轧槽有砂眼、掉块或龟裂。

6) Surface inclusion.

6) 表面夹杂。

The surface inclusions are generally point, block or strip mechanical bonded on the surface of the reinforcement, with a certain depth and irregular size and shape. Inclusions produced in steel making are generally white, gray or gray white; inclusions produced in rolling are generally red or brown, sometimes gray white, but the depth is generally light.

表面夹杂一般呈点状、块状或条状机械黏结在钢筋表面上，具有一定深度，大小形状无规律。炼钢带来的夹杂物一般呈白色、灰色或灰白色；在轧制中产生的夹杂物一般呈红色或褐色，有时也呈灰白色，但深度一般很浅。

Causes: ①Surface non-metallic inclusions brought by slab. ②In the process of heating and rolling, there are some non-metallic inclusions (such as refractory of heating furnace, slag of furnace bottom and ash of fuel) sticking to the surface of rolling piece.

产生原因：①铸坯带来的表面非金属夹杂物。②在加热轧制过程中偶然有非金属夹杂物（如加热炉耐火材料、炉底炉渣、燃料的灰烬）粘在轧件表面。

7) Hairline (also called hairline).

7) 发纹（又称发裂）。

Hairline is a family of fine lines scattered on the surface of the section steel. It is generally consistent with the rolling direction, and its length and depth are less than the crack.

发纹是在型钢表面上分散成族断续分布的细纹，一般与轧制方向一致，其长度、深度比裂纹轻微。

Causes: ①The subcutaneous bubbles or non-metallic inclusions of ingot (Billet) are exposed after rolling. ②Uneven heating, low temperature or improper cooling of the rolled piece. ③The groove of rough rolling is seriously worn.

产生原因：①铸锭（坯）皮下气泡或非金属夹杂物轧后暴露。②加热不均、温度过低或轧件冷却不当。③粗轧孔槽磨损严重。

8) Size out of tolerance.

8) 尺寸超差。

Dimension out of tolerance means that the dimension of each part of the reinforcement exceeds the deviation range specified in the standard.

尺寸超差指钢筋各部位尺寸超过标准规定的偏差范围。

Causes: ①Unreasonable pass design. ②Improper adjustment of rolling mill. ③The bearing bush, rolling groove or guide device are improperly installed and severely worn. ④The uneven heating temperature causes the local dimension out of tolerance. ⑤Tension and looper exist in pulling steel.

产生原因：①孔型设计不合理。②轧机调整操作不当。③轴瓦、轧槽或导卫装置安装不当，磨损严重。④加热温度不均造成局部尺寸超差。⑤张力及活套存在拉钢。

9) Cross rib size out of tolerance.

9) 横肋尺寸超差。

Transverse rib size out of tolerance (transverse rib thin) refers to the deviation value that the height and volume of transverse rib are less than the standard requirements.

横肋尺寸超差（横肋瘦）是指横肋高度及体积均小于标准要求的偏差值。

Causes: ①The pass design is unreasonable, and the size of the red blank before the finished product is too small. ②Tension and looper exist in pulling steel.

产生原因：①孔型设计不合理，成品前的红坯尺寸偏小。②张力及活套存在拉钢。

10) Torsion.

10) 扭转。

Torsion means that the steel bar is twisted into a spiral shape around its longitudinal axis.

扭转是指钢筋绕其纵轴扭成螺旋状。

Causes: ①The center line of roller intersects and is not in the same vertical plane, and the center line is not parallel or axial staggered. ②Improper installation or serious wear of guide and guard device. ③Improper adjustment of rolling mill.

产生原因：①轧辊中心线相交且不在同一垂直平面内，中心线不平行或轴向错动。②导卫装置安装不当或磨损严重。③轧机调整不当。

11) Bending.

11) 弯曲。

Bending refers to the phenomenon that the reinforcement is not straight along the vertical or horizontal direction. Generally, it is a wave bend, sometimes there are repeated wave bends or only at the end.

弯曲是指钢筋沿垂直方向或水平方向不平直现象。一般为波浪弯，有时也出现反复的波浪弯或仅在端部出现弯曲。

Causes: ①Poor installation of finished hole guide device. ②Uneven rolling temperature, improper pass design or improper rolling operation. ③Uneven cooling bed, uneven shifting steel rack and uneven cooling of finished products. ④In the hot state, the finished products are lifted or stacked unevenly, resulting in lifting and bending. ⑤If the guide plate at the outlet of the finished hole is too short or the rolling piece runs too fast, the end will be bent easily after colliding with the baffle plate. ⑥The gap between the blades of cold shear is too large or the number of cutting branches is too large, resulting in head bending.

产生原因：①成品孔导卫装置安装不良。②轧制温度不均、孔型设计不当或轧机操作不当。③冷床不平、移钢齿条不齐、成品冷却不均。④热状态下成品吊运或堆放不平整，造成吊弯、压弯等。⑤成品孔出口导板过短或轧件运行速度过快，撞挡板后容易出现端部弯曲。⑥冷剪机剪刀间隙过大或剪切枝数过多，造成头部弯曲。

12) Head deformation.

12) 切头变形。

Cutting head deformation refers to that the head of the steel bar after cold shear is in the shape of horseshoe or triangle, often associated with the bending of the head.

切头变形是指经冷剪剪切后钢筋头部呈马蹄形或三角形，常与头部弯曲伴生。

Causes: ①Too much clearance between cutting edges. ②The cutting edge is blunt. ③Too much shear.

产生原因：①剪刃间隙过大。②剪刃磨钝。③剪切量过大。

13) Weight out of tolerance.

13) 重量超差。

Weight out of tolerance refers to that the weight per meter of rebar is lower than the lower limit specified in the standard, which is often associated with size out of tolerance.

重量超差是指螺纹钢筋每米重量低于标准规定的下限值，常与尺寸超差伴生。

Causes: ①Unreasonable pass design. ②In the process of negative tolerance rolling, when the finished hole is replaced with new groove, the negative difference rate is too large. ③Improper adjustment of steel rolling. ④Pull steel before finished product.

产生原因：①孔型设计不合理。②负公差轧制过程中，当成品孔换新槽时，负差率过大。③轧钢调整不当。④成品前拉钢。

References
参 考 文 献

[1] 赵松筠,唐文林,等. 型钢孔型设计 [M]. 北京:冶金工业出版社,2005.
[2] 袁志学,马水明. 中型型钢生产 [M]. 北京:冶金工业出版社,2005.
[3] 王子亮. 螺纹钢生产工艺与技术 [M]. 北京:冶金工业出版社,2008.
[4] 苏世怀,等. 热轧钢筋 [M]. 北京:冶金工业出版社,2009.
[5] 新疆八一钢铁(集团)公司. 小型连轧机的工艺与电气控制 [M]. 北京:冶金工业出版社,2000.
[6] 王有铭. 钢材的控制轧制和控制冷却 [M]. 北京:冶金工业出版社,2009.
[7] 袁志学,王淑平. 塑性变形与轧制原理 [M]. 北京:冶金工业出版社,2008.
[8] 《小型型钢连轧生产工艺与设备》编写组. 小型型钢连轧生产工艺与设备 [M]. 北京:冶金工业出版社,1999.
[9] 崔艳. 国内棒材生产线与工艺综述 [J]. 重型机械科技,2004 (4):37~49.
[10] 吕立华. 金属塑性加工力学 [M]. 北京:冶金工业出版社,2007.
[11] 赵志业. 金属塑性加工力学 [M]. 北京:冶金工业出版社,1987.
[12] 王占学. 塑性加工金属学 [M]. 北京:冶金工业出版社,2003.
[13] 王廷溥,齐克敏. 金属塑性加工学——轧制理论与工艺 [M]. 北京:冶金工业出版社,2001.
[14] 杨忠民. The Hot Rolling High Strength Reinforcing Steel Bar and Manufacturing Practicing in China [C]. 2007 年建筑用钢会议论文集. 天津:2007.
[15] 李曼云. 钢的控制轧制和控制冷却技术手册 [M]. 北京:冶金工业出版社,1998.
[16] 苏世怀,孙维,汪开忠,等. 高效节约型建筑用钢——热轧钢筋 [M]. 北京:冶金工业出版社,2010.
[17] 黄肇信. HRB400 小规格热轧带肋钢筋的试制与开发 [J]. 轧钢,2001,18 (5):25~27.
[18] 东涛,付俊岩. 钢筋的微合金化原理及生产要领 [J]. 微合金化技术,2004,4 (1):60~62.
[19] 闻雷. 大规格钢筋余热处理工艺及温度场有限元分析 [D]. 沈阳:东北大学,2000.
[20] 张晓香,魏国增. HRB400 钢中钒微合金化工艺改进 [J]. 河北冶金,2003 (6):44~46.